Praise for Max Tegmark's

Our Mathematical Universe

"This is a valuable book, written in a deceptively simple style but not afraid to make significant demands on its readers, especially once the multiverse level gets turned up to four. It is impressive how far Tegmark can carry you until, like a cartoon character running off a cliff, you wonder whether there is anything holding you up."   —*Nature*

"Max Tegmark is arguably the nearest we have to a successor to Richard Feynman.... Half of [*Our Mathematical Universe*] is a lucid tour d'horizon of what we know about the Universe. The rest is an exhilarating expedition far beyond conventional thinking, in search of the true meaning of reality.... His insights and conclusions are staggering—and perhaps even crazy enough to be true."

—*BBC Focus magazine*

"Today multiple universes are scientifically respectable, thanks to the work of Tegmark as much as anyone.... Physics could do with more characters like Tegmark. He combines an imaginative intellect and a charismatic presence with a determination to promote his subject. Written in a lively and slightly quirky style, it should engage any reader interested in the infinite variety of nature."

—*Financial Times* (UK)

"Mind-bending book about the cosmos. . . . Tegmark's achievement is to explain what on earth he is talking about in language any reasonably attentive reader will understand. He is a professor at MIT, and clearly a fine teacher as well as thinker. He tackles the big, interrelated questions of cosmology and subatomic physics much more intelligibly than, say, Stephen Hawking." —*The Times* (London)

"Tegmark writes at the cutting edge of cosmology and quantum theory in friendly and relaxed prose, full of entertaining anecdotes and down-to-earth analogies."
—*The Guardian* (London)

"*Our Mathematical Universe* is a delightful book in which the Swedish-born author, now at MIT, takes readers on a roller coaster ride through cosmology, quantum mechanics, parallel universes, sub-atomic particles and the future of humanity. It is quite an adventure with many time-outs along the way. . . . *Our Mathematical Universe* gives keen insight into someone who asks questions for the pure joy of answering them." —*Pittsburgh Post-Gazette*

"Lively and lucid, the narrative invites general readers into debates over computer models for brain function, over scientific explanations of consciousness, and over prospects for finding advanced life in other galaxies. Though he reflects soberly on the perils of nuclear war and of hostile artificial intelligence, Tegmark concludes with a bracingly upbeat call for scientifically minded activists who recognize a rare opportunity to make our special planet a force for cosmic progress. An exhilarating adventure for bold readers."
—*Booklist* (starred review)

"Daring, radical, innovative. A game changer. If Dr. Tegmark is correct, this represents a paradigm shift in the relationship between physics and mathematics, forcing us to rewrite our textbooks. A must read for anyone deeply concerned about our universe."
—Michio Kaku, author of *Physics of the Future*

MAX TEGMARK

Our Mathematical Universe

Max Tegmark is author or coauthor of more than two hundred technical papers, twelve of which have been cited more than five hundred times. He holds a Ph.D. from the University of California, Berkeley, and is a physics professor at MIT.

Our Mathematical Universe

My Quest for the
Ultimate Nature of Reality

Max Tegmark

VINTAGE BOOKS
A DIVISION OF RANDOM HOUSE LLC
NEW YORK

To Meia,
WHO INSPIRED ME TO WRITE THIS BOOK

Contents

Preface

I'm truly grateful to everyone who has encouraged and helped me write this book, including:

my family, friends, teachers, colleagues and collaborators for support and inspiration over the years,

Mom for sharing her passion and curiosity for life's big questions,

Dad for sharing his fascination and wisdom about mathematics and its meaning,

my sons, Philip and Alexander, for asking such great questions about the world and for unwittingly providing anecdotes for the book,

all the science enthusiasts around the world who've contacted me over the years with questions, comments and encouragement to pursue and publish my ideas,

my agents, John and Max Brockman, for convincing me to write this book and setting everything in motion,

those who gave me feedback on parts of the manuscript, including Mom, my brother Per, Josh Dillon, Marty Asher, David Deutsch, Louis Helm, Andrei Linde, Jonathan Lindström, Roy Link, David Raub, Shevaun Mizrahi, Mary New, Sandra Simpson, Carl Shulman and Jaan Tallinn.

the superheroes who commented on drafts of the entire book, namely Meia, Dad, Paul Almond, Julian Barbour, Phillip Helbig, Adrian Liu, Howard Messing, Dan Roberts, Edward Witten and my editor, Dan Frank, and most of all,

my beloved wife, Meia, my muse and fellow traveler, who's given me more encouragement, support and inspiration than I ever dreamed of.

Our Mathematical Universe

1

What Is Reality?

. . . trees are made of air, primarily. When they are burned, they go back to air, and in the flaming heat is released the flaming heat of the Sun which was bound in to convert the air into tree. And in the ash is the small remnant of the part which did not come from air, that came from the solid earth, instead.

—Richard Feynman

There are more things in heaven and earth, Horatio, than are dreamt of in your philosophy.

—William Shakespeare, *Hamlet*, act 1, scene 5

Not What It Seems

A second later, I died. I stopped pedaling and hit the brakes, but it was too late. Headlights. Grille. Forty tons of steel, honking furiously, like a modern-day dragon. I saw the panic in the truck driver's eyes. I felt time slow down and my life pass before me, and my very last thought in life was "I hope this is just a nightmare." Alas, I felt in my gut that this was for real.

But how could I be completely sure that I wasn't dreaming? What if, just before impact, I'd perceived something that didn't happen except in dreamland, say my dead teacher Ingrid alive and well, sitting behind me on my bike rack? Or what if, five seconds earlier, a pop-up window had appeared in the upper left corner of my visual field, with the text line "Are you sure it's a good idea to zoom out of this underpass without looking to the right?" materializing above clickable Continue and Cancel buttons? Had I watched enough movies like *The Matrix* and *The Thirteenth Floor*, I might have started wondering whether my whole life had been a computer simulation, questioning some of my most basic assumptions about the nature of reality. But I experienced no such things, and died certain that my problem was real. After all, how much more solid and real than a forty-ton truck can something get?

However, not everything is the way it first seems, and this goes even for trucks and reality itself. Such suggestions come not only from philosophers and science-fiction authors, but also from physics experiments. Physicists have known for a century that solid steel is really mostly empty space, because the atomic nuclei that make up 99.95% of the mass are tiny balls that fill up merely 0.0000000000001% of the volume, and that this near-vacuum only feels solid because the electrical forces that hold these nuclei in place are very strong. Moreover, careful measurements of subatomic particles have revealed that they appear able to be in different places at the same time, a well-known puzzle at the heart of quantum physics (we'll explore this in Chapter 7). But I'm made of such particles, so if they can be in two places at once, then can't I as well? Indeed, about three seconds before the accident, I was subconsciously deciding whether to simply look to the left where I'd always turned on my way to Blackebergs Gymnasium, my Swedish high school, since there was never any traffic on this cross street, or whether to also look to the right just in case. My ill-fated snap decision that morning in 1985 ended up being a very close call. It all came down to whether a single calcium atom would enter a particular synaptic junction in my prefrontal cortex, causing a particular neuron to fire an electrical signal that would trigger a whole cascade of activity by other neurons in my brain, which collectively encoded "Don't bother." So if that calcium atom started in two slightly different places at once, then half a second later, my pupils would have been pointing in two opposite directions at once, two seconds later my bike would have been in two different places at once, and before long, I'd have been dead and alive at once. The world's leading quantum physicists argue passionately about whether this really happens, effectively splitting our world into parallel universes with different histories, or whether the so-called Schrödinger equation, the supreme law of quantum motion, needs to be somehow amended. So did I really die? I just barely made it in this particular universe, but did I die in another equally real universe where this book never got written? If I'm both dead and alive, then can we somehow amend our notion of what reality is so that it all makes sense?

If you feel that what I've just put forth sounds absurd and that physics has muddied the waters, it gets even worse if we consider how I personally perceived this. If I'm in these two different places in two parallel universes, then one version of me will survive. If you apply the same argument to all other ways I can die in the future, it seems there will always be at least one parallel universe where I never die. Since my consciousness exists only where I'm alive, does that mean that I'll subjectively feel immortal? If so, will you, too, find yourself subjectively immortal, eventually the oldest person on Earth? We'll answer these questions in Chapter 8.

Are you surprised that physics has uncovered our reality to be much stranger than we'd imagined? Well, it's actually not surprising if we take Darwinian evolution seriously! Evolution endowed us with intuition only for those aspects of physics that had survival value for our distant ancestors, such as the parabolic orbits of flying rocks (explaining our penchant for baseball). A cavewoman thinking too hard about what matter is ultimately made of might fail to notice the tiger sneaking up behind her and get cleaned right out of the gene pool. Darwin's theory thus makes the testable prediction that whenever we use technology to glimpse reality beyond the human scale, our evolved intuition should break down. We've repeatedly tested this prediction, and the results overwhelmingly support Darwin. At high speeds, Einstein realized that time slows down, and curmudgeons on the Swedish Nobel committee found this so weird that they refused to give him the Nobel Prize for his relativity theory. At low temperatures, liquid helium can flow upward. At high temperatures, colliding particles change identity; to me, an electron colliding with a positron and turning into a Z-boson feels about as intuitive as two colliding cars turning into a cruise ship. On microscopic scales, particles schizophrenically appear in two places at once, leading to the quantum conundrums mentioned above. On astronomically large scales—surprise!—weirdness strikes again: if you intuitively understand all aspects of black holes, I think you're in a minority of one, and should immediately put down this book and publish your findings before someone scoops you on the Nobel Prize for quantum gravity. Zoom out to still larger scales, and more weirdness awaits, with a reality vastly grander than everything we can see with our best telescopes. As we'll explore in Chapter 5, the leading theory for what happened early on is called *cosmological inflation*, and it suggests that space isn't merely really, really big, but actually infinite, containing infinitely many exact copies of you, and even more near-copies living out every possible variant of your life in two different types of parallel universes. If this theory proves to be true, it means that even if there's something wrong with the quantum physics argument I gave above for a copy of me never making it to school, there will be infinitely many other Maxes on solar systems far out there in space who lived identical lives up until that fateful moment, and then decided not to look to the right.

In other words, discoveries in physics challenge some of our most basic ideas about reality *both* when we zoom into the microcosm *and* when we zoom out to the macrocosm. As we'll explore in Chapter 11, many ideas about reality get challenged even on the intermediate scale of us humans if we use neuroscience to delve into how our brains work.

Last but not least, we know that mathematical equations offer a window into the workings of nature, as metaphorically illustrated in Figure 1.1. But why has our physical world revealed such extreme mathematical regularity that astronomy superhero Galileo Galilei proclaimed nature to be "a book written in the language of mathematics," and Nobel Laureate Eugene Wigner stressed the "unreasonable effectiveness of mathematics in the physical sciences" as a mystery demanding an explanation? Answering this question is the main goal of this book, as its title suggests. In Chapters 10–12, we'll explore the fascinating relations between computation, mathematics, physics and mind, and explore a crazy-sounding belief of mine that our physical world not only is *described* by mathematics, but that it *is* mathematics, making us self-aware parts of a giant mathematical object. We'll see that this leads to a new and ultimate collection of parallel universes so vast and exotic

Figure 1.1: When we look at reality through the equations of physics, we find that they describe patterns and regularities. But to me, mathematics is more than a window on the outside world: in this book, I'm going to argue that our physical world not only is *described* by mathematics, but that it *is* mathematics: a mathematical structure, to be precise.

that all the above-mentioned bizarreness pales in comparison, forcing us to relinquish many of our most deeply ingrained notions of reality.

What's the Ultimate Question?

For as long as our human ancestors have walked the Earth, they've undoubtedly wondered what reality is all about, pondering deep existential questions. *Where did everything come from? How will it all end? How big is it all?* These questions are so captivating that virtually all human cultures across the globe have grappled with them and passed on answers from generation to generation, in the form of elaborate creation myths, legends and religious doctrines. As Figure 1.2 illustrates, these questions are so difficult that no global consensus has emerged on the answers. Instead of all cultures converging on a unique worldview that could potentially be the ultimate truth, their answers have differed greatly, and at least some of these differences appear to reflect their differences in lifestyle. For example, creation myths from ancient Egypt, where the Nile River kept the land fertile, all have our world emerge

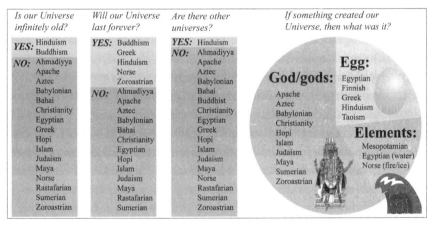

Figure 1.2: Many cosmological questions that we'll tackle in this book have fascinated thinkers throughout the ages, but no global consensus has emerged. The classification above is based on a 2011 presentation by MIT grad student David Hernandez for my cosmology class. Because such simplistic taxonomies are strictly impossible, they should be taken with a large grain of salt: many religions have multiple branches and interpretations, and some fall into multiple categories. For example, Hinduism contains aspects of all three creation options shown: according to one legend, both the creation god Brahma (depicted) and our Universe emerged from an egg, which in turn may have originated from water.

from water. In my native Sweden, on the other hand, where fire and ice used to strongly affect survival, Norse mythology held that life originated from (surprise!) fire and ice.

Other big questions tackled by ancient cultures are at least as radical. *What is real? Is there more to reality than meets the eye? Yes!* was Plato's answer over two millennia ago. In his famous cave analogy, he likened us to people who'd lived their entire lives shackled in a cave, facing a blank wall, watching the shadows cast by things passing behind them, and eventually coming to mistakenly believe that these shadows were the full reality. Plato argued that what we humans call our everyday reality is similarly just a limited and distorted representation of the true reality, and that we must free ourselves from our mental shackles to begin comprehending it.

If my life as a physicist has taught me anything at all, it's that Plato was right: modern physics has made abundantly clear that the ultimate nature of reality isn't what it seems. But if reality isn't what we thought it was, then what is it? What's the relation between the internal reality of our mind and the external reality? What's everything ultimately made of? How does it all work? Why? Is there a meaning to it all, and if so, what? As Douglas Adams put it in his sci-fi spoof *The Hitchhiker's Guide to the Galaxy:* "What's the answer to the ultimate question of life, the universe, and everything?"

Thinkers throughout the ages have offered a fascinating spectrum of responses to the question "What is reality?"—either attempting to answer it or attempting to dismiss it. Here are some examples (this list makes no claims to be complete, and not all alternatives are mutually exclusive).

This book (and indeed my scientific career) is my personal attempt to tackle this question. Part of the reason that thinkers have offered such a broad spectrum of answers is clearly that they've chosen to interpret the question in different ways, so I owe you an explanation of how I interpret it and how I approach it. The word *reality* can have many different connotations. I use it to mean the ultimate nature of the outside physical world that we're part of, and I'm fascinated by the quest to understand it better. So what's my approach?

One evening back in high school, I started reading Agatha Christie's detective novel *Death on the Nile*. Although I was painfully aware that my alarm clock would go off at seven a.m., I couldn't for the life of me

Some Responses to "What Is Reality?"	
The question has a meaningful answer.	Elementary particles in motion Earth, wind, fire, air and quintessence Atoms in motion Elementary particles in motion Strings in motion Quantum fields in curved spacetime M-theory (substitute your favorite capital letter . . .) A divine creation A social construct A neurophysiological construct A dream Information A simulation (à la *The Matrix*) A mathematical structure The Level IV Multiverse
The question lacks a meaningful answer.	There is a reality, but we humans can't fully know it: we have no access to what Immanuel Kant called "das Ding an sich." Reality is fundamentally unknowable. Not only don't we know it, but we couldn't express it if we did. Science is nothing but a story (postmodern answer by Jacques Derrida and others). Reality is all in our head (constructivist answer). Reality doesn't exist (solipsism).

put it down until the mystery had been solved, around four a.m. I've been inexorably drawn to detective stories ever since I was a kid, and when I was around twelve, I started a detective club with my classmates Andreas Bette, Matthias Bothner and Ola Hansson. We never captured any criminals, but the idea of solving mysteries captured my imagination. To me, the question "What is reality?" represents the ultimate detective story, and I consider myself incredibly fortunate to be able to spend so much of my time pursuing it. In the chapters ahead, I'll tell you about other occasions when my curiosity has kept me up in the wee hours of the morning, totally unable to stop reading until the mystery was resolved. Except that I wasn't reading a book, but what my hand was writing, and what I was writing was a trail of mathematical equations that I knew would ultimately lead me to an answer.

I'm a physicist, and I'm taking a physics approach to the mysteries of reality. To me, this means starting with great questions such as "How big is our Universe?" and "What's everything made of?" and treating them exactly like detective mysteries: combining clever observations and reasoning and persistently following each clue wherever it leads.

The Journey Begins

A physics approach? Isn't that a great way to turn something exciting into something boring? When the person sitting next to me on a plane asks what I do, I have two options. If I feel like talking, I'll say, "Astronomy," which infallibly triggers an interesting conversation.* If I don't, I'll say, "Physics," at which point they typically say something like "Oh, that was my worst subject in high school," and leave me alone for the rest of the flight.

Indeed, physics was also *my* least-favorite subject in high school. I still remember my very first physics class. With a monotonous and sedative voice, our teacher announced that we were going to learn about density. That density was mass divided by volume. So if the mass was blah and the volume was blah, then we could calculate that the density was blah blah. After that point, all I remember is a big blur. And that whenever his experiments failed, he'd blame humidity and say, "It worked this morning." And that some friends of mine couldn't figure out why their experiment wasn't working until they discovered that I'd mischievously attached a magnet underneath their oscilloscope. . . .

When the time came to apply for college, I decided against physics and other technical fields, and ended up at the Stockholm School of Economics, focusing on environmental issues. I wanted to do my small part to make our planet a better place, and felt that the main problem wasn't that we lacked technical solutions, but that we didn't properly use the technology we had. I figured that the best way to affect people's behavior was through their wallets, and was intrigued by the idea of creating economic incentives that aligned individual egoism with the common good. Alas, I soon grew disillusioned, concluding that economics was largely a form of intellectual prostitution where you got rewarded for saying what the powers that be wanted to hear. Whatever a politician wanted to do, he or she could find an economist as advisor who had argued for doing precisely that. Franklin D. Roosevelt wanted to increase government spending, so he listened to John Maynard Keynes,

---

* This conversation sometimes begins: "Oh, astrology! I'm a Virgo." When I've instead given the more precise answer "Cosmology," I've gotten answers such as "Oh, cosmetology!"—with follow-up questions about eyeliner and mascara.

whereas Ronald Reagan wanted to decrease government spending, so he listened to Milton Friedman.

Then my classmate Johan Oldhoff gave me the book that changed everything: *Surely You're Joking, Mr. Feynman!* I never got to meet Richard Feynman, but he's the reason I switched to physics. Although the book wasn't really about physics, dwelling more on topics such as how to pick locks and how to pick up women, I could read between the lines that this guy just loved physics. Which really intrigued me. If you see a mediocre-looking guy walking arm in arm with a gorgeous woman, you probably wonder if you're missing something. Presumably, she's seen some hidden quality in him. Suddenly I felt the same way about physics: what did Feynman see that I'd missed in high school?

I just had to solve this mystery, so I sat down with volume 1 of *The Feynman Lectures on Physics*, which I found in Dad's bookcase, and started reading: "If, in some cataclysm, all of scientific knowledge were to be destroyed, and only one sentence passed on to the next generation of creatures, what statement would contain the most information in the fewest words?"

Whoa—this guy was *nothing* like my high-school physics teacher! Feynman continued: "I believe it is that [. . .] all things are made of atoms—little particles that move around in perpetual motion, attracting each other when they are a little distance apart but repelling upon being squeezed into one another."

A lightbulb went off in my head. I read on and on and on, spell-bound. I felt like I was having a religious experience. I finally got it! I had the epiphany that explained what I'd been missing all along, and what Feynman had realized: physics is the ultimate intellectual adventure, the quest to understand the deepest mysteries of our Universe. Physics doesn't take something fascinating and make it boring. Rather, it helps us see more clearly, adding to the beauty and wonder of the world around us. When I bike to work in the fall, I see beauty in the trees tinged with red, orange and gold. But seeing these trees through the lens of physics reveals even more beauty, captured by the Feynman quote that opens this chapter. And the deeper I look, the more elegance I glimpse: we'll see in Chapter 3 how the trees ultimately come from stars, and we'll see in Chapter 8 how studying their building blocks suggests their existence in parallel universes.

At this time, I had a girlfriend studying physics at the Royal Institute of Technology, and her textbooks seemed so much more interesting than mine. Our relationship didn't last, but my love for physics did. Since college was free in Sweden, I enrolled in her university without telling the Stockholm School of Economics administrators about my secret double life. My detective investigation had officially begun, and this book is my report a quarter of a century later.

So what is reality, then? My goal with this audaciously titled chapter isn't to arrogantly try to sell you on an ultimate answer (although we'll explore intriguing possibilities in the last part of the book), but rather to invite you along on my personal journey of exploration, and to share with you my excitement and reflections about these mind-expanding mysteries. Like me, I think you'll conclude that whatever reality is, it's wildly different from what we once thought, and a fascinating enigma at the very heart of our everyday lives. I hope you will, like me, find that this places everyday problems such as parking tickets and heartaches in a refreshing perspective, making it easier to take them in stride and focus on enjoying life and its mysteries to the fullest.

When I first discussed my ideas for this book with John Brockman, now my book agent, he gave me clear marching orders: "I don't want a textbook—I want your book." So this book is a scientific autobiography of sorts: although it's more about physics than it's about me, it's certainly not your standard popular science book that attempts to survey physics in an objective way, reflecting the community consensus and giving equal space to all opposing viewpoints. Rather, it's my personal quest for the ultimate nature of reality, which I hope you'll enjoy seeing through my eyes. Together, we'll explore the clues that I personally find the most fascinating, and try to figure out what it all means.

We'll begin our journey by surveying how the whole context of the question "What is reality?" has been transformed by recent scientific breakthroughs, with physics shedding new light on our external reality from the largest (Chapters 2–6) to the smallest (Chapters 7–8) scales. In Part I of the book, we'll pursue the question "How big is our Universe?" and seek its ultimate conclusion by traveling out to ever-larger cosmic scales, exploring both our cosmic origins and two types of parallel universes, finding hints that space is in a sense mathematical. In Part II of the book, we'll relentlessly pursue the question "What's everything made of?" by journeying into the subatomic microcosm, examining a third kind of parallel universe and finding hints that the ultimate

How to read this book:	Science-curious reader	Hard-core reader of popular science Physi-cist		Chapter Title	Focus	Status
	1	1	1	What Is Reality?	Introduction	
Zooming Out (What is reality on the largest scales?)	2			Our Place in Space	How big is space?	
	3	skip	skip	Our Place in Time	History of our Universe	Mainstream
	4			Our Universe by Numbers	Precision cosmology	
	5	5		Our Cosmic Origins	Cosmological inflation	
Zooming In (What is reality on the smallest scales?)	6	6	6	Welcome to the Multiverse	Level I and II parallel universes	Controversial
	7	skip	skip	Cosmic Legos	Quantum mechanics	Mainstream
	8	8	8	The Level III Multiverse	Quantum parallel universes	Controversial
Stepping Back (Is reality math?)	9	9	9	Internal and External Reality	The role of consciousness	
	10	10	10	Physical and Mathematical Reality	The "reality is math" idea	Extremely Controversial
	11	11	11	Is Time an Illusion?	Making sense of it	
	12	12	12	The Level IV Multiverse	The ultimate multiverse	
	13	13	13	Life, Our Universe and Everything	Future of Universe and humanity	Controversial

Figure 1.3: How to read this book. If you've read lots of modern popular-science books and feel that you already understand curved space, our Big Bang, the cosmic microwave background, dark energy, quantum mechanics etc., then you may consider skipping Chapters 2, 3, 4 and 7 after reviewing the "Bottom Line" boxes that follow them, and if you're a professional physicist, you might consider skipping Chapter 5 as well. But many concepts that may sound familiar are startlingly subtle, and if you can't answer all of questions 1–16 in Chapter 2, I hope you'll learn from the early material as well and see how the later chapters logically build on it.

building blocks of matter are also in a sense mathematical. In Part III of the book, we'll take a step back and consider what all this might mean for the ultimate nature of reality. We'll begin by arguing that our failure to understand consciousness doesn't stand in the way of a complete understanding of the external physical reality. We'll then delve into my most radical and controversial idea: that the ultimate reality is purely mathematical, demoting familiar notions such as randomness, complexity, and even change to the status of illusions, and implying that there's a fourth and ultimate level of parallel universes. We'll wrap up our journey in Chapter 13 by returning home, exploring what this all means for the future prospects of life in our Universe, for us humans, and for you personally. You'll find our travel planner in Figure 1.3 with my reading tips. A fascinating journey awaits us. Let's begin!

THE BOTTOM LINE

- I feel that the most important lesson physics has taught us about the ultimate nature of reality is that, whatever it is, it's very different from how it seems.
- In Part I of this book, we'll zoom out and explore physical reality on the largest scales, from planets to stars, galaxies, superclusters, our Universe and two possible levels of parallel universes.
- In Part II of the book, we'll zoom in and explore physical reality on the smallest scales, from atoms to their even more fundamental building blocks, encountering a third level of parallel universes.
- In Part III, we'll take a step back and examine the ultimate nature of this strange physical reality, investigating the possibility that it's ultimately purely mathematical, specifically a mathematical structure that's part of a fourth and ultimate level of parallel universes.
- *Reality* means very different things to different people. I use the word to mean the ultimate nature of the outside physical world that we're part of, and ever since I was a kid, I've been inspired and fascinated by the quest to understand it better.
- This book is about my personal journey to explore the nature of reality—please join me!

Part One

ZOOMING OUT

2

Our Place in Space

Space . . . is big. Really big. You just won't believe how vastly hugely mind-bogglingly big it is.
 —Douglas Adams, in *The Hitchhiker's Guide to the Galaxy*

Cosmic Questions

He raises his hand, and I gesture to him that it's okay to ask his question. "Does space go on forever?" he asks.

My jaw drops. Wow. I've just finished a little astronomy presentation at Kids' Corner, my kids' after-school program in Winchester, and this extremely cute group of kindergartners is sitting on the floor, looking at me with big inquisitive eyes, awaiting a response. And this five-year-old boy just asked me a question I can't answer! Indeed, a question that nobody on our planet can answer. Yet it's not a hopelessly metaphysical question, but a serious scientific question for which theories I'll soon tell you about make definite predictions, and one on which ongoing experiments are shedding further light. In fact, I think it's a truly great question about the fundamental nature of our physical reality—as we'll see in Chapter 5, this question will lead us to two different kinds of parallel universes.

I'd been growing progressively more misanthropic over the years by following world news, but in just a few seconds, this kindergartner managed to give a major boost to my faith in the potential of humankind. If a five-year-old can say such profound things, then imagine what we grown-ups have the potential to accomplish together in the right circumstances! He also reminded me of the importance of good teaching. We're all born with curiosity, but at some point, school usually manages to knock that out of us. I feel that my main responsibility as a teacher isn't to convey facts, but to rekindle that lost enthusiasm for asking questions.

I love questions. Especially big ones. I feel so fortunate to be able to spend much of my time tackling interesting questions. That I can call this activity work and make a living from it is just luck beyond my wildest expectations. Here's my top-sixteen list of questions that I often get asked:

1. *How could space not be infinite?*
2. *How could an infinite space get created in a finite time?*
3. *What's our Universe expanding into?*
4. *Where in space did our Big Bang explosion happen?*
5. *Did our Big Bang happen at a single point?*
6. *If our Universe is only 14 billion years old, how can we see objects that are 30 billion light-years away?*
7. *Don't galaxies receding faster than the speed of light violate relativity theory?*
8. *Are galaxies really moving away from us, or is space just expanding?*
9. *Is the Milky Way expanding?*
10. *Do we have evidence for a Big Bang singularity?*
11. *Doesn't creation of the matter around us from almost nothing by inflation violate energy conservation?*
12. *What caused our Big Bang?*
13. *What came before our Big Bang?*
14. *What's the ultimate fate of our Universe?*
15. *What are dark matter and dark energy?*
16. *Are we insignificant?*

Let's tackle these questions together. We'll answer eleven of them in the next four chapters, and find interesting twists on the remaining five. But first, let's return to that kindergartner's question, which will form a central theme of this entire first part of the book: *Does space go on forever?*

How Big Is Space?

My dad once gave me the following advice: "If you have a tough question that you can't answer, first tackle a simpler question that you can't answer." In this spirit, let's begin by asking what the minimum size is that space must have without contradicting our observations. Figure 2.1 illustrates that the answer to this question has increased dramatically

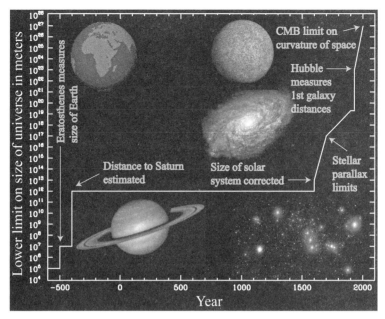

Figure 2.1: Our lower bound on the size of our Universe has kept growing, as we'll describe in this chapter. Note that the vertical scale is extreme, increasing tenfold with every tick mark.

over the centuries: we now know our space to be at least a billion trillion (10^{21}) times bigger than the largest distances our hunter-gatherer ancestors knew about—which was essentially how far they walked in a lifetime. Moreover, the figure shows that this expansion of our horizons wasn't a one-shot deal, but something that recurred repeatedly. Every time we humans have managed to zoom out and map our Universe on larger scales, we've discovered that everything we previously knew about was part of something greater. As illustrated in Figure 2.2, our homeland is part of a planet, which is part of a solar system, which is part of a galaxy, which is part of a cosmic pattern of galactic clustering, which is part of our observable Universe, which we'll argue is part of one or more levels of parallel universes.

Like an ostrich with its head in the sand, we humans have repeatedly assumed that all we could see was all that existed, hubristically imagining ourselves at the center of everything. In our quest to understand the cosmos, underestimation has thus been a persistent theme. However, the insights illustrated in Figure 2.1 reflect also a second theme, which I find inspiring: *we've repeatedly underestimated not only the size of our cosmos, but also the power of our human mind to understand it.* Our

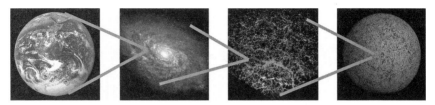

Figure 2.2: Every time we humans have managed to zoom out to larger scales, we've discovered that everything we knew was part of something greater: our homeland is part of a planet (left), which is part of a solar system, which is part of a galaxy (middle left), which is part of a cosmic pattern of galactic clustering (middle right), which is part of our observable Universe (right), which may be part of one or more levels of parallel universes.

cave-dwelling ancestors had just as big brains as we have, and since they didn't spend their evenings watching TV, I'm sure they asked questions like "What's all that stuff up there in the sky?" and "Where does it all come from?" They'd been told beautiful myths and stories, but little did they realize that they had it in them to actually figure out the answers to these questions for themselves. And that the secret lay not in learning to fly into space to examine the celestial objects, but in letting their human minds fly.

There's no better guarantee of failure than convincing yourself that success is impossible, and therefore never even trying. In hindsight, many of the great breakthroughs in physics could have happened earlier, because the necessary tools already existed. The ice-hockey equivalent would be missing an open goal because you mistakenly think your stick is broken. In the chapters ahead, I'm going to share with you striking examples of how such confidence failures were finally overcome by Isaac Newton, Alexander Friedmann, George Gamow and Hugh Everett. In that spirit, this quote by physics Nobel laureate Steven Weinberg resonates with me: "This is often the way it is in physics—our mistake is not that we take our theories too seriously, but that we do not take them seriously enough."

Let's first explore how to figure out the size of the Earth and the distances to the Moon, the Sun, stars and galaxies. I personally find it to be one of the most flavorful detective stories ever, and arguably the birth of modern science, so I'm eager to share it with you as an appetizer before the main course: the latest breakthroughs in cosmology. As you'll see, the first four examples involve nothing more complicated than some measurements of angles. They also illustrate the importance of letting

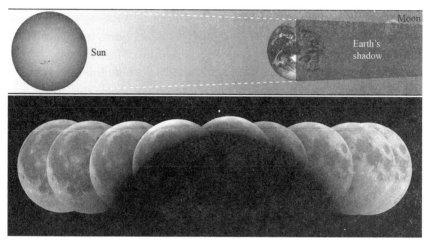

Figure 2.3: During a lunar eclipse, the Moon passes through the shadow cast by Earth (as seen above). Over two millennia ago, Aristarchos of Samos compared the size of the Moon to the size of the Earth's shadow during a lunar eclipse to correctly deduce that the Moon is about four times smaller than the Earth. (*Time-lapse photography by Anthony Ayiomamitis*)

yourself be puzzled by seemingly everyday observations, since they may turn out to be crucial clues.

The Size of Earth

As soon as sailing caught on, people noticed that when ships departed over the horizon, their hulls disappeared before their sails. This gave them the idea that the surface of the ocean was curved and that Earth was spherical, just as the Sun and Moon appeared to be. Ancient Greeks also found direct evidence of this by noticing that Earth cast a rounded shadow on the Moon during a lunar eclipse, as you can see in Figure 2.3. Although it's easy to estimate the size of Earth from the ship-sail business.* Eratosthenes obtained a much more accurate measurement over 2,200 years ago by making clever use of angles. He knew that the Sun was straight overhead in the Egyptian city of Syene at noon on the summer solstice, but that it was 7.2 degrees south of straight overhead in Alexandria, located 794 kilometers farther north. He therefore concluded that traveling 794 kilometers corresponded to going 7.2 degrees

* Earth's radius is approximately $d^2/2h$, where d is the greatest distance at which you can see a sail of height h from sea level.

out of the 360 degrees all around Earth's circumference, so that the circumference must be about 794 km × 360°/7.2° ≈ 39,700 km, which is remarkably close to the modern value of 40,000 km.

Amusingly, Christopher Columbus totally bungled this by relying on subsequent less-accurate calculations and confusing Arabic miles with Italian miles, concluding that he needed to sail only 3,700 km to reach the Orient when the true value was 19,600 km. He clearly wouldn't have gotten his trip funded if he'd done his math right, and he clearly wouldn't have survived if America hadn't existed, so sometimes being lucky is more important than being right.

Distance to the Moon

Eclipses have inspired fear, awe and myths throughout the ages. Indeed, while stranded on Jamaica, Columbus managed to intimidate natives by predicting the lunar eclipse of February 29, 1504. However, lunar eclipses also reveal a beautiful clue to the size of our cosmos. Over two millennia ago, Aristarchos of Samos noticed what you can see for yourself in Figure 2.3: when Earth gets between the Sun and the Moon and causes a lunar eclipse, the shadow that Earth casts on the Moon has a curved edge—and Earth's round shadow is a few times larger than the Moon. Aristarchos also realized that this shadow is slightly smaller than Earth itself, because Earth is smaller than the Sun, but correctly accounted for this complication and concluded that the Moon is about 3.7 times smaller than Earth. Since Erathostenes had already figured out the size of Earth, Aristarchos simply divided it by 3.7 and got the size of the Moon! To me, this was the moment when our human imagination finally got off the ground and started conquering space. Countless people before Aristarchos had looked at the Moon and wondered how big it was, but he was the first to figure it out. And he managed to do it with mental power rather than rocket power.

One scientific breakthrough often enables another, and in this case, the size of the Moon immediately revealed its distance. Please hold your hand up at arm's length and check which things around you can be blocked from view by your pinkie. Your little finger covers an angle of about one degree, which is about double what you need to cover the Moon—make sure to check this for yourself the next time you do some lunar observing. For an object to cover half a degree, its distance from you needs to be about 115 times its size, so if you're looking out your airplane window and can cover a 50-meter (Olympic-size) swimming pool with half your pinkie, you'll know that your altitude is 115 × 50 m = 6 km. In the exact same way, Aristarchos calculated the distance to the Moon to be 115 times its size, which came out to be about 30 times the diameter of Earth.

Distance to the Sun and the Planets

So what about the Sun? Try blocking it with your pinky and you'll see that it covers about the same angle as the Moon, about half a degree. It's clearly farther away than the Moon, since the Moon (just barely) blocks it from view during a total solar eclipse, but how much farther away? That depends on its size: for example, if it were three times the size of the Moon, it would need to be three times as far away to cover the same angle.

Aristarchos of Samos was on a roll back in his time, and cleverly answered this question as well. He realized that the Sun, the Moon and Earth formed the three corners of a right triangle during "quarter Moon," when we see exactly half the Earth-facing lunar surface illuminated by sunlight (see Figure 2.4), and he estimated that the angle between the Moon and the Sun was about 87 degrees at this time. So he

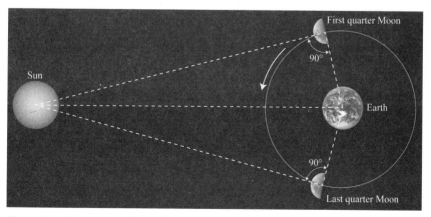

Figure 2.4: By measuring the angle between the quarter moon and the Sun, Aristarchos was able to estimate our distance from the Sun. (This drawing isn't to scale; the Sun is over one hundred times larger than Earth and about four hundred times as distant as the Moon.)

knew both the shape of the triangle and the length of the Earth–Moon edge, and was able to use trigonometry to figure out the length of the Earth–Sun edge, that is, the distance between the Earth and the Sun. His conclusion was that the Sun was about twenty times farther away than the Moon and therefore twenty times bigger than the Moon. In other words, the Sun was *huge*: over five times bigger than Earth in diameter. This insight prompted Aristarchos to propose the heliocentric hypothesis long before Nicolaus Copernicus: he felt that it made more sense for Earth to be orbiting the much larger Sun than vice versa.

This tale is both inspiring and cautionary, teaching us about both the importance of cleverness and the importance of quantifying uncertainties in our measurements. The ancient Greeks were less adept at the second part, and Aristarchos was unfortunately no exception. It turned out to be quite difficult to tell precisely when the Moon was 50% illuminated, and the correct Moon–Sun angle at that time isn't 87 degrees but about 89.85 degrees, extremely close to a right angle. This makes the triangle in Figure 2.4 very long and skinny: in fact, the Sun is almost 20 times farther away than Aristarchos estimated, and about 109 times larger than Earth in diameter—so you could fit over a million Earths inside the volume of the Sun. Unfortunately, this glaring mistake wasn't corrected until almost two thousand years later, so when Copernicus came along and figured out the size and shape of our Solar System with further geometric ingenuity, he got the shapes and relative sizes right for all the planetary orbits, but the overall scale of his Solar System model was about twenty times too small—that's like confusing a real house with a doll house.

Distance to the Stars

But what about the stars? How far away are they? And what are they? Personally, I think this is one of the greatest "cold case" detective stories ever. Figuring out the distances to the Moon and the Sun was impressive, but at least there was some information to use as clues: they change their sky positions in interesting ways, and they have shapes and angular sizes that we can measure. But a star seems totally hopeless! It looks like a faint white dot. You look at it more carefully and see . . . still just a faint white dot, with no discernible shape or size, merely a point of light. And the stars never seem to move across the sky, except for the apparent overall rotation of whole patterns of stars, which we know to be a mere illusion caused by the fact that Earth is rotating.

Some ancients speculated that the stars were small holes in a black sphere through which distant light shone through. The Italian astronomer Giordano Bruno suggested that they were instead objects like our Sun, just much farther away, perhaps with their own planets and civilizations—this didn't go down too well with the Catholic Church, which had him burned at the stake in 1600.

In 1608, a sudden glimmer of hope: the telescope is invented! Galileo Galilei quickly improves the design, looks at stars through his ever-improving telescopes, and sees . . . just white dots again. Back to square one. I have fond memories of playing "Twinkle, Twinkle, Little Star" on my grandma Signe's piano as a kid. As recently as 1806, when this song was first published, the line "How I wonder what you are" *still* resonated with many people, and nobody could honestly claim to really know the answers.

If stars are really distant suns as Bruno suggested, then they must be dramatically farther away than our Sun to look so faint. But how much farther? That depends on how luminous they really are, which we'd also like to know. Thirty-two years after the song was published, the German mathematician and astronomer Friedrich Bessel finally achieved a breakthrough in this detective case. Please hold your thumb up at arm's length and alternate closing your left and right eyes a few times. Do you see how your thumb appears to jump left and right by a certain angle relative to background objects? Now move your thumb closer to your eyes, and you'll see this jump angle growing. Astronomers call this jump angle the *parallax*, and you can clearly use it to figure out how far away your thumb is. In fact, you needn't worry about doing the math, since your brain does it for you so effortlessly that you don't even notice—this fact that your two eyes measure different angles to objects depending on their distance is the very essence of how your brain's depth-perception system works to provide you with three-dimensional vision.

If your eyes were farther apart, you'd have better depth perception at large distances. In astronomy, we can use this same parallax trick and pretend that we're giants with eyes 300 billion meters apart, which is the diameter of Earth's orbit around the Sun. We can do this because we can compare telescopic photographs taken six months apart, when Earth is on opposite sides of the Sun. Doing this, Bessel noticed that while most stars appeared in the exact same positions in both of his pictures, one particular star didn't: a star that went by the obscure name 61 Cygni. Instead, it had moved by a tiny angle, revealing its distance to be almost a million times that to the Sun—a distance so huge that it would take eleven years for its starlight to reach us, whereas sunlight gets here in just eight minutes.

Before long, many more stars had their parallax measured, so many of these mysterious white dots now had distances! If you watch a car drive away at night, the brightness of its taillights drops as the inverse square of its distance (twice as far means four times dimmer). Now that Bessel knew the distance to 61 Cygni, he used this inverse-square law to figure out how luminous it was. His answer was a luminosity in the same ballpark as that of the Sun, suggesting that the late Giordano Bruno had been right after all!

Around the same time, there was a second major break in the case using a totally different approach. In 1814, the German optician Joseph von Fraunhofer invented a device called a *spectrograph*, which let him separate white light into the rainbow of colors from which it's made up, and measure them in exquisite detail. He discovered mysterious dark lines in the rainbow (see Figure 2.5), and that the detailed positions of these lines within the spectrum of colors depended on what the light source was made of, constituting a type of spectral fingerprint. During the following decades, such spectra were measured and cataloged for many common substances. You can use this information to pull a great party trick, impressing your friends by telling them what's glowing in their lantern just from analyzing its light, without ever going near

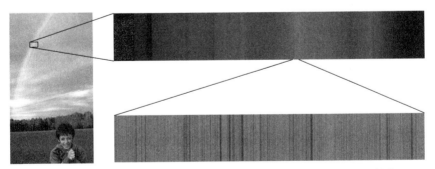

Figure 2.5: The rainbow spotted by my son Alexander leads not to a pot of gold, but to a goldmine of information about how atoms and stars work. As we'll explore in Chapter 7, the relative intensities of the various colors are explained by light being made of particles (photons), and the positions and strengths of the many dark lines can all be calculated from the Schrödinger equation of quantum mechanics.

it. Sensationally, the spectrum of sunlight revealed that the Sun, this mysterious fiery orb in the sky, contained elements well known from Earth, such as hydrogen. Moreover, when starlight from a telescope was observed through a spectroscope, it revealed that stars are made of roughly the same mixture of gases as the Sun! This clinched it in favor of Bruno: stars are distant suns, similar in both their energy output and contents. So in a brief few decades, stars had gone from being inscrutable white dots to being giant balls of hot gas whose chemical composition we could measure.

A spectrum is a goldmine of astronomical information, and every time you think you've milked it for all it's worth, you find more clues encoded in it. For starters, a spectrum lets you measure the temperature of an object without touching it with a thermometer. You know without touching that a piece of metal glowing white is hotter than one glowing red, and similarly that a whitish star is hotter than a reddish star; with a spectrograph, you can determine the temperatures quite accurately. As a surprise bonus, this information now reveals the star's size, much like figuring out a word in a crossword puzzle can reveal another word. The trick is that the temperature tells you how much light emerges from each square meter of the star's surface. Since you can calculate the total amount of light radiated by the star (from its distance and apparent brightness), you now know how many square meters of surface area the star must have, and therefore how big it is.

As if this weren't enough, the spectrum of a star also contains hidden clues about its motion, which slightly shifts the frequency (color) of the light through the so-called Doppler effect, the effect that makes the pitch fall in the *vroooooooom* of a passing car: the frequency is higher when the car is moving toward you, then lower as it moves away from you. Unlike our Sun, most stars are in stable pair relationships with a companion star, and the two partners dance around each other in a regular orbit. We can often detect this dance through the Doppler effect, which causes the spectral lines of the stars to move back and forth once per orbit. The magnitude of the shift reveals the speed of motion, and by looking at the two stars, we can sometimes measure how far apart they are. Combining this information allows us to pull another major stunt: we can weigh the stars without putting them on a gigantic bathroom scale, using Newton's laws of motion and gravitation to calculate how massive they must be to have the observed orbits. In some cases, such Doppler shifts have also revealed that planets orbit a star. If the planet moves in front of the star, the slight dip in the star's brightness reveals the size of the planet, and the slight change in spectral lines can reveal whether the planet has an atmosphere and what it's made of. And spectra are the gift that just keeps giving. For example, by measuring the width of spectral lines for a star of a given temperature, we can measure its gas pressure. And by measuring the extent to which spectral lines split into two or more nearby lines, we can measure how strong the magnetism is at the star's surface.

In conclusion, the only information we have about stars is in their faint light that reaches us, but through clever detective work, we can decode this light into information about their distance, size, mass, composition, temperature, pressure, magnetism and any solar system they may host. That our human minds have deduced all this from seemingly inscrutable white dots is a feat that I think would have made even the great detectives Sherlock Holmes and Hercule Poirot proud!

Distance to the Galaxies

When my grandma Signe passed away at age 102, I spent some time reflecting on her life, and it struck me that she grew up in a different universe. When she went off to college, our known Universe was simply our Solar System and a swarm of stars around it. She and her friends probably thought of these stars as incredibly distant, with light taking several years to arrive from the closest ones and thousands of years from the farthest ones known. All of which we nowadays consider merely our cosmic backyard.

If there were astronomers at her college, they'd have argued about the so-called nebulae, diffuse cloudlike objects in the night sky, some with beautiful spiral shapes like those in van Gogh's famous painting *Starry Night*. What were these things? Many astronomers dismissed them as boring gas clouds between the stars, but some had a more radical idea: that they were "island universes," which we today call *galaxies*: enormous groups of stars so far away that they couldn't be seen individually with our telescopes, appearing instead as a nebulous haze. To settle this controversy, astronomers needed to measure the distance to some nebulae. But how?

The parallax technique, which had worked so well for nearby stars, failed for the nebulae: they were so far away that their parallax angles were too small to detect. How else can you measure large distances? If you look at a distant lightbulb with a telescope and notice that it has "100 watts" printed on it, you're all set: you simply use the inverse-square law to calculate how far away it must be to look as bright as it does. Astronomers call such useful objects of known luminosity *standard candles*. Using the above-mentioned detective methods, astronomers had unfortunately discovered that stars are anything but standard, some shining a million times more brightly than the Sun and others a thousandfold more faintly. However, if you could observe a star and see that it had "4 $\times 10^{26}$ watts" written on it (that would be the correct label for our Sun), you'd have your standard candle and could calculate its distance just as for the lightbulb. Fortunately, nature has provided us with a particular type of stars this helpful, called *Cepheid variables*. Their luminosity oscillates over time as they pulsate in size, and Harvard astronomer Henrietta Swan Leavitt discovered in 1912 that their pulsation rate acts like a watt meter: the more days there are between successive pulses, the more watts of light are radiated.

These Cepheid stars also have the advantage of being bright enough to see at vast distances (some can shine 100,000 times brighter than our Sun), and the American astronomer Edwin Hubble discovered several of them in the so-called Andromeda nebula—a Moon-size haze that you can see with your naked eye if you're far from city lights. Using the recently completed Hooker telescope in California (its 2.5-meter mirror was the largest in the world), he measured their pulsation rates, used Leavitt's formula to figure out how luminous they were, compared that with how bright they appeared, and calculated their distances. When he announced his answer at a 1925 conference, jaws dropped: he argued that Andromeda was a galaxy about a million light-years away, a thousandfold farther than most stars my grandma saw in her night sky! We now know that the Andromeda galaxy is even more distant than Hubble estimated, about three million light-years from us, so Hubble inadvertently continued the tradition of accidental underestimation from Aristarchos and Copernicus.

In the years that followed, Hubble and other astronomers went on to discover ever more distant galaxies, expanding our horizons from millions to billions of light-years away from us, and we'll push into the trillions and beyond in Chapter 5.

What Is Space?

So, as that kindergartner asked: does space go on forever? We can approach this question in two ways: observationally and theoretically. So far in this chapter, we've done the former, exploring how clever measurements have gradually revealed ever more distant regions of space, with no end in sight. However, lots of progress has been made on the theoretical front as well. First of all, how could space *not* go on forever? As I discussed with those kindergartners, it would be pretty weird if we reached a sign like the one in Figure 2.6, warning that we'd reached the end of space. I remember thinking about this when I was a kid: what would there be beyond the sign? To me, worrying about reaching the end of space sounded as silly as ancient seafarers worrying about falling off the end of the Earth. I therefore concluded that space simply had to go on forever and be infinite, based on pure logic. Indeed, using logical reasoning back in ancient Greece, Euclid realized that geometry was really mathematics, and that infinite 3-D space could be described with the same rigor as other mathematical structures such as sets of numbers. He developed this beautiful mathematical theory of infinite 3-D space and its geometric properties, and this was widely viewed as the only logically possible way that our physical space could be.

In the 1800s, however, the mathematicians Carl Friedrich Gauss, János Bolyai and Nikolai Lobachevsky all discovered that there were other logical possibilities for uniform 3-D space, and Bolyai excitedly wrote to his father: "Out of nothing I have created a strange new uni-

Figure 2.6: It's hard to imagine how space could be finite. If it could end, then what would lie beyond?

verse." These new spaces obey different rules: for example, they no longer have to be infinite like the space Euclid envisioned, and the angles in a triangle no longer have to add up to 180 degrees as Euclid's formula stipulates. Imagine drawing a triangle on each of the 2-D surfaces of the 3-D shapes in Figure 2.7: its three angles will add up to more than 180 degrees for the sphere (left), exactly 180 degrees for the cylinder (middle), and less than 180 degrees for the hyperboloid (right). Moreover, the 2-D surface of the sphere is finite even though it lacks any sort of edge.

This example shows that surfaces can break Euclid's geometry rules if they aren't flat. However, Gauss and the others had a more radical insight: a space can be curved all by itself, even if it isn't the surface of anything! Suppose you're a blind ant and want to figure out which one of the three surfaces in Figure 2.7 you're walking around on. You feel like you're effectively living in a 2-D space, because you have no access to the third dimension (away from your surface), but this won't thwart your detective work: you can still define a straight line (as the shortest path between two points), so you simply sum the three angles of a triangle. For example, if you get 270 degrees, you exclaim: "Aha! It's more than 180 degrees so I'm on the sphere!" To further impress your ant friends, you can even figure out how far you'll need to walk in a straight line before returning to where you started. In other words, all the usual geometry business of points, lines, angles, curvature and so on can be rigorously defined by referring only to what's in your 2-D space, without making any reference to a third dimension. This means that mathematicians can rigorously define a curved 2-D surface even if no third dimension exists: a curved 2-D space all by itself, which isn't the surface of anything.

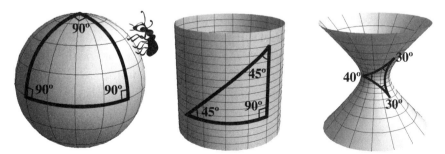

Figure 2.7: If you draw triangles on these surfaces, their angles will add up to more than 180 degrees (left), exactly 180 degrees (middle) and less than 180 degrees (right), respectively. Einstein taught us that these three options are possible for triangles in our 3-D physical space as well.

To most people, this mathematical discovery of non-Euclidean spaces probably seemed like little more than esoteric mathematical abstraction, of no practical relevance to our physical world. But then Einstein came along with his theory of general relativity, which effectively says: "We're the ants!" Einstein's theory allows our 3-D space to be curved—even without it having any hidden fourth dimension for it to curve within. So the question of what kind of space we inhabit *can't* be settled from pure logic alone, as some Euclid fans had hoped. It can only be resolved by performing measurements—such as making a huge triangle in space (with light rays as edges, say) and checking whether the angles add up to 180 degrees. In Chapter 4, I'll tell you about how my colleagues and I have had fun doing precisely this; the answer turns out to be about 180 degrees for universe-sized triangles, but significantly more than 180 degrees if a neutron star or a black hole fills up much of the triangle, so the shape of our physical space is more complicated than the three simple options illustrated in Figure 2.7.

Returning to that kindergartner's question, we see that Einstein's theory allows space to be finite in a way that isn't silly as in Figure 2.6: it can be finite by being curved. For example, if our 3-D space is curved like the surface of a 4-D hypersphere, then if we could travel as far as we wanted in a straight line, we'd eventually return home from the opposite direction. We wouldn't fall off the edge of our 3-D space because it has no edge, just as the ant in Figure 2.7 encounters no edge when crawling around the sphere.

Indeed, Einstein allows our 3-D space to be finite even if it isn't curved! The cylinder in Figure 2.7 is flat rather than curved in the mathematical sense: if you draw a triangle on a paper cylinder, its angles will sum to 180 degrees. To see this, simply cut the triangle out with a pair of scissors, and note that you can lay it flat on a table; you couldn't do this with a paper sphere or hyperboloid without the paper tearing or crumpling. However, although the cylinder in Figure 2.7 therefore looks flat to an ant walking on a small patch of it, the cylinder nonetheless connects back on itself: the ant can return home after walking in a horizontal straight line. Mathematicians call the connectedness of a space its *topology*. They've defined flat spaces that connect back on themselves in *all* their dimensions, and call such a space a *torus*. A 2-D torus has the same topology as the surface of a bagel or a traditional donut (the kind with a hole in it). Einstein allows the possibility that the physical space we inhabit is a 3-D torus, in which case it's both flat and finite. Or it could be infinite.

In summary, the space we live in might go on forever and it might not—both possibilities are perfectly reasonable according the best theory we have for the nature of space, Einstein's general relativity. So which way is it? We'll return to this fascinating question in Chapters 4 and 5, finding evidence that space is truly infinite after all. But our pursuit of the kindergartner's deep question raises another one: what *is* space, really? Although we all start our lives thinking about space as something *physical*, forming the very fabric of our material world, we've now seen how mathematicians talk of spaces as being *mathematical* things. To them, studying space is the same as studying geometry, and geometry is just part of mathematics. One could indeed argue that space is a mathematical object, in the sense that its only intrinsic properties are mathematical properties—properties such as dimensionality, curvature and topology. We'll push this argument much further in Chapter 10, arguing that in a well-defined sense, our entire physical reality is a purely mathematical object.

We've spent this chapter exploring our place in space, revealing a vastly larger Universe than our ancestors were aware of. To really understand what's going on at the greatest distances we can observe with our telescopes, however, it's not enough to explore only our place in space. We also need to explore our place in time. That's our battle cry for the next chapter.

THE BOTTOM LINE

- Over and over again, we humans have realized that our physical reality is vastly larger than we'd imagined, and that everything we knew of was part of an even grander structure: a planet, a solar system, a galaxy, a galaxy supercluster, etc.
- Einstein's theory of general relativity allows for the possibility that space goes on forever.
- It also allows the alternative option where space is finite without having an end, so that if you could travel far and fast enough, you'd return home from the opposite direction.
- The very fabric of our physical world, space itself, could be a purely mathematical object in the sense that its only intrinsic properties are mathematical properties—numbers such as dimensionality, curvature and topology.

3

Our Place in Time

Real knowledge is to know the extent of one's ignorance.

—Confucius

The highest form of ignorance is when you reject something you don't know anything about.

—Wayne Dyer

Where does our Solar System come from? My son Philip got into a heated discussion about this question when he was in second grade, which went something like this:

"I think it was made by God," a girl in his class said.

"But my dad said it was made by a giant molecular cloud," Philip interjected.

"But where did the giant molecular cloud come from?" another boy asked.

"Maybe God made the giant molecular cloud, and then the giant molecular cloud made our Solar System," said the girl.

I bet that as long as people have walked the Earth, they've gazed into the night sky and wondered where everything comes from. Just as in times past, there are things we know and things we don't. We know lots about here and now, and also quite a bit about events close in space and time, such as what's right behind us and what we ate for breakfast. Farther away and longer ago, we eventually hit the frontier of our knowledge, where our ignorance begins. In the last chapter, we saw how human ingenuity gradually pushed this knowledge frontier outward in *space*, expanding our realm of the known to incorporate our entire planet, our Solar System, our Galaxy, and even billions of light-years of space in all directions. Let's now launch a second intellectual expedition, and explore how we humans have gradually pushed this frontier backward in *time*.

Why doesn't the Moon fall down? The answer to this question triggered our first push.

Where Did Our Solar System Come From?

As recently as four hundred years ago, this question still seemed rather hopeless. We just saw how ingenious detective work revealed the locations of the key parts visible to the naked eye: the Sun, the Moon, Mercury, Venus, Mars, Saturn and Jupiter. Diligent sleuthing by Nicolaus Copernicus, Tycho Brahe, Johannes Kepler and others also revealed the motions of these objects: our Solar System was found to be reminiscent of a clockwork, with its parts moving in precise orbits over and over again, seemingly forever. There was no indication whatsoever that the clockwork would stop one day, or that it had started at any particular time in the past. But was it really eternal? If not, where did it come from? We were still clueless.

For the man-made clockworks for sale at the time, the laws that governed the motion of their cogwheels, springs and other parts were so well understood that one could make predictions about both the future and the past. One could predict that a clock would keep ticking at a steady rate, and also that, because of friction, it would eventually stop unless it was wound up. By studying it carefully, you might conclude that it must have been wound up within the last month, say. If there were similarly precise laws that described and explained celestial motions, then might they, too, involve some frictionlike effects that would eventually alter our Solar System, and that might also give clues to when and how it formed?

The answer seemed to be a resounding *no*. Down on the ground, we'd developed a fairly good understanding of how things move through space, from hurled stones to rocks launched by Roman catapults to iron balls fired from cannons. But whatever laws governed heavenly objects seemed to be different from the laws governing things down here on the ground. For example, what about the Moon? If it's some kind of giant rock in the sky, why doesn't it fall down like ordinary rocks do? The classic answer was that the Moon was a heavenly object, and heavenly objects simply play by different rules. Like being immune to gravity and not falling down. Some went further and offered an explanation: heavenly objects are this way because they're perfect. They have perfectly spherical shapes because the sphere is the perfect shape; they move in circular orbits because circles are also perfect; and falling down would be about as far from perfect as it gets. On Earth, imperfection abounds: friction slows things down, fires burn out and people die. In the heavens, on the other hand, the motions appear frictionless, the Sun doesn't burn out and there's no end in sight.

This perfect reputation of the heavens didn't hold up to closer scrutiny, however. By analyzing the measurements of Tycho Brahe, Johannes Kepler established that planetary motions weren't circles but ellipses, which are elongated and arguably less-perfect versions of circles. Through his telescopes, Galileo saw that the Sun had its perfection tarnished by ugly black spots. And that the Moon wasn't a perfect sphere but what looked like a *place*, complete with mountains and giant craters. So why didn't it fall down?

Isaac Newton finally answered this question by exploring an idea that was as simple as it was radical: that heavenly objects obey the *same* laws as objects here on Earth. Sure, the Moon doesn't fall down like a dropped rock, but might it be possible to throw an ordinary rock in such a way that it doesn't fall down either? Newton knew that Earth rocks fall toward Earth rather than toward the much more massive Sun, and concluded that this must be because the Sun was much farther away and the gravitational attraction of an object weakens with distance. So could you hurl a rock upward so fast that it escaped Earth's gravitational pull before this pull had time to reverse the rock's motion? Newton personally couldn't do it, but realized that a hypothetical supercannon should do the trick, provided that it could give the rock enough speed. As you can see in Figure 3.1, this means that the fate of a horizontally fired cannon ball depends on its speed: it crashes into the ground only if its speed is below some magic value. If you keep firing balls with ever higher speeds, they'll travel farther and farther before landing, until you reach the magical speed where they keep their height over the ground exactly constant and never land, merely orbiting Earth in a circle—just like the Moon! Since he knew the strength of gravity near Earth's surface from experiments with falling rocks, apples, etc., he was able to calculate what this magic speed was: a roaring 7.9 kilometers per second. Assuming that the Moon really was obeying the same laws as a cannon ball, he could similarly predict what speed it needed to have to be in a circular orbit—all that was missing was a rule for how much

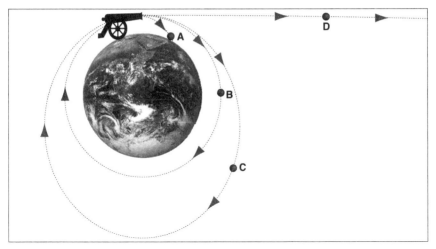

Figure 3.1: A cannon ball (D) fired faster than 11.2 kilometers per second escapes from Earth never to return (ignoring air resistance). If fired slightly more slowly (C), it instead enters an elliptical orbit around Earth. If fired horizontally at 7.9 kilometers per second (B), its orbit will be perfectly circular, and if fired at lower speeds (A), it eventually crashes into the ground.

weaker Earth's gravity was out there where the Moon was. Moreover, since the Moon took one month to travel around a circle whose circumference Aristarchos had figured out, Newton already knew its speed: about 1 kilometer per second, the same as for an M16 rifle bullet. Now he made a remarkable discovery: if he assumed that the force of gravity weakened like the inverse square of the distance from the center of Earth, then this magical speed that would give the Moon a circular orbit exactly matched its measured speed! He had discovered the law of gravity and found it to be universal, applying not merely here on Earth, but in the heavens as well.

Suddenly, the pieces of the puzzle started falling into place. By combining his law of gravity with mathematical laws of motion he formulated, Newton was able to explain not merely the motion of the Moon, but also the motions of the planets around the Sun: Newton was even able to mathematically derive the fact that general orbits are ellipses rather than circles, which to Kepler had been just a mysterious, unexplained fact.

Like most great breakthroughs in physics, Newton's laws answered way more questions than those that prompted the discovery. For example, they explained tides: the gravitational pull from the Moon and the Sun is greater on the seawater closer to them, causing water to slosh around as Earth rotates. Newton's laws also showed that the total amount of energy is conserved (in physics, we use the word *conserved* to mean "preserved" and "unchanging"), so if energy appears somewhere, it can't have been created from nothing, but must have come from somewhere else. Tides dissipate lots of energy (some of which can be recovered by tidal power plants), but where is all this energy coming from? In large part from Earth's rotation, which is slowed down by tidal friction: if you ever feel that there aren't enough hours in a day, just wait 200 million years, and days will be twenty-five hours long!

This means that friction affects even planetary motion, which kills the idea of an eternal solar system: Earth must have spun faster in the past, and you can calculate that the Earth-Moon system in its present form can't be more than 4 to 5 billion years old, or else Earth would once have spun so fast that centrifugal forces would have torn it apart. Finally, a first clue to the origin of our Solar System: we have an estimate for the time of the crime!

Newton's breakthrough empowered our human minds to conquer space: he showed that we could first discover physical laws by making experiments down here on the ground, then extrapolate these laws to explain what was happening in the heavens. Although Newton first applied this idea only to motion and gravity, the concept spread like wildfire and was gradually applied to other topics such as light, gases, liquids, solids, electricity and magnetism. People boldly extrapolated not only to the macrocosmos of space, but also to the microcosmos, finding that many properties of gases and other substances could be explained by applying Newton's laws of motion to the atoms that they were made of. The scientific revolution had begun. It ushered in both the Industrial Revolution and the information age. This progress in turn enabled us to create powerful computers that could help further advance science, solving our equations of physics and calculating answers to many interesting physics questions that had previously stumped us.

We can make use of the laws of physics in several different ways. Often we wish to use knowledge of the present to predict the future, as with weather forecasts. However, the equations can be solved equally well in reverse, using knowledge of the present to reveal the past—such as reconstructing the exact details of the eclipse Columbus witnessed on Jamaica. A third is to imagine a hypothetical situation and use our physics equations to calculate how it will change over time—as when we simulate the launch of a rocket to Mars and figure out whether it will arrive at the desired destination. This third approach has produced new clues about the origin of our Solar System.

Imagine a large cloud of gas in outer space: what will happen to it over time? The laws of physics predict a battle between two forces that will seal its fate: its gravity will try to crush it while its pressure will try to blow it apart. If gravity starts gaining the upper hand, compressing the cloud, it will get hotter (this is why my bike pump heats up with use), which in turn boosts its pressure, halting gravity's advance. The cloud can remain stable for a long time while gravity and pressure balance each other out, but this uneasy truce is eventually upset. Because it's hot, the gas cloud glows, radiating away some of the heat energy that gave it pressure. This allows gravity to compress the cloud further, and so on. By plugging the laws of gravity and gas physics into our computers, we can simulate this hypothetical battle in detail to see what happens. Eventually, the densest part of the cloud gets so hot and dense that it turns into a fusion reactor: hydrogen atoms are fused into helium, while intense gravity prevents it all from blowing apart. A star has been born. The outer parts of the nascent star are hot enough to shine intensely, and this starlight begins to blow away the rest of the gas cloud, bringing the newborn star into sight of our telescopes.

Rewind. Replay. As the gas cloud gradually contracts, any slight rotation of the cloud gets amplified, just as a figure skater spins faster when she pulls her arms closer to her body. The centrifugal forces from this ever-faster rotation prevents gravity from crushing the gas cloud down to a point—instead, it's crushed into a pizza shape, just as when the pizza baker near my old elementary school spun his dough to flatten it out. The main ingredients of all such cosmic pizzas are hydrogen and helium gas, but if the ingredient list also contains heavier atoms such as carbon, oxygen, and silicon, then while the center of this gas pizza forms a star, the outer parts may clump into other colder objects, *planets*, which become revealed once the newborn star blows away the rest of the pizza dough. Since all the spin (or *angular momentum*, as we physicists call it) comes from the rotation of the original cloud, it's no surprise that all planets in our Solar System are orbiting around the Sun in the same direction (counterclockwise if you're looking down at the North Pole), which is also the same direction that the Sun itself rotates roughly once per month.

This explanation of our Solar System's origins is now supported not only by theoretical calculations, but also by telescope observations of many other solar systems "caught in the act" of the birth process in various stages. Our Galaxy contains vast numbers of giant molecular clouds, gas clouds containing molecules that help them radiate away their heat, cool and contract, and we can see new stars being born in many of them. In some cases, we can even see baby stars with their pizzalike protoplanetary discs of gas still largely intact around them. The recent discovery of vast numbers of solar systems around other stars has given astronomers a wealth of new clues with which to refine our understanding of how our Solar System formed.

If this birth process is *what* happened to form our Solar System, then *when* exactly did it happen? Just over a century ago, it was still widely believed that the Sun may have formed as recently as 20 million years ago, because if you waited much longer, the loss of energy radiated away as sunshine would have caused gravity to compress it to a much smaller size than we observe. Similarly, it was calculated that if one waited much longer than that, most of Earth's inner heat (manifested as volcanoes and geothermal vents) would cool away.

The mystery of what keeps the Sun warm wasn't solved until the 1930s when nuclear fusion was discovered. But before then, the 1896 discovery of radioactivity demolished the old estimates of Earth's age and also provided a great method for making better ones. The most common isotope of uranium atoms spontaneously decays into thorium and other lighter atoms at such a rate that half of the atoms have fallen apart after 4.47 billion years. Such radioactive decays generate enough heat to keep Earth's core nice and toasty for billions of years, explaining why Earth is so warm even if it's way older than 20 million years. Moreover, by measuring what fraction of the uranium atoms in a rock have decayed, you can determine the age of the rock, and in this way, some rocks from the Jack Hills of western Australia have been found to be over 4.404 billion years old. The record age for meteorites is 4.56 billion years, suggesting that both our planet and the rest of our Solar System formed in the ballpark of 4.5 billion years ago—in good agreement with those much cruder estimates from tides.

In summary, discovering and using laws of physics has given us humans a qualitative and quantitative answer to one of our ancestors' greatest questions: *How and when was our Solar System created?*

Where Did the Galaxies Come From?

So we've pushed the frontier of our knowledge back to 4.5 billion years ago, when our Solar System was formed by the gravitational collapse of a giant molecular cloud. But as Philip's classmate asked: *Where did the giant molecular cloud come from?*

Galaxy Formation

Armed with telescopes, pencils and computers, astronomers have discovered a convincing resolution to this mystery as well, although important details still remain to be filled in. Basically, the same battle between gravity and pressure that formed our pizza-shaped Solar System repeats itself on a vastly larger scale, compressing a much larger region of gas into a pizza shape millions to trillions of times heavier than the Sun. This collapse turns out to be quite unstable, so it doesn't lead to a solar system on steroids with a single mega-star surrounded by mega-planets. Instead, it fragments into countless smaller gas clouds that form separate solar systems: thus, a galaxy has been born. Our Solar System is one of hundreds of billions in one of these pizza-shaped galaxies, the Milky Way, and we orbit around it once every couple of hundred million years, about halfway from the center (see Figure 2.2).

Galaxies sometimes collide with one another in huge cosmic traffic accidents. This isn't quite as bad as it sounds, as their stars mostly miss each other; in the end, gravity merges most of the stars into a new, larger galaxy. Both the Milky Way and our nearest big neighbor, Andromeda, are pizza-shaped galaxies, usually called spiral galaxies because of their beautiful spiral arm structure, which you can see in Figure 2.2. When two spiral galaxies collide, the result looks really messy at first, then settles into a roundish blob of stars known as an elliptical galaxy. This is our fate, since we're heading for a collision with Andromeda in a few billion years—we don't know if our descendants will call their home "Milkomeda," but we're pretty sure it will be an elliptical galaxy, because telescopes have imaged many other similar collisions in various stages, and the results roughly match our theoretical predictions.

If today's galaxies have been built up by mergers of smaller ones, then how small were the first ones? This quest to push our knowledge frontier backward in time was the topic of the very first research project I ever got really stuck on. A key part of my calculation was to figure out how chemical reactions in the gas produced molecules that could in turn reduce the gas pressure by radiating away heat energy. But every time I thought my calculations were done, I discovered that the molecule formulas I'd been using were wrong in some major way, invalidating all my conclusions and forcing me to start over. Four years after my grad school thesis advisor, Joe Silk, first got me started on this, I was so frustrated that I considered printing a custom-designed T-shirt saying I HATE MOLECULES, with my nemesis, the hydrogen molecule, crossed out by a big red stripe as on a no-smoking sign. Then luck intervened: after moving to Munich to do a postdoc, I met a friendly undergrad named Tom Abel who'd just completed a truly encyclopedic calculation of all the molecule formulas I needed. He joined our team of coauthors, and twenty-four hours later, we were done. We predicted that the very first galaxies weighed "only" about a million times as much as our Sun; we lucked out, since this finding basically agrees with the much more sophisticated computer simulations that Tom is making nowadays as a professor at Stanford.

Our Universe *Could Be* Expanding

We've seen that Earth's grand drama—generation upon generation of organisms being born, interacting and dying—had a beginning about 4.5 billion years ago. Moreover, we've discovered that this is all part of a much grander drama, where generation upon generation of galaxies are born, interact and eventually die in a cosmic ecosystem of sorts. So could there be a third level in this dramaturgy, whereby even universes are created and die? In particular, is there any indication that our Universe itself had some sort of beginning? If so, what happened, and when?

Why don't the galaxies fall down? The answer to this question triggered the next push of our knowledge frontier backward in time. We saw that the Moon doesn't fall down because it's orbiting us at high speed. Our Universe is teeming with galaxies in all directions, and it's pretty obvious that this same explanation doesn't work for them. They're not all orbiting around us. If our Universe has been eternal and essentially static, so that distant galaxies aren't moving much, then why don't they eventually fall toward us just as the Moon would do if you stopped it in its orbit, held it still and dropped it?

Back in Newton's day, people of course didn't know about galaxies. But if they, as Giordano Bruno, contemplated an infinite static universe uniformly filled with stars, then they had at least a half-baked excuse not to worry about why it didn't come crashing down on us: Newton's laws showed that each star would feel a strong (in fact, infinite) force pulling on it equally hard in each and every direction, so you could argue that these opposing forces would cancel each other out and the stars would all stay put.

In 1915, this excuse was refuted by Albert Einstein's new theory of gravity, the general theory of relativity. Einstein himself realized that a static infinite universe uniformly filled with matter didn't obey his new gravity equations. So what did he do? Surely, he'd learned the key lesson from Newton to boldly extrapolate, figuring out what sort of universe *did* obey his equations, and then asking whether there were observations that could test whether we inhabit such a universe. I find it ironic that even Einstein, one of the most creative scientists ever, whose trademark was questioning unquestioned assumptions and authorities, failed to question the most important authority of all: himself, and his prejudice that we live in an eternal unchanging universe. Instead, in what he later described as his greatest blunder, he changed his equations by adding an extra term that allowed our Universe to be static and eternal. In a double irony, it now seems as if this extra term is really there in the form of the cosmic dark energy we'll discuss later, but with a different value that doesn't make our Universe static.

The person who finally had the confidence and ability to really listen to Einstein's equations was the Russian physicist and mathematician Alexander Friedmann. He solved them for the most general case of a universe uniformly filled with matter, and discovered something shocking: most of the solutions were *not* static, but changing over time! Einstein's static solution wasn't merely an exception to typical behavior, but it was unstable, so that an almost static universe couldn't remain that way for long. Just as Newton's work showed that the natural state of the Solar System is to be in motion (Earth and the Moon can't just sit still forever), Friedmann's work showed that the natural state of our entire Universe is to be in motion.

But what sort of motion, precisely? Friedmann discovered that the most natural state of affairs was to find yourself in a universe that's either *expanding* or *contracting*. If it's expanding, that means that all distant objects are moving away from one another, like chocolate chips in a rising muffin (Figure 3.2). In that case, everything must have been closer together in the past. Indeed, in Friedmann's simplest solutions for an expanding universe, there was a particular time in the past when everything we can observe today was in the same place, creating an infinite density. In other words, our Universe had a beginning, and this cosmic birth was a cataclysmic explosion of something infinitely dense. The Big Bang was born.

The response to Friedmann's Big Bang was a deafening silence. Although his paper was published in one of Germany's most prestigious physics journals and was discussed by Einstein and others, it ended up largely ignored and had essentially no impact whatsoever on the prevailing worldview at the time. Ignoring great insights is a venerable tradition in cosmology (and indeed in science more generally): we've already discussed the heliocentrism of Aristarchos and the distant solar systems of Bruno, and we will encounter many more examples in the pages and chapters ahead. In Friedmann's case, I think part of the reason he was ignored was that he was ahead of his time: in 1922 the known Universe was essentially our Milky Way Galaxy (actually, just the limited part of it that we could see), and our Galaxy is *not* expanding, with its hundreds of billions of stars bound into orbits by its gravitational attraction. This

Galaxies moving, space not expanding: **Galaxies not moving, space expanding:**

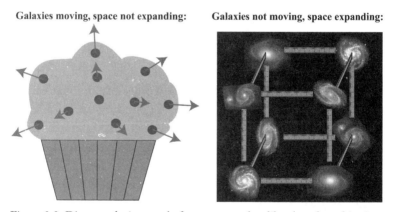

Figure 3.2: Distant galaxies recede from one another like chocolate chips in a rising muffin (left): from the vantage point of any one of them, all others are moving straight away with a speed proportional to their distance. But if we think of space as stretching as the muffin dough does, then the galaxies aren't moving relative to space, and space simply has all its distances stretched uniformly (right), as if we relabeled the tick marks on our rulers from millimeters to centimeters.

is the answer to *question 9* on our list from the last chapter: **Is the Milky Way expanding?** Friedmann's expansion applies only on scales so large that we can ignore the clumping of matter into galaxies and galaxy clusters. We can see in Figure 2.2 that the distribution of galaxies gets rather smooth and uniform on huge scales such as 100 million light-years, implying that Friedmann's homogeneous-universe solutions apply and that all galaxies separated by such large distances should be moving away from each other. But as we discussed earlier, Hubble didn't establish that galaxies even *existed* until 1925, three years later! Now time was finally ripe for Friedmann. Unfortunately time was also up for him: he died of typhoid fever that same year, only thirty-seven years old.

To me, Friedmann is one of the great unsung heroes of cosmology. While writing this, I couldn't resist reading his original 1922 paper, and noticed that it ends by giving an intriguing example of a vast universe containing five billion trillion suns' worth of mass, from which he calculates a lifetime of about ten billion years—in the same ballpark as the accepted modern value for the age of our Universe. He doesn't explain where he got this from, years before galaxies were discovered, but it was certainly a fitting ending to a remarkable paper by a remarkable person.

Our Universe *Is* Expanding

Five years later, history repeated itself: an MIT graduate student, the Belgian priest and astrophysicist Georges Lemaître, again published Friedmann's Big Bang solution, which he had been unaware of and had rediscovered. And once again, it was largely ignored by the scientific community.

What finally made people take note of the Big Bang idea wasn't new theoretical work, but new measurements. Now that Edwin Hubble had established that galaxies existed, an obvious next step for him was to start mapping out how they were distributed in space and how they moved. As I mentioned in the previous chapter, it's often easy to measure how fast something is moving toward or away from us, since this motion shifts the lines of its spectrum. Red light has the lowest frequency of all the colors in the rainbow, so if a galaxy is moving away from us, the colors of all its spectral lines will be *redshifted*, shifted toward redder colors, and the higher its speed, the greater its redshift. If the galaxy is moving toward us, its colors will instead be *blueshifted* toward higher frequencies.

If galaxies were just moving around at random, we'd expect about half of them to be redshifted and the rest blueshifted. Surprisingly, almost all the galaxies that Hubble studied were redshifted. Why were they all receding from us? Didn't they like us? Did we say something wrong? Moreover, Hubble discovered that the greater the distance d to the galaxy, the higher the velocity v with which it receded from us, according to the formula:

$$v = Hd$$

which we now know as *Hubble's law*. Here H is the so-called Hubble parameter, which Hubble modestly called K in his seminal 1929 paper on the subject, so as not to appear too conceited. Interestingly, Georges Lemaître had shown in his ignored 1927 paper that the expanding universe solution *predicted* Hubble's law: if everything was expanding away from everything else, then we'd see the distant galaxies expand away from us like this.

If a galaxy is moving straight away from us, this suggests that it was very close to us in the past. How long ago? If you see a car speeding away after a bank robbery, you can estimate how long ago it left the bank by dividing its distance by its speed. If we do this for the receding galaxies, Hubble's law gives the same answer $d/v = 1/H$ for all of them! This answer is $1/H \approx 14$ billion years, using modern measurements, so Hubble's discovery suggests that something rather dramatic happened about 14 billion years ago, involving lots of matter squeezed together here at high density. To get a more exact answer, we need to factor in the extent to which the car/universe has been accelerating/decelerating/cruising at constant speed since leaving the crime scene. When we do this today, using Friedmann's equations and modern measurements, we find that the required correction is quite small, at the percent level: after its Big Bang, our Universe spent about the first half of its time decelerating, then the rest of the time accelerating, so the corrections roughly cancel out.

Making Sense of an Expanding Universe

After Hubble's measurements were announced, even Einstein was convinced, and now our Universe was expanding even officially. But what does it *mean* that our Universe is expanding? We're now ready to tackle four more of the questions from the beginning of Chapter 2.

First of all, ***are galaxies really moving away from us, or is space just expanding?*** Conveniently, Einstein's theory of gravity (general relativity) says that these are two equivalent viewpoints that are equally valid, as illustrated in Figure 3.2, so you're free to think about it in whichever way you find more intuitive.* From the first viewpoint (left), space isn't changing but the galaxies are moving through space like the chocolate chips in a muffin that's rising because of the baking powder you put in the batter. All galaxies/chocolate chips move farther apart from all others, and more widely separated pairs get separated faster. In particular, if you're standing on a specific chocolate chip/galaxy, you'll see that the motion of all the others relative to you obeys Hubble's law: they're all receding straight away from you, and one twice as far recedes at twice the speed. Remarkably, things will look the same whichever chocolate chip or galaxy you're observing from, so if the distribution of galaxies has no end, then the expansion has no center—it looks the same from everywhere.

From the second viewpoint, space is like the muffin dough: it expands, so just as the chocolate chips aren't moving relative to the dough, the galaxies aren't moving through space. Instead, we can think of the galaxies as being at rest in space (Figure 3.2, right) while all the distances between them get redefined. It's as if the tick marks on imaginary rulers connecting the galaxies get relabeled so that their spacing corresponds not to a millimeter but to a centimeter—now all intergalactic distances are ten times larger than they used to be.

This answers another one of our questions: ***Don't galaxies receding faster than the speed of light violate relativity theory?*** Hubble's law $v = Hd$ implies that galaxies will move away from us faster than the speed of light c if their distance from us is greater than $c/H \approx 14$ billion light-years, and we have no reason to doubt that such galaxies exist, so doesn't this violate Einstein's claim that nothing can go faster than light? The answer is yes and no: it violates Einstein's special relativity theory from 1905 but not his general relativity theory from 1915, and the latter is Einstein's final word on the subject, so we're okay. General

* Mathematically, the different viewpoints correspond to different choices of space coordinates, and Einstein's theory allows you to pick whichever coordinate system you want for space and time.

relativity liberalizes the speed limit: whereas special relativity says that no two objects can move faster than light relative to one another *under any circumstances*, general relativity merely insists that they can't move faster than light relative to one another *when they're in the same place*—in contrast, the galaxies speeding away from us superluminally are all very far from us. If we think of space as expanding, then we can rephrase this by saying that nothing is allowed to move faster than light *through space*, but space itself is free to stretch however fast it wants to.

Speaking of distant galaxies, I've seen newspaper articles talking about ones as far as about 30 billion light-years away from us. ***If our Universe is only 14 billion years old, how can we see objects that are 30 billion light-years away?*** How did their light have time to reach us? Moreover, we just figured out that they're receding from us faster than the speed of light, which makes the notion that we can see them sound even weirder. Here the answer is that we're not seeing these distant galaxies where they are now, but where they were when they emitted the light that reaches us now. Just as we see the Sun the way it looked eight minutes ago at the position where it was eight minutes ago, we might see a distant galaxy the way it looked 13 billion years ago, at the position where it was back then—which was about eight times closer to Earth than it is now! So the light from this galaxy never needed to travel more than 13 billion light-years through space to reach us, because the stretching of space made up for the difference—it's as if you walk up an escalator and move twenty meters while taking only ten one-meter steps.

What's Our Universe Expanding Into?

Won't there be a cosmic traffic accident somewhere far away where galaxies expanding away from us crash into whatever they're expanding into? If our Universe expands according to Friedmann's equations, there are no such problems: as Figure 3.2 illustrated, the expansion looks the same from everywhere in space, so there can't be any such trouble spots. If we take the viewpoint that distant galaxies really are receding through a static space, then the reason they never collide with more distant galaxies is that those are receding even faster: you can't rear-end a speeding Porsche if you're driving a Model T Ford. If you instead take the viewpoint that space is expanding, the explanation is simply that volume isn't conserved. From reading about the Middle East, we're used to the idea that you can't get more space without taking it away from someone else. However, general relativity says the exact opposite: more volume can be created in a particular region between some galaxies without this new volume expanding into other regions—the new volume simply stays between those same galaxies (Figure 3.2, right).

The Cosmic Classroom

In other words, as crazy and counterintuitive as it sounds, the expanding universe is both logical and supported by astronomical observations. In fact, the observational evidence has grown dramatically stronger since the days of Edwin Hubble, thanks to modern technology and new discoveries that we'll explore below. The most basic conclusion is simply that even our Universe itself is changing: when we push our knowledge frontier back many billions of years, we discover a universe that hadn't expanded as much, and was therefore denser and more crowded. This means that the space we inhabit isn't the boring static space once axiomatized by Euclid, but a dynamic evolving space that once had some sort of childhood—and perhaps some sort of birth about 14 billion years ago.

Dramatically better telescopes have now improved our vision to the point that we can see our evolving cosmos quite directly. Imagine that you're giving a presentation in a large auditorium. Suddenly you notice something funny about the audience. The rows of chairs closest to you are all occupied by people around your own age. But about ten rows back, you see only teenagers. Behind them are a bunch of younger kids, and behind them a row of toddlers. Behind them, near the very back of the room, you see only babies. The very last row is completely empty, as far as you can see. When we gaze out into our Universe with our best telescopes, we see something similar: nearby are lots of large and mature galaxies like our own, but very far away, we see mostly small baby galaxies that don't yet look fully developed. Beyond them we see no galaxies at all, merely darkness. Since it takes light longer to reach us from farther away, gazing into the distance is equivalent to observing the past. The darkness behind the galaxies is the epoch before the first galaxies had time to form. Back then, space was filled with hydrogen and helium gas that gravity hadn't yet had time to clump into galaxies, and since this gas is transparent like helium in balloons at birthday parties, it's invisible to our telescopes.

But there's a mystery: during your presentation, you suddenly realize that there's energy coming from beyond that empty last row: the rear wall of the auditorium isn't completely dark, but gives off a faint glow of microwaves! Why? Bizarre as it sounds, this is what we see when we peer into the most distant depths of our Universe. To understand this, we need to continue our quest to push our knowledge frontier even farther back in time.

Where Did the Mysterious Microwaves Come From?

To me, a key lesson from both Newton and Friedmann is this simple mantra: "Dare to extrapolate!" Specifically, take your current understanding of the laws of physics, apply them in a new uncharted situation, and ask whether they predict something interesting that we can observe. Newton took the laws of motion that Galileo had established on Earth and extrapolated them to the Moon and beyond. Friedmann took the laws of motion and gravity that Einstein had established in our Solar System and extrapolated them to our entire Universe. Given how successful this mantra was, you might think that it would catch on as a meme in the scientific community. In particular, you might think that after 1929 when Friedmann's expanding-universe idea gained acceptance, scientists around the world would race against each other to systematically explore what happened if you extrapolated it backward in time. Well, if you'd have thought this, you'd have been wrong. . . . No matter how emphatically we scientists claim to be rational seekers of truth, we're as prone as anyone to human foibles such as prejudice, peer pressure and herd mentality. Overcoming these shortcomings clearly takes more than just talent for calculating.

To me, the next cosmological superhero who had what it took was another Russian: George Gamow. His Ph.D. advisor in Leningrad was none less than Alexander Friedmann, and although Friedmann died two years into Gamow's studies, both his ideas and his intellectual boldness lived on in Gamow.

The Cosmic Plasma Screen

Given that our Universe is currently expanding, it must have been denser and more crowded in the past. But has it always been expanding? Perhaps not: Friedmann's work allows for the possibility that our Universe was once contracting, and that all the material moving toward us gently slowed down, stopped and started accelerating away from us. Such a cosmic bounce could only have happened if the density of matter were much lower than we now know it to be. Gamow decided to systematically explore the other option, which was more generic and more radical: expansion ever since the beginning. As he explained in a 1946 book, this proposition implies that if we imagine the cosmic drama to be a movie and rewind it, playing it backward, we'll see the density of our Universe increase without limit. Since intergalactic space is filled with hydrogen, this gas will get more and more compressed and therefore hotter and hotter the farther back in time we look. If you keep heating an ice cube, it melts. If you keep heating liquid water, it transforms into gas: steam. Similarly, if you keep heating hydrogen gas, it turns into a fourth phase: plasma. Why? Well, a hydrogen atom is simply an electron orbiting around a proton, and hydrogen gas is just a bunch of such atoms bouncing against each other. If the temperature rises, the atoms move faster and bump each other harder. If it gets hot enough, the bumps get so violent that the atoms break apart and the electrons and protons go their separate ways—a hydrogen plasma is simply such a soup of free electrons and protons.

In other words, Gamow predicted that our Universe began with a hot Big Bang, and that plasma once filled all of space. What's exceptionally interesting about this is that the prediction is testable: whereas cold hydrogen gas is transparent and invisible, hot hydrogen plasma is opaque and glows brightly, like the surface of the Sun. This means that when we gaze ever farther into space as in Figure 3.3, we should encounter old galaxies nearby, then young galaxies beyond them, then transparent hydrogen gas, then a wall of glowing hydrogen plasma. We can't see beyond this wall, because it's opaque and therefore obstructs what came before it like a cosmic censor. Moreover, as illustrated in Figure 3.4, this is what we should see in *all* directions, since wherever we look, we're also looking back in time. It therefore looks to us like we're surrounded by a gigantic plasma sphere.

In his 1946 book, Gamow's Big Bang theory predicted that we should be able to observe this plasma sphere. He got his students Ralph Alpher and Robert Herman to work things out in more detail, and a few years later, they published a paper predicting that it would glow with a temperature of about five degrees above absolute zero, meaning that it

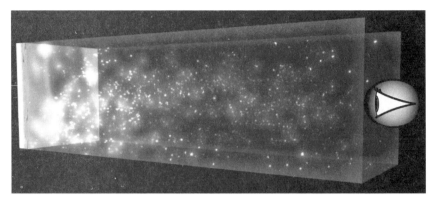

Figure 3.3: Since it takes time for distant light to reach us, looking farther away means looking farther back in time. Beyond the most distant galaxies, we see an opaque wall of glowing hydrogen plasma, whose glow has taken about 14 billion years to reach us. This is because the same hydrogen that fills space today was hot enough to be plasma about 14 billion years ago, when our Universe was only about 400,000 years old. *(Credit: Adapted from NASA/WMAP team)*

would mainly give off microwaves rather than visible light. They unfortunately failed to convince any astronomers to search for this cosmic microwave–background radiation in the sky, and their work was largely forgotten just as Friedmann's expanding-universe discovery was.

Seeing the Afterglow

By 1964, a group at Princeton University had realized that this observable microwave signal should exist and planned an observational search for it, but they were beaten to the punch. The same year, Arno Penzias and Robert Wilson were testing a new state-of-the-art microwave telescope at Bell Labs in New Jersey and discovered something puzzling:

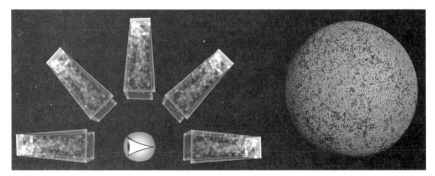

Figure 3.4: It looks like we're in the middle of a giant plasma sphere, because we see this plasma wall from the previous figure in every direction that we look.

their telescope detected a signal they couldn't explain, and this signal remained the same regardless of where they pointed the telescope. Weird! They were expecting to detect signals only when they pointed at particular objects in the sky, such as the Sun or a satellite transmitting microwaves. Instead, it was as if the whole sky was glowing, with a temperature of three degrees above absolute zero—close to the five degrees that Gamow's group had predicted. They carefully checked for local sources of noise, and for a while, suspicion fell on pigeons that were nesting in the telescope and leaving droppings there. I got to have lunch with Arno a while back, and he told me that they put the pigeons in a wooden box with food and sent it to another Bell Labs campus far away with instructions to release the birds. Unfortunately, they were homing pigeons. . . . Although his book merely states that they "eliminated" the pigeons when they returned, I got him to reveal the grim truth after some wine: it involved a shotgun. . . . Although the pigeons were gone, the mysterious signal remained: they had discovered the cosmic microwave background, the faint afterglow of our Big Bang.

The discovery was a sensation, and earned them the 1978 Nobel Prize in physics. From the calculations of Gamow and his students, it followed that the plasma sphere in Figure 3.4 must have been about half as hot as the Sun's surface, and that as the radiation from its hot glow traveled through space for 14 billion years to reach us, it cooled down thousandfold to the observed three degrees above absolute zero as space expanded thousandfold. In other words, our entire Universe was once as hot as a star, and the wild thousandfold extrapolation of Gamow's hot Big Bang theory had been tested and vindicated.

Baby Pictures of Our Universe

Now that the plasma sphere had been detected, the race was on to take the first photos of it. Because the temperature of the radiation was basically the same in all directions, the images Penzias and Wilson could make looked like one of those joke postcards labeled "San Francisco in the Fog," where all you see is uniform white. To get interesting photos that would qualify as the first baby pictures of our Universe, one would need to increase the contrast to detect slight variations from place to place. These variations had to exist, because if the conditions had been identical everywhere in the past, then the laws of physics would have kept them identical everywhere today, in stark contrast to the clumpy Universe we now observe with galaxies in some places and not in others.

However, taking these cosmic baby pictures proved so difficult that it took almost three decades of technological development. To suppress measurement noise, Penzias and Wilson had to use liquid helium to cool their detector down to near the temperature of the cosmic microwave background. The temperature fluctuations from place to place in the sky turned out to be tiny, around a thousandth of a percent, so producing baby pictures required a hundred thousand times more sensitivity than that of Penzias and Wilson's measurements. Experimentalists around the world took on the challenge and failed. Some said it was hopeless, but others refused to give up. On May 1, 1992, when I was halfway through grad school, the fledgling Internet was abuzz with rumors: George Smoot was going to announce the results from the most ambitious microwave-background experiment to date, performed from the cold darkness of space by a NASA satellite called COBE, the Cosmic Background Explorer. My Ph.D.-thesis advisor, Joe Silk, was scheduled to introduce George's talk, and before he flew out to Washington, D.C., I asked him what he thought the odds were of a discovery. Joe guessed that they hadn't seen the cosmic fluctuations, just a radio noise from our own Galaxy.

But instead of delivering an anticlimactic lecture, George Smoot dropped a bombshell that transformed not only my own career, but the entire field of cosmology: he and his team members had found the fluctuations! Stephen Hawking hailed this as "the most important discovery of the century, if not of all time," because as we'll see below, these baby pictures of when our Universe was "only" 400,000 years old contained crucial clues to our cosmic origins.

The Gold Rush

Now that COBE had found gold, there was a wild rush to mine more of it. As you can see in Figure 3.5, the COBE sky map was pretty fuzzy, because of low-resolution imaging that smoothed out features smaller than about 7 degrees—the natural next step was therefore to zoom in on a small part of the sky with higher resolution and less noise. As I'll explain below, such high-resolution maps encode the answer to some key cosmological questions. I'd loved photography ever since I saved up for my first camera at age twelve by delivering junk mail in Stockholm,

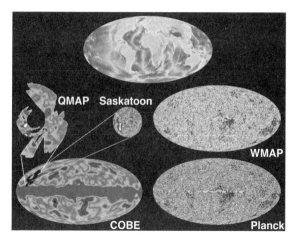

Figure 3.5: When showing maps of the whole sky, it's convenient to project them onto a flat page just as we do with Earth maps (top), simply interpreting them as looking up toward the sky rather than down toward the ground. The "baby picture of our Universe" from COBE (bottom left) was quite fuzzy, motivating many experiments to zoom in on small sky patches with higher resolution (middle left) before the WMAP and Planck satellites delivered high-resolution maps of the entire sky (right) with three megapixels and fifty megapixels, respectively. These sky maps are rotated relative to the Earth map so that map midplane corresponds not to the plane of Earth's equator but to the plane of our Galaxy (bottom left gray stripe); Earth's North Pole points toward the center of the Saskatoon map. (*Earth map credit: Patrick Dineen*)

so imaging our Universe instinctively appealed to me. I'd also enjoyed messing with images and computer graphics, whether it was for my high school's newspaper, *Curare*, or for the shareware computer game FRAC, a 3-D Tetris clone that paid for my 1991 around-the-world trek. I therefore felt very fortunate when various experimentalists let me team up with them on converting their data into sky maps.

My first stroke of luck was meeting Lyman Page, a young professor from Princeton. I liked his playful, boyish smile, and worked up the courage to ask him about possible collaboration after a conference talk he gave. I liked him even more after learning that he'd spent years sailing the Atlantic before grad school. He ended up entrusting me with data from a microwave telescope in the Canadian town Saskatoon, with which he and his group had spent three years scanning the sky patch directly above the North Pole.

Converting this into a map was surprisingly hard because the data didn't consist of sky photos, merely of long tables of numbers encoding how many volts had been measured by adding and subtracting different parts of the sky in various complicated ways. But I also found it surprisingly exciting, requiring my utmost efforts with information theory and computational number-crunching, and after many müsli-fueled evenings in the Munich office where I was doing my postdoc, I was able to finish the Saskatoon map in Figure 3.5 just in time for my talk at a big cosmology conference in the French Alps. Although I've given hundreds of talks by now, there are a few that stand out in my memory as magic moments that infallibly make me smile every time I remember them. This was one of those magic ones. My heart pounded as I walked up to the podium and looked around the room. It was packed with people, many of whom I knew from reading their work and most of whom had no idea who I was. They'd come to the conference more for the great skiing than for hearing total beginners like me. But I didn't just feel my heart race—I also felt a great energy in the room. People were really excited about all the new cosmic microwave–background developments, and I felt honored and thrilled to get to be a small part of this. The year 1996 was back in the pre-Cambrian era when we still gave our talks using plastic transparencies, and I ended with the ace in my deck: a slide showing the Saskatoon map just as in Figure 3.5, as a zoom of the COBE map. I could feel a ripple of excitement spreading through the room, and a bunch of people stood around the overhead projector for most of the ensuing coffee break requesting to look at it again and asking questions. Dick Bond, one of the founders of cosmic microwave–background cosmology, came over and said, smiling: "I can't believe Lyman gave you the data!"

I felt that cosmology had entered a golden age, where new discoveries were bringing new people and new funding into the field, which led to new discoveries in a virtuous circle. The very next month, in April 1996, funding was approved for two new satellites with radically better resolution and sensitivity than COBE. One was the NASA mission WMAP, spearheaded by Lyman Page and a close-knit group of colleagues, and the other was the European-led mission Planck, for which I'd had lots of fun making calculations and forecasts for the grant proposal. Since space missions involve many years of planning, smaller teams around the world raced to steal the thunder from WMAP and Planck, or at least grab some of the lowest-hanging fruit before they launched. As a result, the Saskatoon project ended up being just the first of many fun data collaborations for me. I got to work with the builders of experiments with exotic names such as HACME, QMAP, Tenerife, POLAR, PIQ and Boomerang to make baby pictures of our Universe from their data or figure out what they taught us about our cosmos. My basic game plan was to be the middleman between theory and experiment: I felt that cosmology was transforming from a data-starved field into one with more data than people knew how to handle, so I decided to develop tools for taking full advantage of this data avalanche. Specifically, my strategy was to use a branch of mathematics known as information theory to figure out how much relevant information about our Universe was contained in a given data set. Typically, the megabytes, gigabytes or terabytes measured would contain only a modest number of bits of cosmological information, scrambled and hidden in some complicated way among vast amounts of noise from detector electronics, atmospheric emission, Galactic radiation, and other sources. Although there was a known mathematically perfect method for extracting this needle from the haystack, it was usually too complex to do in practice, requiring millions of years of computer calculations. I published various data-analysis methods that weren't necessarily perfect, but were able to extract almost all the information quickly enough to be useful in practice.

I love the cosmic microwave background for many reasons. For example, I can thank it for my first marriage and for the existence of my sons, Philip and Alexander: I got together with my ex-wife, Angélica de Oliveira Costa, because she came from Brazil to Berkeley as a grad student to work with George Smoot, and we ended up collaborating closely not only on diaper changing, but also on many of the data-analysis projects I mentioned. One such project was QMAP, a telescope flown by Lyman Page, Mark Devlin and collaborators on a high-altitude balloon to avoid most of the microwave noise caused by Earth's atmosphere.

. . .

Oh, no! It's about two a.m. on May 1, 1998, and things look grim. There are only seven hours left before our flight will depart to Chicago, where I'm supposed to present the new QMAP results at a cosmology conference, but Angélica and I are still in my office at the Institute for Advanced Study in Princeton, shaking our heads. So far, all cosmic microwave–background experiments had required you to believe that no mistakes had been made and nothing important had been overlooked. A key to believability in science is having an independent experiment confirm your results, but because people had looked in different directions with different resolutions, it was never possible to compare the sky images made by two different experiments and check whether they agreed with each other. Up until this moment, that is: the Saskatoon and QMAP sky maps have a major overlap in a banana-shaped sky patch that you can see in Figure 3.5. Angélica and I are staring with dismay at my computer screen and feeling our hearts sink: there are the Saskatoon and QMAP maps side by side, and they don't agree at all! We squint and try to imagine that the discrepancies are just due to instrumental noise. No, wishful thinking goes only so far. All this work just to realize that at least one of the maps is all wrong. And how can I possibly give a talk about this? It would be total humiliation not just for us, but for all the people who built and ran the experiments.

Suddenly Angélica, who's been poring over our computer program, discovers a suspicious minus sign, which would, crudely speaking, cause the QMAP map to come out upside down. We fix it, rerun the code, and look at each other with disbelief as the new map appears on the screen: now the agreement between the two maps is stunning! A clutch play! We sleep for a few hours, fly to Chicago, I whip my talk together on pure adrenaline, and I run all the way from our rental car to the Fermilab auditorium to arrive just in time for my talk. I'm so excited that I don't even realize my transgression until the evening, when our car is mysteriously missing.

"Where did you park it?" the guard asks.

"Oh, right outside, in front of the fire hydrant," I reply—and suddenly find myself thinking *Doh!!!* for the second time that day. . . .

The Cosmic Beach Ball

The great gold rush to mine the microwave sky continued for years, with over twenty different experiments spurring each other on—I'll tell you more about some of them below. And then there was WMAP. At two p.m. on March 11, 2003, the room was packed: we were all glued to the screen where the WMAP team members were announcing their results live on NASA-TV. Whereas ground- and balloon-based experiments could only map parts of the sky, the WMAP satellite had mapped the whole sky just as COBE had, with dramatically better sensitivity and resolution. I felt like when I was a little kid on Christmas Eve and Santa Claus finally arrived—except that I'd been eagerly awaiting this moment not for months but for years. It was worth the wait: the resulting images were stunning. As was their work ethic and sleep deprivation: they'd gone from funding to construction, launch, data analysis, and results in under six years, three times faster than COBE. Indeed, the WMAP project leader, Chuck Bennett, almost killed himself keeping it on schedule: David Spergel, another key contributor to the project, told me that Chuck collapsed and had to be hospitalized for three weeks after launch.

Moreover, they made all their data publicly available online, so that cosmologists around the world could take a crack at reanalyzing it themselves. Cosmologists like me. Now it was my turn to work like crazy while they caught up on sleep. Their measurements were superb, but contaminated with radio noise from our own Galaxy, which you can see in Figure 3.5 as the horizontal band in the COBE map. The bad news is that such microwave contamination from our Galaxy and others exists everywhere in the sky, even if the level is too low to be easily seen. The good news is that the contamination has a different color than the signal (it depends on frequency in a different way), and that WMAP imaged the sky at five separate frequencies. The WMAP team had used this extra information to clean out the contamination, but I was excited about an even better method for doing this, based on information theory, which produced a cleaner map with higher-resolution (Figure 3.5, bottom right). After working all out on this for a month with Angélica and my old friend Andrew Hamilton, we submitted our paper and my life started returning to normal. I had fun making the ball-like image of the microwave background in Figure 3.4 and on the front cover of this book, and the WMAP team liked it so much that they made their own version and printed it on a plastic beach ball, which to this day adorns my office. I call it my "universe," because it's the iconic image of what bounds everything we can in principle observe.

The Axis of Evil

As I'll explain further on, key cosmic clues lie encoded in the sizes of the spots you see in the cosmic microwave background. Just as we can decompose sounds and colors into different frequencies, we can decompose two-dimensional microwave-background maps as a sum of many different component maps (see Figure 3.6) that go by the geeky-sounding name of *multipoles*. These multipole maps, in essence, contain the contributions from spots of different sizes, and ever since COBE, something had seemed to be fishy with the second multipole, called the *quadrupole*:

the largest spots in the map appeared weaker than expected. Yet nobody had ever been able to make a *map* of the quadrupole to see what was going on with it: this required a map of the entire sky, but microwaves from our Galaxy contaminated part of the sky beyond repair.

Until now: our map appeared so clean that perhaps it was usable across the whole sky. It was late at night, shortly before we submitted our map paper. Angélica and the kids were asleep, and I was tempted to hit the sack. But I was really curious to see what that pesky quadrupole looked like, and decided to write a computer program making a picture of it. When it finally popped up on my screen (Figure 3.6, left), I got intrigued: it wasn't just weak as expected (the temperature fluctuations in the hot and cold spots were really close to zero), but the pattern formed a funny-looking one-dimensional band across the sky rather than being a random mess as theory predicted. I was really sleepy now, but decided to reward myself for all this late-night programming and debugging with one more image, so I changed 2 to 3 in my program and reran it to get a plot of the third multipole, known as the *octupole*. *Whoa! What the . . . ?* Up popped another one-dimensional band (Figure 3.6, middle), seemingly aligned with the quadrupole. This was *not* how our Universe was supposed to be! As opposed to photos of you, photos of our Universe weren't supposed to have any special direction, such as "up": they should look similar no matter how you rotate them. Yet these baby-universe images on my computer screen had these bands of zebra-like stripes aligned in only one particular direction. Suspecting a bug in my code, I changed 3 to 4 and reran, but the plot of the fourth multipole (Figure 3.6, right) looked just as expected: a random mess with no special direction.

After Angélica had double-checked everything, we mentioned this surprising discovery in our map paper. I was amazed by how it caught

−34μK ▬▬▬▬▬▬▬▬▬▬▬▬▬▬▬▬▬▬▬▬▬▬▬ 34μK

Figure 3.6: When decomposing the WMAP map from Figure 3.5 into a sum of *multipoles* showing spots of progressively smaller sizes, the first two (left and middle maps) show a mysterious alignment around what's been dubbed "the axis of evil." The different colors show how much warmer or colder than average the sky is in different directions; the bar shows the scale in μK, millionths of degrees.

on. It got mentioned in the *New York Times*, which sent a photographer to take mug shots of us. We and many other groups looked into it in more detail, one of which dubbed the special direction "the axis of evil." Some argued that it was a statistical fluke or galaxy contamination, while others argued that it was even more puzzling than we'd claimed, finding additional anomalies even for multipoles 4 and 5 using a different method. Some exotic explanations, such as our living in a small "bagel universe" where space connects back on itself (see page 32), were ruled out by further analysis, and to this day, I'm as puzzled by the axis of evil as I was that first night.

A Microwave Background Comes of Age

In 2006 Angélica and I were invited to Stockholm to help celebrate that the COBE discovery had been awarded the Nobel Prize in Physics. As is so common in science, there had been acrimony within the COBE team about credit attribution. The prize was shared between George Smoot and John Mather, and I was relieved to see them both take a conciliatory approach. They were able to invite the entire COBE team to come and bask in well-deserved glory, and I felt that the unending stream of elegant parties helped bring closure to the rifts by emphasizing the obvious—they'd all accomplished something much more important than helping two guys get prizes: their first baby pictures of our Universe had created a vibrant new research field and ushered in a whole new era in cosmology. I just wish Gamow, Alpher and Herman could have been there, too.

On March 21, 2013, I got up at five a.m. full of anticipation and tuned in to a live webcast from Paris, where the Planck satellite team were releasing their first microwave background images. ACBAR, ACT, the South Pole Telescope and other experiments had improved our microwave background knowledge over the past decade, but this was the greatest milestone since WMAP. While I was shaving, George Efstathiou described the results, and I felt a wave of nostalgia and excitement sweep over me. I had a flashback to March 1995, when George had invited me to Oxford to work with him on new methods for analyzing Planck data. It was the first time I'd ever been invited anywhere for research collaboration, and I felt most grateful for the opportunity. We developed a novel technique for cleaning out contaminating symbols, which helped bolster the case that the European Space Agency should fund Planck. Now the results would finally be revealed to the eighteen-years-older Max I saw in the bathroom mirror!

When George showed the new Planck sky map, I just had to put my razor down so I could place our foreground-cleaned WMAP map next to George's map on my laptop screen. *Wow—they agreed beautifully!* I thought to myself. *And the axis of evil is still there!* I've placed the two maps together in Figure 3.5 so that you can compare them. As you can see, all the large-scale patterns match up exquisitely, but the Planck map has much more tiny spots. This is because of its greatly superior sensitivity and resolution, which allows it to image tiny patterns that the WMAP satellite blurred out. The Planck map was definitely worth the wait! I've projected it as a sphere for you so that you can enjoy it in high-quality color on the front cover of this book. Because of its superb quality, Planck effectively provides the answer sheet for grading the performance of WMAP, and after carefully digesting the Planck results, it's clear to me that the WMAP team deserves an *A+*. As does the Planck team. However, I think the greatest surprise with Planck was that there was no surprise: it basically confirmed the cosmological picture we'd already come to believe, with much better precision. The cosmic microwave background had come of age.

In summary, we've now pushed the frontier of knowledge back from about 14 billion to about 400,000 years after our Big Bang, and seen that everything around us came from a hot plasma that filled all space. Back then, there were no people, planets, stars or galaxies—just atoms bouncing around and radiating light. But we still haven't explored the mystery of where these atoms came from.

Where Did the Atoms Come From?

The Cosmic Fusion Reactor

We saw that Gamow's audacious extrapolation backward in time successfully predicted the cosmic microwave background, which has now given us stunning baby pictures of our Universe. As if this smashing success wasn't enough, he pushed his extrapolation even farther back in time and worked out the consequences. The longer ago it was, the hotter it was. We saw that 400,000 years after our Big Bang, the hydrogen that filled space was thousands of degrees, about half as hot as the *sur-*

face of the Sun, so it did what hydrogen in the Sun's surface does: glows, producing the cosmic microwave–background radiation. Gamow also realized that a minute after our Big Bang, the hydrogen temperature was about a billion degrees, even hotter than the *core* of the Sun, so it must have done what hydrogen in the Sun's core does: fusion, converting hydrogen into helium. However, the expansion and cooling of our Universe soon switched off this cosmic fusion reactor, by making it too cold to function, so it didn't have time to turn everything into helium. Encouraged by Gamow, his students Alpher and Herman made a detailed calculation of what would happen with the fusion, although since they were working in the late 1940s, their calculations were limited by the lack of modern computers.

But how can this prediction be tested, given that our Universe wasn't transparent during its first 400,000 years, with everything that happened back then hidden from view, censored by the cosmic microwave–background plasma screen? Gamow realized that the situation was the same as with the dinosaur theory: you can't see what happened directly, but you can find fossil evidence! Repeating their calculations with modern data and computers, you predict that back when our entire Universe was a fusion reactor, it fused about 25% of its mass into helium. When you measure the helium fraction of distant intergalactic gas by studying its spectrum with a telescope, you find . . . 25%! To me, this finding is just as impressive as discovering a fossilized Tyrannosaurus rex femur: direct evidence that crazy things happened in the past, in this case everything being crazy hot like the center of the Sun. And helium isn't the only fossil. Big Bang nucleosynthesis, as Gamow's theory became known, also predicts that about one in every 300,000 atoms out there should be deuterium* and about one in every five billion atoms should be lithium—both of these fractions have now been measured and agree beautifully with the theoretical prediction.

Big Bang in Trouble

However, success didn't come easy: Gamow's hot Big Bang got a cool reception. Indeed, the name *Big Bang* was coined by one of its detractors, Fred Hoyle, in an attempt to ridicule it. According to the 1950

---

* Deuterium is hydrogen's big brother, weighing twice as much because it contains not only a proton but also a neutron.

scorecard, the theory had made two major predictions, both wrong: the age of our Universe and the abundance of atoms. Hubble's initial measurement of the cosmic expansion predicted that our Universe was less than two billion years old, and geologists were underwhelmed by the idea that our Universe was younger than some of their rocks. Moreover, Gamow, Alpher and Herman had hoped that Big Bang nucleosynthesis would produce essentially *all* the atoms around us in the right proportions, but found that it failed to produce even remotely enough carbon, oxygen and other everyday atoms—making merely helium, deuterium and puny amounts of lithium.

We now know that Hubble had grossly underestimated the distances to his galaxies. Because of this, he mistakenly concluded that our Universe expands seven times faster than it really does, suggesting that our Universe is seven times younger than it really is. When better distance measurements started correcting this error during the 1950s, the unhappy geologists were vindicated and appeased.

The second Big Bang "failure" also melted away around the same time. Gamow had done pioneering research on fusion reactions in stars, and the work by him and others suggested that stars produce helium and little else, just as our Sun is doing right now. This is why he hoped that Big Bang nucleosynthesis could explain where the rest of the atoms came from. In the 1950s, however, a seemingly surprising nuclear-physics coincidence was discovered that linked nuclear-energy levels of helium, beryllium, carbon and oxygen, facilitating fusion. As Fred Hoyle was the first to realize, this coincidence enabled stars in the late stages of their lives to turn helium into carbon, oxygen and most of the other atoms that you and I are made of. Moreover, it became clear that stars end their lives by blowing apart, recycling many of the atoms that they've made into gas clouds that can later form new stars, planets and ultimately you and me. In other words, we're more connected to the heavens than our ancestors realized: we're made of star stuff. Just as we are in our Universe, our Universe is in us. This insight transformed Gamow's Big Bang nucleosynthesis from failure to smashing success: our Universe made helium and a smidgen of deuterium and lithium during its first few minutes, and stars made the rest of our atoms later on.*

---

* The stars add further to the 25% of helium made by Big Bang nucleosynthesis. We can tell the two sources of helium apart with our telescopes: the farther back in time we look, the less helium we see, bottoming out at 25% when we look back to times before most stars had formed.

The mystery of where the atoms came from had been solved. And when it rains, it pours: just as the hot Big Bang theory was finally coming in from the cold, the cosmology world was electrified by the 1964 discovery of Gamow's other prediction: the Big Bang afterglow, the cosmic microwave–background radiation.

What Is a Big Bang, Really?

We've now pushed the frontier of our knowledge back about 14 billion years, to a time when our entire Universe was a blazing hot fusion reactor. When I say I believe in the Big Bang Hypothesis, I mean that I'm convinced of this but nothing more.

> **Big Bang Hypothesis:** *Everything we can observe was once hotter than the core of the Sun, expanding so fast that it doubled its size in under a second.*

That's definitely big enough a bang that I feel we can call it Big with a capital *B*. However, take note that my definition is quite conservative, saying nothing whatsoever about what happened before that. For example, this hypothesis does *not* imply that our Universe was one second old at the time, or that it was ever infinitely dense or came from some sort of singularity where our math breaks down. The question ***Do we have evidence for a Big Bang singularity?*** from the last chapter has a very simple answer: *No!* Sure, if we extrapolate Friedmann's equations as far back in time as they'll go, they break down in an infinitely dense singularity about a second before Big Bang nucleosynthesis, but the theory of quantum mechanics that we'll explore in Chapter 7 tells us that this extrapolation breaks down before reaching a singularity. I think it's crucial to distinguish between what we have solid evidence for and what's highly speculative, and the truth is that although we have some exciting theories and hints about what happened earlier, which we'll explore in Chapter 5, we frankly don't yet know. This is the current frontier of our knowledge. Indeed, we don't even know for sure that our Universe really had a beginning at all, as opposed to spending an eternity doing something we don't understand prior to Big Bang nucleosynthesis.

In summary, we humans have now pushed our knowledge frontier remarkably far back in time, revealing the cosmic storyline I've tried to illustrate in Figure 3.7. A million years after our Big Bang, space

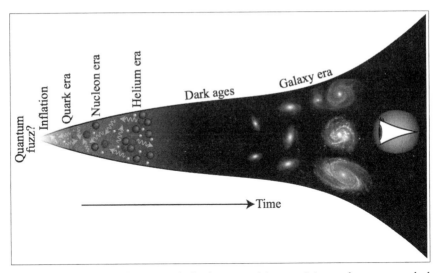

Figure 3.7: Although we know very little about our ultimate origins, we know a great deal about what happened during the subsequent 14 billion years. As our Universe expanded and cooled, quarks assembled into protons (hydrogen nuclei) and neutrons, which in turn fused into helium nuclei. Then these nuclei formed atoms by capturing electrons, and gravity clumped these atoms into the galaxies, stars and planets that we observe today.

was filled with nearly uniform transparent gas. If we were to watch the cosmic drama running backward in time, we'd see this gas get gradually hotter, with its atoms smashing into each other progressively harder until they break apart into atomic nuclei and free electrons—a plasma. Then we'd see the helium atoms get smashed apart into protons and neutrons. Then these get smashed into their building blocks: quarks. Then we cross our knowledge frontier and enter the realm of scientific speculation—in Chapter 5, we'll explore what's labeled "inflation" and "quantum fuzz" in Figure 3.7. If we jump back to a million years after our Big Bang and instead let time run forward, we see gravity amplify the slight clumping of the gas into galaxies, stars and the rich cosmic structure we observe around us today.

But gravity can only amplify small fluctuations into larger fluctuations—it can't create fluctuations out of nowhere. If something is perfectly smooth and uniform, gravity will keep it that way forever, unable to create any dense clumps, let alone galaxies. This means that, early on, there must have been small seed fluctuations for gravity to amplify, acting like a form of cosmic blueprints that determined where galaxies would form. Where did these seed fluctuations come from? In other words, we've seen where the atoms in our Universe came from, but what about the grand galactic patterns into which they got arranged? Where did the cosmic large-scale structure come from? Of the many questions we've asked in cosmology, I feel that this one has turned out to be the most fruitful of all. In the next two chapters, let's explore why.

THE BOTTOM LINE

- Because distant light takes time to reach us, telescopes let us see cosmic history unfold.
- About 14 billion years ago, everything that we can now observe was hotter than the core of the Sun, expanding so fast that it doubled its size in under a second. This is what I call our Big Bang.
- Although we don't know what happened earlier, we know a great deal about what's happened since then: expansion and clustering.
- Our Universe spent a few minutes being a giant fusion reactor, like the core of our Sun, converting hydrogen into helium and other light elements, until the cosmic expansion diluted and cooled our Universe enough to stop the fusion.
- Doing the math, we predict that about 25% of the hydrogen turned into helium; measurements beautifully agree with this prediction and also match the predictions for other light elements.
- After another 400,000 years of expansion and dilution, this hydrogen-helium plasma had cooled into transparent gas. We see this transition as a distant plasma wall whose faint glow has become known as the cosmic microwave background, triggering two Nobel Prizes.
- Over the billions of years that followed, gravity transformed our Universe from uniform and boring to clumpy and interesting, amplifying the tiny density fluctuations that we see in the cosmic microwave background into planets, stars, galaxies and the cosmic large-scale structure that we see around us today.
- The cosmic expansion predicts that distant galaxies should be receding from us according to a simple formula, in good agreement with what we actually observe.
- This entire history of our Universe is accurately described by simple physical laws that let us predict the future from the past, and the past from the future.
- These physical laws that govern the history of our Universe are all cast in terms of mathematical equations, so our most accurate description of our cosmic history is a mathematical description.

4

Our Universe by Numbers

Cosmologists are often wrong, but never in doubt.

—Lev Landau

In theory, theory and practice are the same. In practice, they are not.

—Albert Einstein

"Wow!" My jaw dropped, and I stood there by the roadside utterly speechless. I'd looked at it every day of my life, yet I'd never really seen it before. It was about five a.m., and I'd pulled off the highway through the Arizona desert to check the map. When all of a sudden, it hit me: the sky! This wasn't the lame light-polluted Stockholm firmament under which I'd grown up, with the Big Dipper and a sparse smattering of other dim stars. It was spectacular and absolutely overwhelming, with thousands of brilliant points of light forming beautiful and intricate patterns, and the Milky Way glowing like a magnificent Galactic highway across the sky.

My view was enhanced by the dry desert air and being more than two kilometers above sea level, but I suspect that you, too, have at some point gotten far enough from city lights to be awed by the sky. So what exactly was it that we marveled at? Partly the stars themselves, no doubt, and the vastness of it all. But also something else: the *patterns*. Our ancestors were so intrigued by them that they created myths to explain them, and some cultures imagined them grouped into constellations depicting mythological figures. The stars clearly aren't uniformly spread across the sky like polka dots, but clustered. The largest stellar clustering pattern I saw that night was our Milky Way Galaxy, and our telescopes have revealed that galaxies, too, are clustered into intricate patterns, forming groups, galaxy clusters, and enormous filamentary patterns spanning hundreds of millions of light-years. Where did these patterns come from? What's the origin of this grand cosmic structure?

At the end of the last chapter, our exploration of the destabilizing

effects of gravity also led us to wonder about the origin of the cosmic large-scale structure. In other words, we're led intellectually to
the same question that we're led to emotionally when awestruck by the
sky: whence the structure? This is the key question that we're going to
explore in this chapter.

Wanted: Precision Cosmology

As we saw in the last chapter, we humans still don't understand the
ultimate origins of our Universe, specifically what happened before the
epoch when our Universe was a giant fusion reactor and doubled its
size in under a second. However, we now understand a great deal about
what's happened during the 14 billion years since then: *expansion* and
clustering. These two basic processes, both controlled by gravity, have
transformed hot, smooth quark soup into today's star-studded cosmos.
In the last chapter's fast-forward history of our Universe, we saw that the
gradual expansion diluted and cooled the elementary particles, enabling
them to cluster into ever-larger structures such as atomic nuclei, atoms,
molecules, stars and galaxies. We know of four fundamental forces of
nature, and three of them have taken turns driving this clustering process: first the strong nuclear force pulled the nuclei together, then the
electromagnetic force made the atoms and the molecules, and finally
gravity built the grand structures that adorn our night sky.

How exactly did gravity do this? If you stop your bike at a red light,
you quickly realize that gravity can be destabilizing: you inevitably start
tipping sideways and need to put your foot down on the asphalt to avoid
falling. The essence of instability is that small fluctuations get amplified. In the stopped-bike example, the farther you get from balance,
the stronger gravity will push you in the wrong direction. In the cosmic example, the farther our Universe gets from perfect uniformity, the
more forcefully gravity amplifies its clumpiness. If a region of space is
slightly denser than its surroundings, then its gravity will pull in neighboring material and make it even denser. Now its gravity is even stronger, making it accrete mass even faster. Just as it's easier to make money
when you have lots of it, it's easier to accrete mass when you have lots
of it. Fourteen billion years was ample time for this gravitational instability to transform our Universe from boring to interesting, amplifying
even tiny density fluctuations into huge dense clumps such as galaxies.

Although this basic picture of expansion and clustering had been worked out during the preceding decades, the details remained sketchy when I started grad school in 1990 and got my first exposure to cosmology. People were still arguing about whether our Universe was 10 billion years old or 20 billion years old, reflecting a long-standing dispute about how fast it was currently expanding, and the harder question of how fast it had expanded in the past was wide open. The clustering story was on even shakier grounds: attempts to get detailed agreement between theory and observation gradually revealed that we had no clue what 96% of our Universe was made of! After the COBE experiment measured 0.002% clumping 400,000 years after our Big Bang, it became clear that gravity wouldn't have had time to amplify this faint clustering into today's cosmic large-scale structure unless some invisible form of matter contributed extra gravitational pull.

This mysterious stuff is known as *dark matter,* which is really little more than a name for our ignorance. The name *invisible matter* would be more apt, since it looks transparent rather than dark, and can pass through your hand without your noticing. Indeed, dark matter from space that strikes Earth appears to typically pass unaffected through our entire planet, emerging unscathed on the other side. As if dark matter wasn't crazy enough, a second mystery substance dubbed *dark energy* was introduced to make the theoretical predictions match the observed expansion and clustering. It was assumed to affect the cosmic expansion without clustering at all, remaining perfectly uniform at all times.

Both dark matter and dark energy had had a long and controversial history. The simplest candidate for dark energy was the so-called cosmological constant, the above-mentioned fudge factor that Einstein added to his gravity theory and later called his greatest blunder. Fritz Zwicky postulated dark matter in 1934 to explain the extra gravitational pull that kept galaxy clusters from flying apart, and Vera Rubin dis-

What dark matter looks like:	What dark energy looks like:

Figure 4.1: Both dark matter and dark energy are invisible, which means that they refuse to interact with light and other electromagnetic phenomena. We know of their existence only through their gravitational effects.

covered in the 1960s that spiral galaxies rotated so fast that they, too, would fly apart unless they contained invisible mass gravitating enough to hold them together. These ideas met with considerable skepticism: if we're willing to blame unexplained phenomena on entities that are both invisible and can pass through walls, shouldn't we also start believing in ghosts while we're at it? Moreover, there was a disturbing precedent: in ancient Greece, when Ptolemy realized that planetary orbits weren't perfect circles, he cooked up a complicated theory in which they moved on smaller circles (called epicycles) that in turn moved in circles. As we saw earlier, the subsequent discovery of a more accurate law of gravity killed the epicycles, predicting that the orbits aren't circles but ellipses. Perhaps the need for dark matter and dark energy could be eliminated just as the epicycles were, by discovering a still more accurate law of gravity? Could modern cosmology really be taken seriously?

These were the sorts of questions we asked while I was a grad student. Answering them would require much more accurate measurements, to transform cosmology from the data-starved and speculative field that it was into a precision science. Fortunately, that's exactly what happened.

Precision Microwave-Background Fluctuations

As we saw in Figure 3.6, the baby picture of our Universe produced by a cosmic microwave–background experiment can be decomposed as a sum of many different component maps called *multipoles* which, in essence, contain the contributions from spots of different sizes. Figure 4.2 plots the total amount of fluctuation in each of these multipoles; this curve is called the *power spectrum* of the microwave background, and encodes the key cosmological information from the map. When you look at a sky map like the one in Figure 3.4 and on the book cover, you see spots of many different sizes just as on a Dalmatian: some spots are about 1 degree across in the sky, others are 2 degrees, and so on. The power spectrum encodes information about how many spots there are of each size.

What's so great about the power spectrum is that not only can we measure it, but we can also predict it: for any mathematically defined model of how our Universe expanded and clustered, we can calculate exactly what the power spectrum should be. As you can see in Figure 4.2, the predictions differ wildly between models: indeed, the measure-

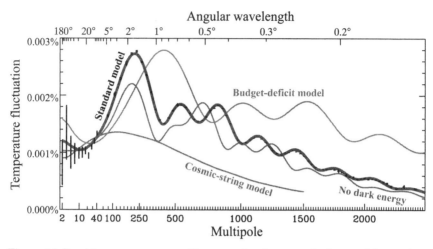

Figure 4.2: Precision measurements of how cosmic microwave–background fluctuations depend on angular scale have totally ruled out many previously popular theoretical models, but agree beautifully with the curve predicted by the current standard model. You can appreciate the most remarkable aspect of modern cosmology here without worrying about any of the details: highly accurate measurements now exist, and they agree with theoretical prediction.

ments have now ruled out all but one of the theoretical models in Figure 4.2 beyond any reasonable doubt, even though for each of the killed models, I have at least one respected colleague who used to believe that it was the right one back when I was in grad school. The predicted shape of the power spectrum depends in complicated ways on all the things that affect cosmic clustering (including the density of atoms, the density of dark matter, the density of dark energy and the nature of the seed fluctuations), so if we can adjust our assumptions about all these things so that the prediction matches what we measure, then we've not only found a model that works, but also measured these important physical quantities.

Telescopes and Computers

When I first learned about the cosmic microwave background in grad school, there were no measurements whatsoever of the power spectrum. Then the COBE team gave us our first grip on this elusive wiggly curve, measuring that its height on the leftmost side was about 0.001% and that its slope around there was roughly horizontal. There was more information about the power spectrum in the COBE data, but nobody had squeezed it out because this would involve tedious manipulations of a table of numbers called a matrix, which took up 31 megabytes. Although this quantity sounds like a joke these days, being the size of a short video clip on your phone, it was daunting in 1992. So my classmate Ted Bunn and I hatched a sneaky plan: Professor Marc Davis in our department owned a computer called "magicbean" that had more than 32 megabytes of memory, and night after night, I logged on to it in the wee hours of the morning when nobody was paying attention, and let it work on our data analysis. After a few weeks of this clandestine moonshine number crunching, we published a paper with the most accurate measurement so far of the power spectrum shape.

This project made me realize that, just as telescopes had once transformed astronomy, dramatic improvements in computer technology had the power to take it to yet another level. Indeed, your own computer today is so much better that it could repeat all my calculations with Ted in minutes. I decided that if experimentalists were putting so much hard work into collecting this data about our Universe, people like me owed it to them to milk their data for all it was worth. This became a central theme of my work for the next decade.

One question I obsessed about was how to best measure the power spectrum. There were quick methods that suffered from inaccuracies and other problems. Then there was the optimal method that my friend Andrew Hamilton had worked out, which unfortunately required an amount of computer time that grew as the sixth power of the number of pixels in the sky map, so measuring the power spectrum from the COBE map would take longer than the age of our Universe.

It's November 21, 1996, and it's dark and quiet at the Institute for Advanced Study in Princeton, New Jersey, where I'm having another crazy night in the office. I'm excited about an idea for replacing the sixth power in Andrew Hamilton's method by the third power, enabling me to optimally measure the COBE power spectrum in under an hour, and I'm scrambling to finish my paper in time for a Princeton conference the next day. In the physics community, we post all our papers to this free website http://arXiv.org as soon as we've finished them, so that our colleagues can read them before they get bogged down in the refereeing and publishing process. The problem was that I had this terrible habit of submitting my papers *before* I'd finished them, right after the day's submission deadline ended. This way I could be first in the next daily paper listing. The downside was that, if I failed to finish within twenty-four hours, I'd be publicly humiliated by having an unfinished draft displayed to the world as a permanent monument to my stupidity. This time, my strategy finally backfired, with early birds in Europe getting access to the incomplete mess that was my discussion section before I finally finished it around four a.m. At the conference, my friend Lloyd Knox presented a similar method that he'd developed with Andrew Jaffe and Dick Bond in Toronto, but hadn't yet written up for publication. When I presented my results, Lloyd smiled and said to Dick: "Fast fingers Tegmark!" Our methods turned out to be quite useful, and have been used for essentially all microwave-background power-spectrum measurements since then. Lloyd and I seem to follow parallel paths through life: we have the same ideas at the same time (indeed, he'd scooped me earlier on a cool formula for noise in microwave-background maps), we got two sons at the same time, and we even got divorced at the same time.

Gold in the Hills

With improved experiments, computers and methods, the measurements of the power-spectrum curve in Figure 4.2 kept getting better. As you can see in the figure, the curve was predicted to look a bit like the rolling hills of California, with a series of distinct peaks. If you measure lots of Great Danes, poodles and Chihuahuas and plot their distribution of sizes, you'll get a curve with three peaks. Similarly, if you measure lots of cosmic microwave–background spots as shown in Figure 3.4 and plot their distribution of sizes, you'll find that there are certain characteristic spot sizes that are particularly common. The most prominent peak in Figure 4.2 corresponds to spots that are about one degree in angular size. Why? Well, these spots were caused by sound waves rippling through the cosmic plasma near the speed of light, so since the plasma existed for 400,000 years after our Big Bang, the spots grew to be about 400,000 light-years in size. If you calculate how large an angle such a 400,000 light-year blob will cover in the sky today, 14 billion years later, you get about a degree. Unless space is curved, that is. . . .

As we discussed in Chapter 2, there is more than one kind of uniform three-dimensional space: in addition to the flat kind that Euclid axiomatized and we all learned about in school, there are curved spaces where the angles obey different rules. I learned in school that the angles in a triangle on a flat sheet of paper add up to 180 degrees. But if you draw it on the curved surface of an orange, they'll add up to more than 180 degrees, and if you draw it on a saddle, they'll add up to less (see Figure 2.7 for examples). Similarly, if our physical space were curved like a spherical surface, the angle covered by each microwave-background spot would be bigger, shifting the peaks in the power-spectrum curve to the left, and if space had saddlelike curvature, the spots would look smaller and shift the peaks to the right.

To me, one of the most beautiful ideas in Einstein's gravity theory is that geometry isn't just mathematics: it's also physics. Specifically, Einstein's equations show that the more matter space contains, the more curved space gets. This curvature of space causes things to move not in straight lines, but in a motion that curves toward massive objects—thus explaining gravity as a manifestation of geometry. This opens up a totally new way of weighing our Universe: just measure the first peak in the cosmic microwave–background power spectrum! If its position shows that space is flat, then Einstein's equations tell us that our average cosmic density is about 10^{-26}kg/m^3, corresponding to about ten milligrams per Earth volume or about six hydrogen atoms per cubic meter. If the peak is farther to the left, the density is higher, and vice versa. Given all the confusion about dark matter and dark energy, measuring this total density was hugely important, and experimental teams around the world raced for the first peak—which was expected to be the easiest peak to detect because bigger spots are easier to measure than smaller ones.

I caught my first glimpse of the peak in 1996, in a paper spearheaded by Lyman Page's student Barth Netterfield using Saskatoon data. "*Wow!*" I thought, and had to put down my spoonful of Munich müsli to really take it in. At the cerebral level, the theory behind the power-spectrum peaks was very elegant and all, but in my gut, I still felt that our human extrapolations couldn't work this well. Three years later, Lyman Page's student Amber Miller spearheaded a more accurate measurement of the first peak, and found it to be roughly in the right place for a flat universe, but somehow, it all still felt too good to be true. Finally, in April 2000, I just had to accept it. A microwave telescope called Boomerang, which had spent eleven days circumnavigating Antarctica, dangling under a high-altitude balloon the size of a football field, had produced by far the most accurate power-spectrum measurements to date, showing a beautiful first peak at exactly the place for a flat universe. So now we knew the total density of our Universe (averaged over space).

Dark Energy

This measurement presents an interesting situation when accounting for the cosmic-matter budget. As you can see in Figure 4.3, we know the size of the total budget from the first peak position, but we also know the density of ordinary matter, and of dark matter from measuring its gravitational effects on cosmic clustering. But all this matter makes up only 30% of the total budget, which means that the remaining 70% must be some form of matter that doesn't cluster—so-called dark energy.

The most impressive thing I just said is what I didn't say: the word *supernova*. Because completely independent evidence for dark energy suggested exactly the same 70% number, based on cosmic expansion rather than clustering. Earlier, we talked about using Cepheid variable stars as standard candles to measure cosmic distances. We cosmologists now have another standard candle in our toolbox that's even more luminous, so that it can be seen not only millions but even billions of light-years away. These are huge cosmic explosions known as Type Ia supernovae, which during a few seconds can release more energy than a hundred million billion suns.

Do you remember the rest of the first verse of "Twinkle, Twinkle, Little Star"? When Jane Taylor wrote the lines "Up above the world so high, / Like a diamond in the sky", she had no idea how right she was: when our Sun eventually dies in about 5 billion years, it will end its days as a so-called white dwarf, which is a giant ball that—like a diamond—is made mostly of carbon atoms. Our Universe is teeming with white dwarfs today, created by stars past. Many of them are con-

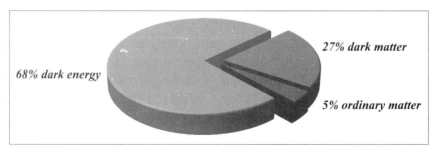

68% dark energy

27% dark matter

5% ordinary matter

Figure 4.3: The cosmic matter budget. The horizontal positions of the microwave-background power-spectrum peaks tell us that space is flat and the total cosmic density is about a million trillion trillion (10^{30}) times lower than that of water (averaged across our Universe). The peak heights tell us that ordinary and dark matter make up only about 30% of this density, so there must be 70% of something else—dark energy.

tinually gaining weight by gobbling up gas from dying companion stars that they're orbiting. Once they become officially overweight (which happens when they reach 1.4 times the mass of the Sun), they suffer the stellar equivalent of a heart attack: they become unstable and detonate in a gigantic thermonuclear explosion—a Type Ia supernova. Since all these cosmic bombs therefore have the same mass, it comes as no surprise that they're roughly equally powerful.

Moreover, the slight variations in explosive power have been shown to be linked both to the spectrum of the explosion and to how fast it brightens and dims, all of which can be measured, allowing astronomers to turn supernovae Ia into accurate standard candles.

This technique was used by Saul Perlmutter, Adam Riess, Brian Schmidt, Robert Kirshner and their collaborators to accurately measure the distances to lots of supernovae Ia and also how fast they were receding from us based on their redshift. From these measurements, they made the most accurate reconstruction to date of how fast our Universe has expanded at various times in the past, and in 1998, they announced a startling discovery that earned them the 2011 Nobel Prize in physics: after spending its first 7 billion years slowing down, the cosmic expansion started speeding up again and has accelerated ever since! If you throw a rock up in the air, Earth's gravitational pull will decelerate its motion away from Earth, so the cosmic acceleration revealed a strange gravitational force that's repulsive rather than attractive. As I'll explain in the next chapter, Einstein's gravity theory predicts that dark energy has exactly this antigravity effect, and the supernova teams found that a cosmic-matter budget with 70% dark energy beautifully explained what they saw.

A 50% Batting Average

One of my favorite things about being a scientist is getting to work with such cool people. The person I've coauthored the most papers with is a friendly Argentinian named Matias Zaldarriaga. My ex-wife and I secretly nicknamed him "the Great Zalda," and we agreed that the only thing that topped his talent was his sense of humor. He'd cowritten the computer program everyone used to predict power-spectrum curves like those in Figure 4.2, and he once bet an air ticket to Argentina that his predictions were all wrong and there were no peaks. In preparation for the Boomerang results, we sped up these calculations and computed a huge database of models against which we could compare measurements. So when the Boomerang data became available, I again posted an unfinished paper to http://arXiv.org, and then we had fun working around the clock to finish it before it went public on Sunday evening. Ordinary matter (atoms) can bump into stuff that dark matter simply sails through, and therefore ends up moving differently through space. This means that ordinary and dark matter affects cosmic clustering and the microwave-background power-spectrum curve (see Figure 4.2) in different ways. In particular, adding more atoms to the matter budget lowers the second peak. The Boomerang team reported a really puny second peak, and Matias and I found that this required atoms to make up at least 6% of the cosmic-matter budget. But Big Bang nucleosynthesis, the cosmic fusion–reactor story we discussed in Chapter 3, only works if atoms make up 5%, so something was wrong! I spent these crazy days in Albuquerque where I'd gone to give a talk, and it felt really exhilarating to get to tell the audience about these brand-new clues that our Universe had revealed. Matias and I just barely made our deadline, and our paper hit the Web even before the Boomerang team's own analysis paper, which was delayed by a pedantic computer on the ridiculous grounds that a figure caption was one word too long.

Cross-checking is a bad thing in ice hockey but a good thing in science, where it can reveal hidden mistakes. Boomerang let us cosmologists make two cross-checks on the cosmic-matter budget:

1. We measured the dark-energy fraction in two different ways (with supernovae Ia and with cosmic microwave–background peaks) and the answers agreed.
2. We measured the ordinary-matter fraction in two different ways (with Big Bang nucleosynthesis and with cosmic microwave–background peaks) and the answers disagreed, so at least one of the two methods was messed up.

The Bump Is Back

A year later, I'm in a swanky press conference room in Washington DC, glued to my seat, feeling like Santa Claus is about to arrive three times over. First up was John Carlstrom, reporting results from a microwave telescope called DASI at the South Pole. After the usual blah blah about technical details I already knew—boom!—the most amazing power-spectrum plot I'd ever seen! With as many as three peaks clearly visible. Then came Santa 2: John Ruhl from the Boomerang team. Blah blah—boom! Another amazing power spectrum with three peaks, in beautiful agreement with the independent DASI measurements. And the once-so-anemic second peak was bigger this time, after they'd improved the modeling of their telescope. Finally, Santa 3: Paul Richards reported measurements from a balloon experiment called MAXIMA, agreeing well with the others' data. I was simply amazed. After all these years of dreaming about these elusive clues encoded in the microwave-background sky, here they were! It had felt so hubristic to assume that we humans knew what our Universe was doing just a few hundred thousand years after our Big Bang, yet we'd been right. That night I quickly reran my model-fitting software with the new microwave-background data, and now that the second peak was higher, my code predicted about 5% atoms, in beautiful agreement with Big Bang nucleosynthesis. The atom cross-check had gone from failure to success, and order was restored in the cosmos. And the order remained: by now, WMAP, Planck and other experiments have measured the power-spectrum curve way more accurately, as you can see in Figure 4.2, showing that these early experiments really got it right.

Precision Galaxy Clustering

By 2003, the cosmic microwave–background radiation had become arguably cosmology's greatest success story ever. It became widely perceived as a panacea that could solve all our problems and let us measure all the key numbers in our cosmological model. But this perception was incorrect. Suppose that you measure my weight to be two hundred pounds. This clearly isn't enough information for you to determine my height and my width, since my weight depends on both: I could be either tall and slim or short and chubby. We have analogous problems when trying to measure key numbers about our Universe. For example, the characteristic microwave background–spot sizes that correspond to the horizontal locations of the cosmic microwave–background peaks in Figure 4.2 depend on both the curvature of space (which magnifies/demagnifies the spots) and on the dark-energy density (which changes the expansion rate of our Universe and therefore the distance to the plasma surface with its spots, also causing them to look larger or smaller). So although many journalists claimed that experiments such as Boomerang and WMAP had shown space to be flat, they hadn't: our Universe could be either flat with about 70% dark energy or curved with a different amount of dark energy. There are other pairs of cosmological parameters that are similarly hard for the microwave background to tease apart, for example the amplitude of clumping in our early Universe and the time when the first stars formed, both of which affect the Figure 4.2 power spectrum in similar ways (in this case changing the peak heights). As we're taught in high-school algebra, we need more than one equation to determine two unknown quantities. In cosmology we want to determine about seven numbers, and the microwave background alone simply doesn't contain enough information to allow this. So we need additional information from other cosmological measurements. Such as 3-D galaxy maps.

Galaxy Redshift Surveys

When we make a 3-D map of where the galaxies are in our Universe, we first analyze 2-D photos of the sky to find galaxies, then make additional measurements to figure out how far away they are. The most ambitious 3-D mapping project to date is called the Sloan Digital Sky Survey, and I had the great fortune to get to join it when I was a postdoc in Princeton even though a small army of people had already spent almost a decade organizing the project, building the telescope hardware, and making things work. It's made the 2-D sky map shown in Figure 4.4 by spending over a decade imaging over a third of the sky with a custom-built 2.5 meter telescope in New Mexico. Jim Gunn, a Princeton professor who reminded me of a friendly bearded wizard, used his magical powers

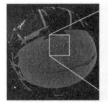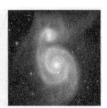

Figure 4.4: The Sloan Digital Sky Survey contains an astonishing amount of information. The left panel, where the sphere represents the whole sky, contains almost a terapixel, a million megapixels. The successive zooms focus behind the Big Dipper on the so-called Whirlpool galaxy, but the same level of detail is available everywhere you zoom. *(Image credit: Mike Blanton and David Hogg/SDSS Collaboration)*

to build this amazing digital camera for the telescope, then the largest camera ever made for astronomical purposes.

If you carefully look through the survey's sky images, like those in Figure 4.5, you'll find lots of stars, galaxies and other objects—more than half a billion of them, in fact. This multiplicity means that if you tell a grad student to find all the objects at a rate of one per second, working eight hours a day without breaks for weekends or holidays, you'll be waiting for fifty years—and get the award for worst thesis advisor ever. This object finding is a surprisingly tricky problem even for a computer: it needs to be able to distinguish between galaxies (which look fuzzy and spread-out), stars (which would look pointlike were it not for atmospheric blurring), comets, satellites and various instrumental artifacts. Worse yet: sometimes objects overlap, as when a nearby star is annoyingly located in front of a distant galaxy. After a large group of people had struggled with this problem for years, it was solved through a heroic programming effort by Robert Lupton, a chipper Englishman who used "Robert Lupton the Good" as his email name and was always barefoot (Figure 4.5).

The next step is to figure out how far away each galaxy is. In Chapter 3 we saw how Edwin Hubble's law $v = Hd$ means that our Universe is expanding, so the greater the distance d to a faraway galaxy, the higher the velocity v with which it recedes from us. Now that Hubble's law has been firmly established, we can use it backwards, as a method to measure distances: by measuring how fast a galaxy recedes using the redshifting

Figure 4.5: A small fraction of the Sloan Digital Sky Survey map has been used to decorate an entire wall in the Princeton University Astronomy Department, which Robert Lupton can be seen scrutinizing with my kids. After Robert's software identifies all objects in the map, the distances to the most interesting galaxies are measured, producing a 3-D map (left) with us at the center and every dot representing a galaxy. You can see the "Sloan Great Wall" about a third of the way from the top in the image.

of its spectral lines, we learn its distance. Basically, measuring redshifts and velocities is easy in astronomy while measuring distances is hard, so Hubble's law can save you work: once you've measured the Hubble parameter H using some nearby galaxies, you just need to measure the velocities v of distant galaxies from their redshifted spectra, then divide by H to get a good estimate of their distance.

From the catalog of objects churned out by Robert Lupton's software, the most interesting million or so were selected to have their spectra measured. The twenty-four galaxy spectra that Edwin Hubble had used to discover our cosmic expansion took weeks to collect at the time. In contrast, the Sloan Digital Sky Survey could mass-produce spectra at a rate of 640 per hour, all measured at the same time. The trick was to position 640 optical fibers at the places in the focal plane of the telescope where Robert's catalog said that the galaxy images would be, and then have these fibers lead the galaxy light to a spectrograph that split it into 640 separate rainbows imaged by a digital camera. Another software package, this one spearheaded by David Schlegel and colleagues, analyzed these rainbows to figure out the distance to each galaxy (from the redshifting of its spectral lines) and other galaxy properties.

On the leftmost side of Figure 4.5, I've plotted a 3D slice of our Universe, with each point representing a galaxy; when I feel I need to get away from it all for a while, I like to fly around among the galaxies with a 3-D cosmological-flight simulator I have. Doing this reveals something I find quite beautiful: we're part of something grander. Not only is our planet a part of a solar system and our Solar System part of a galaxy, but our Galaxy is part of a cosmic web of galaxy groups, clusters, superclusters and gigantic filamentary structures. While poring over this map and noticing what's now become known as the "Sloan Great Wall" (Figure 4.5, left), I was so flabbergasted by its size that I first suspected a bug in my code. But some of my collaborators discovered it independently and it's definitely real: being 1.4 billion light-years long, it's the largest known structure in our Universe. These large-scale clustering patterns are a cosmological treasure trove, encoding the sort of valuable information that the cosmic microwave background is missing.

From Derision Cosmology to Precision Cosmology

These patterns in the galaxy distribution are really the same patterns that we saw manifest themselves in cosmic microwave–background maps—only billions of years later, amplified by gravity. In a region of space where gas was once 0.001% denser than its surroundings, causing a spot in the WMAP map (see Figure 3.4), there might today be a cluster of one hundred galaxies. In this sense, we can think of the cosmic microwave–background fluctuations as the cosmic DNA, the blueprints for what our Universe will grow to become. By comparing the slight past clustering seen in the cosmic microwave background to the strong current clustering seen in a 3-D galaxy map, we can measure the detailed nature of the stuff whose gravity caused the clustering to grow between then and now.

Just as the microwave-background clustering is characterized by a power-spectrum curve (see Figure 4.2), so is the galaxy clustering. However, measuring this curve really accurately turns out to be really

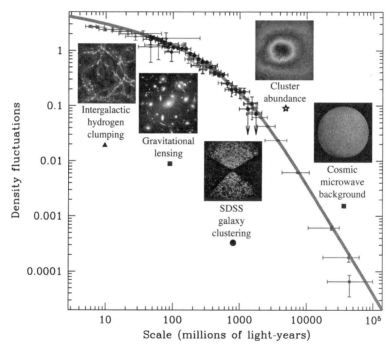

Figure 4.6: The clumping of matter in our Universe is described by the power-spectrum curve shown here. The fact that the curve equals 10% at 1,000 million light-years crudely speaking means that if you measure the amount of mass in a sphere of that radius, then the answer you'll get will vary by 10%, depending on where in space you put the sphere. In contrast to when I started my career, highly accurate measurements now exist, and they agree with theoretical prediction. I find it particularly remarkable that the five different measurements of this curve agree with each other even though the data, the people and the methods involved are totally different.

hard: the Sloan Digital Sky Survey galaxy–power spectrum measurement shown in Figure 4.6 took me six years (six!) to finish, despite lots of help from colleagues, and ended up being my most exhausting project ever. Time and again, I'd think, *Thank goodness I'm finally almost done, since I just can't take this any longer!* just to discover a major new problem with the analysis.

Why was it so hard? Well, it would be easy if we knew the exact position of every galaxy in our Universe and had an infinitely powerful computer with which to analyze them. In practice, we can't see many of the galaxies because of various complications, and some of the ones we do see have a different distance and luminosity than we think. If we ignore these complications, we get an incorrect power spectrum that translates into incorrect conclusions about our Universe.

The first 3-D galaxy maps were so small that it wasn't worth putting lots of time into analyzing them. My colleague Michael Vogeley gave me a nice plot summarizing all the measurements up to 1996 or so, and when I asked him why he hadn't put error bars on them to indicate the measurement uncertainty, he said, "Because I don't believe them." He had good reason for his skepticism: some teams claimed ten times more power than others, so they couldn't all be right.

Groups around the world gradually made bigger 3-D maps and shared them online. I felt that when so many people were putting so much hard work into making these maps, they deserved a really careful analysis. So I teamed up with my friend Andrew Hamilton to go the extra mile, measuring galaxy power spectra with the same sorts of information-theory methods we'd developed for cosmic microwave–background analysis.

Andrew is an incurably cheerful Brit with a mischievous, bright smile, and one of my favorite collaborators. I once showed up late at a restaurant where I was supposed to meet Andrew as well as my friends Wayne Hu and David Hogg, who had recently shaved his head. When I asked a waitress if she'd seen a trio who looked like Robert Redford, Bruce Lee and Kojak, she thought for a moment, then smiled and said: "I can see Robert Redford. . . ." We first analyzed progressively larger 3-D maps with obscure names such as IRAS, PSCz, UZC and 2dF, with about 5,000, 15,000, 20,000 and 100,000 galaxies, respectively. He lived in Colorado, and we had endless conversations about the mathematical intricacies of power-spectrum measurement by email, by phone, and while hiking in the Alps and the Rocky Mountains.

The Sloan Digital Sky Survey map was the largest and cleanest survey of all, based on all-digital imaging and meticulous quality control, so I felt that it also deserved the most painstaking analysis. Because the results would only be as good as the weakest link, I spent years working on many of the dirty-laundry issues that people considered the most boring. Professor Jill Knapp, one of the driving forces in the project and also Jim Gunn's wife, would organize weekly meetings in Princeton where she'd spoil us all with irresistible food while we tried to identify all the skeletons in the analysis closet and figure out what to do about them. For example, how many galaxies we'd map in a particular direction depended on how bad the weather was while it was imaged, how much Galactic dust was in the way, and on the fraction of the visible galaxies that could be covered by optical fibers. Frankly, this stuff truly was boring, so I'll spare you the details, but I nonetheless got huge amounts of help from many people, particularly Professor Michael Strauss and his then grad student Mike Blanton. In parallel, there was the seemingly never-ending cycle of computing terabytes of number tables called matrices during multiweek computer runs, looking at messed-up result plots, debugging my code, and trying again.

After six years of this, I finally submitted two papers with results in 2003, both with over sixty coauthors. I've never in my entire life felt as relieved to finish something, except perhaps for this book. The first paper concerned the galaxy–power spectrum measurement in Figure 4.6, and the second dealt with a measurement of cosmological parameters from combining this with the microwave-background power spectrum. I've listed some of the highlights in Table 4.1; here I've updated the numbers to the most recent measurements by others, but the values haven't changed significantly even though the uncertainties have decreased. I still had in fresh memory the wild debates from my grad-student days about whether our Universe was 10 billion or 20 billion years old, and now we were arguing over whether it was 13.7 or 13.8 billion! Precision cosmology had finally arrived, and I felt excited and honored to have gotten to play a small part in this.

At a personal level, this outcome was quite lucky for me: they evaluated me for tenure at MIT in the fall of 2004, and I'd been told that to get it, I needed "a home run, or at least a couple of doubles." Just as musicians have their top-ten sales charts, we scientists have our citation lists: every time someone cites your paper, it counts as a feather in your hat. The citation business can be rather random and silly, prone to

Parameter name	Parameter symbol	Measured value	Uncertainty
Atom fraction	Ω_b	0.049	2%
Dark-matter fraction	Ω_d	0.27	4%
Dark-energy fraction	Ω_Λ	0.68	1%
Neutrino fraction	Ω_v	0.003	100%
Budget total	Ω_{tot}	1.001	0.7%
Age of Universe in gigayears	t_0	13.80	0.2%
Seed-clustering amplitude	Q	0.0000195	3%
Seed-clustering "tilt"	n	0.96	0.5%

Table 4.1: By combining cosmic microwave–background maps with 3-D galaxy maps, we can measure key cosmic quantities to percent-level precision.

bandwagon effects, since lazy authors tend to copy citations from others without actually reading the papers they cite, but promotion committees nonetheless care as much about citation rates as baseball coaches care about batting averages. And now, just when I could really use some luck, these two papers suddenly became my most-cited ones ever, one even grabbing the spot as the most-cited physics paper of 2004—that distinction didn't last long, but long enough for the tenure decision. My dumb luck continued with the magazine *Science* deciding that the number-one "Breakthrough of the Year: 2003" was that cosmology had finally become believable, mentioning both the WMAP results and our Sloan Digital Sky Survey analysis.

Honestly, though, this data wasn't a breakthrough at all, just a reflection of the slow but steady progress that the worldwide cosmology community had made in recent years. Our work wasn't in any way revolutionary and didn't discover anything surprising—rather, it simply contributed to making cosmology more believable, to its growing up into a more mature science. To me, the most surprising result was that there was no surprise.

The famous Soviet physicist Lev Landau once said that "cosmologists are often wrong, but never in doubt," and we've seen many examples of this, from Aristarchos claiming the Sun was eighteen times too close, to Hubble claiming our Universe was expanding seven times too fast. This Wild West phase is now over: we saw how both Big Bang nucleosynthesis and cosmic clustering gave the same measurement of the atom density, and how both supernovae Ia and cosmic clustering gave the same measurement of the dark-energy density. Of all cross-checks, my favorite is the one in Figure 4.6: here I've plotted five different measurements of the power-spectrum curve, and even though the data, the people, and the methods involved are totally different for all five, you can see that they all agree.

The Ultimate Map of Our Universe

A Lot Left to Explore

So here I am sitting in my bed, typing these words and thinking about how cosmology has changed. Back when I was a postdoc, we used to talk about how cool it was going to be to get all that precision data and finally measure those cosmological parameters accurately. Now we can say, "Been there, done that": the answers are in Table 4.1. So now what? Is cosmology over? Do we cosmologists need to find something else to do?

Here's my answer: "No!" To appreciate how much fun cosmology research remains, let's be honest about how little we cosmologists have accomplished: we've mainly just parameterized our ignorance, in the sense that behind each parameter in Table 4.1 lies an unexplained mystery. For example:

- We've measured the density of dark matter, but what is it?
- We've measured the density of dark energy, but what is it?
- We've measured the density of atoms (there's about one atom for every two billion photons)—but what process produced that amount?
- We've measured that the seed fluctuations were at the level of 0.002%—but what process created them?

As data continue to improve, we'll be able to use it to measure the numbers in Table 4.1 even more accurately, to more decimal places. But I'm a lot more excited about using the better data to measure *new* parameters. For example, we can try to measure other properties of dark matter and dark energy besides their densities. Does the dark matter have a pressure? A velocity? A temperature? This would shed light on its nature. Is the dark-energy density really exactly constant, as it so far seems? If we can measure that it changes even slightly over time or from place to place, this will be a crucial clue about its nature and about how dark energy will affect the future of our Universe. Do the seed fluctuations have any patterns or properties besides their 0.002% amplitude? This would provide clues about the origins of our Universe.

I've thought a lot about what we need to do to tackle these questions, and interestingly, the answer is the same for all of them: map our Universe! Specifically, we need to map as much as possible of our Universe in 3-D. The largest volume we can possibly map is the part of space that light has had time to reach us from so far. This volume is essentially the interior of the plasma sphere (Figure 4.7, left) that we have explored, and as you can see in the middle panel of this figure, over 99.9% of the volume remains unexplored. You can also see that our most ambitious 3D galaxy map from the Sloan Digital Sky Survey covers only our cosmological backyard—our Universe is simply huge! If I added the most distant galaxies ever discovered by astronomers to this figure, they'd be just over halfway to the edge, and way too few and far between to represent a useful 3-D map.

If we could somehow map these unexplored parts of our Universe, it would be terrific for cosmology. Not only would it increase our cosmological information a thousandfold, but because far away equals long ago, it would also reveal in great detail what happened during the first half of our cosmic history. But how? All the techniques we've discussed will continue to improve in various exciting ways, but it unfortunately doesn't look like they'll be able to map a large fraction of that uncharted 99.9% of the volume anytime soon. Cosmic microwave–background experiments map mainly the edge of this volume, since the interior is mostly transparent to microwaves. At such huge distances, most galaxies are very faint and difficult to see even with our best telescopes. Worse still: most of the volume is so far away that it contains almost

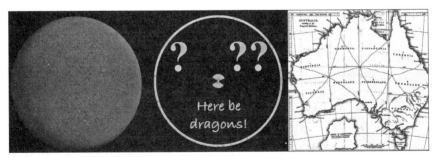

Figure 4.7: The fraction of our observable Universe (left) that has been mapped (center) is tiny, covering less than 0.1% of the volume. Just as for Australia in 1838 (right), we've mapped a thin strip around the perimeter while most of the interior remains unexplored. In the middle panel, the circular region is plasma (the cosmic microwave–background radiation we see comes only from the thin gray inner edge), and the tiny structure near the center is the largest 3-D galaxy map from the Sloan Digital Sky Survey.

no galaxies at all—we're looking so far back in time that most galaxies hadn't formed yet!

Hydrogen Mapping

Fortunately, there's another mapping technique that might work better. As we discussed earlier, what we call empty space isn't really empty: it's filled with hydrogen gas. Moreover, physicists have long known that hydrogen gas emits radio waves with a wavelength of 21 centimeters, which can be detected with a radio telescope. (When my classmate Ted Bunn was teaching this back in Berkeley, a student asked him a question that became an instant classic: "What's the wavelength of the 21-centimeter line?") This means that you can in principle "see" hydrogen with a radio telescope throughout most of our Universe, even long before it's formed stars and galaxies, back while it was invisible to ordinary telescopes. Better still, we can make a 3-D map of the hydrogen gas, using the redshift idea we discussed Chapter 2: since these radio waves are stretched by the expansion of our Universe, the wavelength they have when reaching Earth tells us how far away (and hence how long ago) they come from. For example, waves that arrive with a wavelength of 210 centimeters have been stretched to 10 times their original length, so they were emitted when our Universe was 10 times smaller than it is now. This technique has become known as 21-centimeter tomography, and since it has the potential to become the next big thing in cosmology, it's attracted lots of recent attention. Many teams around the world are currently racing to become the first to convincingly detect this elusive signal from hydrogen halfway across our Universe, but so far, nobody has succeeded.

What Is a Telescope, Really?

Why is it so hard? Because the radio signal is very faint. What do you need to detect a really faint signal? A really big telescope. A square kilometer size would be nice. What do you need to build a really big telescope? A really big budget. But how big exactly? This is where it gets interesting! For a traditional radio telescope like the one in Figure 4.8 (background), its cost more than doubles if you double its size, and gets absurd beyond a certain point. If you ask a friend who's a mechanical engineer to build a square kilometer radio dish with motors that can point it toward arbitrary sky directions, she'll no longer be your friend.

Figure 4.8: Radio astronomy on a big (background) and small (foreground) budget. My grad student Andy Lutomirski is tinkering with our electronics unit, which we put in a tent for rain protection during our expedition to Green Bank, West Virginia.

For this reason, all the experiments racing toward 21-centimeter tomography are using a more modern type of radio telescope called an *interferometer*. Since light and radio waves are electromagnetic phenomena, they create a voltage between different points in space as they fly by. Very faint voltages for sure, vastly lower than the 1.5 volts you have between the two ends of a flashlight battery, but still strong enough that they can be detected with good antennas and amplifiers. The basic idea with interferometry is to measure lots of such voltages using an array of radio antennas, and then have a computer reconstruct what the sky looks like. If all antennas are in a horizontal plane as in Figure 4.8 (foreground), then a wave from straight above will reach them simultaneously. Other waves reach some antennas before others, and the computer uses this fact to figure out what directions they're coming from. Your brain uses the same method to figure our where sound waves are coming from: if your left ear detects the sound before your right ear, the sound is clearly coming from the left, and by measuring the exact time difference, your brain can even estimate if it's coming straight from your left or from an angle. Since you have only two ears, you can't pinpoint the angle very accurately, but you'd do (albeit perhaps not look . . .) much better if you mimicked a large radio interferometer by having hundreds of ears all over your body. This interferometer idea has been enormously successful ever since Martin Ryle pioneered it in 1946, and it earned him a Nobel Prize in 1974.

However, the slowest computing step, which corresponds to measuring these time differences, needs to be done once for every pair of antennas (or ears), and if you increase the number of antennas, the number of pairs grows roughly as the number of antennas squared. This means that if you make the number of antennas a thousand times larger, the computer cost gets a million times larger—ouch! You want the telescope to be astronomical, not the budget! For this reason, interferometers have so far been limited to tens or hundreds of antennas, not the million or so that we really need for 21-centimeter tomography.

When I moved to MIT, I was generously allowed to join an American-Australian 21-centimeter-tomography experiment spearheaded by my colleague Jackie Hewitt. At our project meetings, I'd sometimes daydream about whether there might be a way of building huge telescopes cheaper. And then one afternoon, during one of our meetings at Harvard, it suddenly clicked for me: there is a cheaper way!

The Omniscope

I think of a telescope as a wave-sorting machine. If you look at your hand and measure the intensity of light across it, this won't reveal what your face looks like, because light waves from every part of your face are mixed together at each point on your skin. But if you can somehow sort all these light waves by the direction in which they travel, so that waves going in different directions land on different parts of your hand, then you'll recover an image of your face. This is exactly what a lens does in a camera, in a telescope, or in an eye, and what the curved mirror does in the radio telescope in Figure 4.8. In mathematics, we have a fancy and intimidating name for wave sorting: Fourier transforming. So a telescope is essentially a Fourier transformer. Whereas a traditional telescope does this Fourier transform by analog means, using lenses or curved mirrors, an interferometer does it digitally, using some form of computer. The waves are sorted not only by their travel direction, but also by their wavelength, which for visible light corresponds to its color. My idea that afternoon at Harvard was to design a huge radio interferometer where the antennas were arranged not rather randomly, as for our current project, but in a simple, regular pattern. For a telescope with a million antennas, this would allow the Fourier transform to be computed 25,000 times faster using some clever numerical tricks exploiting the pattern—basically making such a telescope 25,000 times cheaper.

After I'd managed to convince my friend Matias Zaldarriaga that the idea would work, we explored it in detail and published two papers about it, showing that the basic trick worked for a wide range of antenna patterns. We called our proposed telescope an "Omniscope" because it was both omnidirectional (imaging essentially the whole sky at once) and omnichromatic (imaging a broad range of wavelengths/"colors" all at once).

Albert Einstein allegedly said: "In theory, theory and practice are the same. In practice, they are not." We therefore decided to build a small prototype to see if it *really* worked. I discovered that the basic idea for the Omniscope had been tried already twenty years earlier by a Japanese group, for different purposes, but they were limited by the electronics of the time to sixty-four antennas. Thanks to the subsequent cell-phone revolution, the key components needed for our prototype had now dropped dramatically in price, so we could do the whole thing on a shoestring budget. I was also very lucky to get help from an amazing group of MIT students, some of whom came from our Electrical Engineering Department and knew the sort of wizardry needed for electronic circuit-board design and digital signal processing. One of them, Nevada Sanchez, taught me the magic-smoke theory of electronics, which we've subsequently verified in our lab: electronic components work because they contain magic smoke. So if you accidentally do something to them that lets their magic smoke out, they stop working. . . .

Having spent my whole academic career doing merely theory and data analysis, suddenly building an experiment was something completely different, and I loved it. It brought back fond memories of tinkering in the basement as a teenager, except that we were now building something much more exciting, and we were having fun doing it as a team. So far, our fledgling Omniscope is doing well, but it's too early to tell whether we or anyone else will ultimately succeed in making 21-centimeter tomography live up to its full potential.

However, the Omniscope has already taught me something else—something about myself. For me, the most fun part of all has been our team's expeditions: when we load all our gear into a van and drive to a remote location far from radio stations, cell phones and other human sources of radio waves. During those magic days, my normally so-fragmented life of emails, teaching, committees and family obligations gets replaced by a blissful Zen-like state of total focus: there's no cell-phone reception, no Internet, no interruption, and every single one of us in the team is 100% focused on this one common goal of making our experiment work. Sometimes I wonder whether we try to multitask too much in our day and age, and whether I should disappear like this more often for other reasons. Like finishing this book . . .

Where Did Our Big Bang Come From?

In this chapter, we've explored how an avalanche of precision data has transformed cosmology from a speculative, philosophical field into the precision science that it is today, where we've measured the age of our Universe to 1% accuracy. As is usual in science, answering old questions has uncovered new ones, and I predict an exciting decade ahead as cosmologists around the world build new theories and experiments attempting to shed light on the nature of dark matter, dark energy and other mysteries. In Chapter 13, we'll return to this quest and its implications for the ultimate fate of our Universe.

To me, one of the most striking lessons from precision cosmology is that simple mathematical laws govern our Universe all the way back to its fiery origins. For example, the equations that constitute Einstein's theory of general relativity appear to accurately govern the gravitational force over distances ranging from a millimeter up to a hundred trillion trillion (10^{26}) meters, and the equations of atomic and nuclear physics appear to have accurately governed our Universe from the first second after our Big Bang until today, 14 billion years later. And not just crudely, like the equations of economics, but with stunning precision, as illustrated by Figure 4.2. So precision cosmology highlights the mysterious utility of mathematics for understanding our world. We'll return to this mystery in Chapter 10 and explore a radical explanation for it.

Another striking lesson from precision cosmology is that it's incomplete. We saw that everything that we observe in our Universe today evolved from a hot Big Bang where nearly uniform gas as hot as the core of our Sun expanded so fast that it doubled its size in under a second. But who ordered that? I like to think of this as the "Bang Problem": what put the bang into the Big Bang? Where did this hot expanding gas come from? Why was it so uniform? And why was it imprinted with these 0.002%–level seed fluctuations that eventually grew into the galaxies and the large-scale structure that we see around us in our Universe today? In short, how did it all begin? As we'll see, extrapolating Friedmann's expanding-universe equations even further back in time leads to embarrassing problems, suggesting that we need a radical new idea before we can understand our ultimate origins. That's what the next chapter is about.

THE BOTTOM LINE

- A recent avalanche of data about the cosmic microwave background, galaxy clustering, etc., has transformed cosmology into a precision science; for example, we've gone from arguing about whether our Universe is 10 or 20 billion years old to arguing about whether it's 13.7 billion or 13.8 billion years old.
- Einstein's gravity theory arguably broke the record as the most mathematically beautiful theory, explaining gravity as a manifestation of geometry. It shows that the more mass space contains, the more curved space gets. This curvature of space causes things to move not in straight lines, but in a motion that curves toward massive objects.
- By measuring the geometry of universe-sized triangles, Einstein's theory has let us infer the total amount of mass in our Universe. Remarkably, the atoms that were thought to be the building blocks of everything were found to make up only 4% of this mass, leaving 96% unexplained.
- The missing mass is ghostly, being both invisible and able to pass through us undetected. Its gravitational effects suggest that it consists of two separate substances of opposite character, dubbed dark matter and dark energy: Dark matter clusters, dark energy doesn't. Dark matter dilutes as it expands, dark energy doesn't. Dark matter attracts, dark energy repels. Dark matter helps galaxies form, dark energy sabotages.
- Precision cosmology has revealed that simple mathematical laws govern our Universe all the way back to its fiery origins.
- As elegant as it is, the classic Big Bang model fails badly early on, suggesting that to understand our ultimate origins, we need to add another crucial piece to the puzzle.

5

Our Cosmic Origins

In the beginning, the Universe was created. This has made a lot of people very angry and been widely regarded as a bad move.
—Douglas Adams, in *The Restaurant at the End of the Universe*

Oh, no: he's falling asleep! It's 1997, I'm giving a talk at Tufts University, and the legendary Alan Guth has come over from MIT to listen. I'd never met him before, and having such a luminary in the audience made me feel both honored and nervous. Especially nervous. Especially when his head started slumping toward his chest, and his gaze began going blank. In an act of desperation, I tried speaking more enthusiastically and shifting my tone of voice. He jolted back up a few times, but soon my fiasco was complete: he was off in dreamland, and didn't return until my talk was over. I felt deflated.

Only much later, when we became MIT colleagues, did I realize that Alan falls asleep during *all* talks (except his own). In fact, my grad student Adrian Liu pointed out that I've started doing the same myself. And that I've never noticed that he does, too, because we always go in the same order. If Alan, I and Adrian sit next to each other in that order, we'll infallibly replicate a somnolent version of "the wave" that's so popular with soccer spectators.

I've come to really like Alan, who is as warm as he is smart. Tidiness isn't his forte, however: the first time I visited his office, I found most of the floor covered with a thick layer of unopened mail. I pulled up a random envelope as an archaeological sample, and found that it was postmarked over a decade earlier. In 2005, he cemented his legacy by winning the prestigious prize for the messiest office in Boston.

What's Wrong with Our Big Bang?

But this prize isn't Alan's only achievement. Back around 1980, he learned from the physicist Bob Dicke that there are serious problems

Figure 5.1: Andrei Linde (left) and Alan Guth (right) at a Swedish crayfish party, bliss-fully unaware that I'm photographing them, and that they'll need to dress differently to collect the prestigious Gruber and Milner prizes, which recognize them as the two main architects of inflation.

with the earliest stages of Alexander Friedmann's version of the Big Bang model, and proposed a radical solution that he called *inflation*.* As we've seen in the last two chapters, extrapolating Friedmann's expanding-universe equations backward in time was extremely success-ful, accurately explaining why distant galaxies are flying away from us, why the cosmic microwave–background radiation exists, how our light-est atoms originated, and many other observed phenomena.

Let's go back in time to near the frontier of our knowledge, to an instant when our Universe was expanding so fast that it would double its size during the next second. Friedmann's equations tell us that before this event, our Universe was even denser and hotter, without limit. That, in particular, there was a beginning of sorts one third of a second earlier, when the density of our Universe was infinite, and everything was flying away from everything else with infinite speed.

Following in Dicke's footsteps, Alan Guth carefully analyzed this

---

* Few important scientific discoveries are made by one person alone, and the discov-ery and development of inflation is no exception, with important contributions by Alan Guth, Andrei Linde, Alexei Starobinsky, Katsuhiko Sato, Paul Steinhardt, Andy Albrecht, Viatcheslav Mukhanov, Gennady Chibisov, Stephen Hawking, So-Young Pi, James Bardeen, Michael Turner, Alex Vilenkin and others. You'll find interesting histori-cal chronicles of this in many of the books on inflation in the "Suggestions for Further Reading" section at the end of this book.

story of our ultimate origins, and realized that it seemed awfully contrived. For example, it gives the following answers to four of our cosmic questions from the beginning of Chapter 2:

> **Q: What caused our Big Bang?**
> A: *There's no explanation—the equations simply assume it happened.*
> **Q: Did our Big Bang happen at a single point?**
> A: *No.*
> **Q: Where in space did our Big Bang explosion happen?**
> A: *It happened everywhere, at an infinite number of points, all at once.*
> **Q: How could an infinite space get created in a finite time?**
> A: *There's no explanation—the equations simply assume that as soon as there was any space at all, it was infinite in size.*

Do you feel that these answers settle the matter, elegantly laying all your Big Bang questions to rest? If not, then you're in good company! In fact, as we'll see, there's even more that Friedmann's Big Bang model fails to explain.

The Horizon Problem

Let's analyze more carefully the third question from our list above. Figure 5.2 illustrates the fact that the temperature of the cosmic microwave–background radiation is almost identical (agreeing to about five decimal places) in different directions in the sky. If our Big Bang explosion had happened significantly earlier in some regions than in others, then different regions would have had different amounts of time to expand and cool, and the temperature in our observed cosmic micro-

Figure 5.2: Whereas the molecules of hot coffee and cold milk have ample time to interact with each other and reach the same temperature, the plasma in regions A and B have never had time to interact at all: even information traveling at the speed of light couldn't have made it from A to B yet, since light from A is only reaching us coffee drinkers at the halfway point today. The fact that the plasma at A and B nonetheless have the same temperature is therefore an unexplained mystery in Friedmann's Big Bang model.

wave–background maps would vary from place to place not by 0.002% but by closer to 100%.

But couldn't some physical process have made the temperatures equal long after the Big Bang? After all, if you pour cold milk into hot coffee as illustrated in Figure 5.2, you won't be surprised if everything mixes to a uniform lukewarm temperature before you drink it. The catch is that this mixing process takes time: you need to wait long enough for milk and coffee molecules to move through the liquid and mix. In contrast, the distant parts of our Universe that we can see haven't had time for such mixing (Charles Misner and others first pointed this out back in the sixties). As illustrated in Figure 5.2, the regions A and B that we see in opposite directions of the sky haven't had time to interact at all: even information traveling at the speed of light couldn't have made it from A to B yet, since light from A is only now reaching the halfway point (where we're located). This means that Friedmann's Big Bang model offers no explanation whatsoever for why A and B have the same temperature. So regions A and B seem to have had the same amount of time to cool since our Big Bang, which must mean that they independently underwent a Big Bang explosion at almost exactly the same time, without any common cause.

To better understand Alan Guth's puzzlement over this, imagine how you'd feel if you checked your email and found a lunch invitation from a friend. And then realized that every other friend of yours has also sent you a separate email inviting you for lunch. And that every single one of these emails was sent to you at the exact same time. You'd probably conclude that this was some sort of conspiracy, and that all the emails had a common cause. Perhaps your friends had communicated among themselves and decided to throw you a surprise party, say. But to complete the analogy with Alan's Big Bang puzzle, where the regions A, B, etc., correspond to your friends, imagine that you know for a fact that your friends have never met, have never communicated with each other, and have never had access to any common information before they sent you their emails. Then your only explanation would be that it was all a crazy fluke coincidence. Too crazy to be plausible, in fact, so you'd probably conclude that you'd made an incorrect assumption somewhere, and that your friends had somehow managed to communicate after all. This is exactly what Alan concluded: it couldn't just have been a crazy fluke coincidence that infinitely many separate regions of space underwent Big Bang explosions all at once—some physical mechanism must have caused both the exploding and the synchronizing. One unexplained Big Bang is bad enough; an infinite number of unexplained Big Bangs in perfect synchronization strains credulity.

This is known as the *horizon problem*, because it involves what we see on our cosmic horizon, in the most distant regions we can observe. As if this weren't bad enough, Bob Dicke had told Alan of a second problem for Friedmann's Big Bang that he called the *flatness problem*.

The Flatness Problem

As we saw in the last chapter, we've measured our space to be flat to high accuracy. Dicke argued that this is puzzling if Friedmann's Big Bang model is correct, since it's a highly unstable situation, and we shouldn't expect unstable situations to last for long. For example, we discussed in Chapter 3 how a stopped bike is unstable, because any slight departure from perfect balance gets amplified by gravity, so you'd be very puzzled if you saw an unsupported stopped bike remain upright for minutes on end. Figure 5.3 shows three solutions to Friedmann's equation, illustrating the cosmic instability. The middle curve corresponds to a flat universe, which remains perfectly flat and expands forever. The other two curves start out virtually identically on the left side, with space having almost no curvature at all, and after a billionth of a second, their densities differ only in the last of the twenty-four digits.* But gravity amplifies these tiny differences, and over the next 500 million years, this causes our Universe described by the bottom curve to stop expanding and recollapse in a cataclysmic Big Crunch, a sort of Big Bang in reverse. In this ultimately collapsing universe, space gets curved so that triangle angles add up to much more than 180 degrees. In contrast, the top curve describes a universe getting curved so that these angles add up to much less than 180 degrees. It expands much faster than the flat borderline universe, and by the present day, its gas would be way too diluted to form galaxies, rendering its fate a cold and dark "Big Chill."

So why is our Universe so flat? If you change the twenty-four digits in Figure 5.3 to random values and re-solve Friedmann's equation, the probability that you'll get a universe remaining nearly flat for 14 billion years is smaller than the probability that a dart randomly fired into space

---

* We haven't even measured the strength of gravity accurately enough to know what more than the first four of these digits need to be, so the last twenty digits are my guess for illustration.

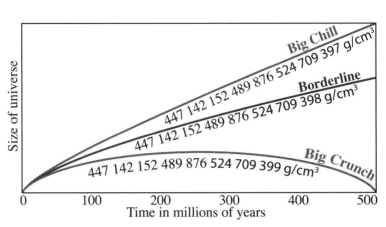

Figure 5.3: Another unexplained mystery in Friedmann's Big Bang model is why our Universe has lasted so long without getting severely curved and undergoing a Big Crunch or Big Chill. Each curve corresponds to a slightly different density when our Universe was a billionth of a second old. The borderline situation we're in is highly unstable: changing merely the very last of the twenty-four digits would have triggered a Big Crunch or a Big Chill before our Universe reached 4% of its current age. *(Figure idea courtesy of Ned Wright)*

from Mars would hit the bull's-eye on a dartboard on Earth. Yet Friedmann's Big Bang model offers no explanation for this coincidence.*

Surely, Alan Guth argued, there must be some mechanism that *caused* our Universe to have exactly the right density required for extreme flatness early on.

How Inflation Works

The Power of Doubling

Alan's radical insight was that by making just one strange-sounding assumption, you can solve both the horizon problem and the flatness problem in one fell swoop—and explain a lot more as well. This assumption is that once upon a time, there was a tiny uniform blob of a substance whose density was very hard to dilute. This means that if one gram of this substance expanded into twice the volume, its density (its mass per volume) would remain basically unchanged, so that you'd now

---

* As pointed out by Phillip Helbig and others, the flatness problem is often misrepresented and overstated, but it remains extremely serious because of the cosmic clumpiness we discussed in the last chapter, which causes random departures from flatness early on.

have about two grams of the stuff. Compare this with a normal substance such as air: if it expands into a larger volume (as when you release compressed air from a tire), then the total number of molecules stays the same, so the total mass remains the same and the density drops.

According to Einstein's theory of gravity, such a tiny nondiluting blob can undergo a most remarkable explosion that Alan called *inflation*, in effect creating a Big Bang! As illustrated in Figure 5.4, Einstein's equations have a solution where each part of this blob doubles its size at regular time intervals, a type of growth that mathematicians refer to as *exponential*. In this scenario, our baby Universe grew very much the way you yourself did right after your conception (see Figure 5.5): each of your cells doubled roughly daily, causing your total number of cells to increase day by day as 1, 2, 4, 8, 16, etc. Repeated doubling is a powerful process, so your Mom would have been in trouble if you'd kept doubling your weight every day until you were born: after nine months (about 274 doublings), you'd have weighed more than all the matter in our observable Universe combined! Crazy as it sounds, this

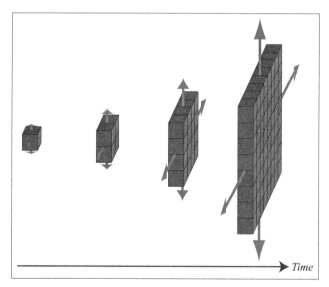

Figure 5.4: According to Einstein's theory of gravity, a substance whose density is undilutable can "inflate," doubling its size at regular intervals, growing from a subatomic scale to a size vastly larger than our observable Universe in a split second and effectively putting the bang into our Big Bang. This repeated doubling occurs in all three dimensions, so that doubling the diameter makes the volume eight times larger. Here, I've drawn only two dimensions just for illustration, where doubling the diameter quadruples the volume.

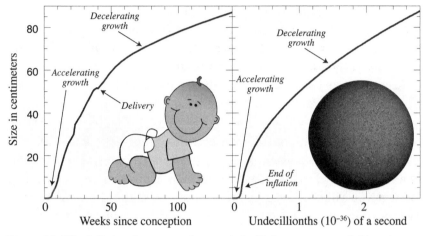

Figure 5.5: The inflation theory says that our baby Universe grew much like a human baby: an accelerating growth phase where the size doubled at regular intervals was followed by a more leisurely decelerating growth phase. Amusingly, the vertical axis is the same for the two plots: in the simplest model, our Universe stopped inflating when it was about the size of an orange (but weighed about 10^{81} times more). Our baby Universe doubled its size about 10^{43} times faster than the first cells of the baby.

is exactly what Alan's inflation process does: starting out with a speck much smaller and lighter than an atom, it repeatedly doubles its size until it's more massive than our entire observable Universe.

Problems Solved

As you can see in Figure 5.4, repeated doubling of the size automatically causes repeated doubling of the expansion speed, which I've indicated by arrows. In other words, it causes accelerated expansion. If you'd really kept doubling your mass daily until birth, then you'd have expanded quite slowly initially (by just a few cell sizes per day). But toward the end of your gestation period when you weighed more than our observable Universe and doubled daily, you'd have expanded with a mind-bogglingly large speed of many billion light-years per day. Whereas you used to double your mass once per day, our inflating baby Universe doubled its mass extremely often—in some of the most popular versions of inflation, one mass doubling occurred about every ten trillionths of a trillionth of a quadrillionth (10^{-38}) of a second, and about 260 mass doublings were required to create all the mass in our observable Universe. This means that the whole inflation process, from beginning to end, could have been almost instantaneous by human standards, requiring less than about 10^{-35} seconds, less time than light takes to travel a trillionth of the size of a proton. In other words, exponential expansion takes something tiny that isn't moving much and turns it into a humongous, fast-expanding explosion. In this way, inflation solves the "Bang problem," explaining what caused our Big Bang: it was caused by this repeated doubling process. It also explains why the expansion is uniform, as Edwin Hubble discovered: Figure 5.4 illustrates that regions that are twice as far from each other move apart twice as fast.

Figure 5.5 illustrates that, just as you eventually replaced your exponential body expansion by more leisurely growth, our baby Universe eventually stopped inflating. The inflating material decayed into ordinary matter which kept expanding at a more relaxed pace, coasting along with the speed it got from the explosive inflationary phase, gradually decelerated by gravity.

Alan Guth realized that inflation also solves the horizon problem. The distant regions A and B in Figure 5.2 were extremely close together during the early stages of inflation, so they had time to interact back then. The explosive expansion of inflation then brought A and B out of contact with each other, and they're only now beginning to come back into contact. A cell in your nose has the same DNA as a cell in your toe because they have a common progenitor: they're both produced by successive doublings of your very first cell. In the same way, distant regions of our cosmos have similar properties because they have a common origin: they're produced by successive doublings of that same tiny speck of inflating matter.

As if this weren't enough of a success, Alan realized that inflation solves the flatness problem as well. Suppose you're the ant on the sphere in Figure 2.7 and can only see a small area of the curved surface that you live on. If inflation suddenly makes the sphere vastly larger, that small area that you can see will look much flatter; a square centimeter on a ping-pong ball is noticeably curved, whereas a square centimeter on the surface of Earth is almost perfectly flat. Similarly, when inflation dramatically expands our own 3-D space, the space within any given cubic centimeter becomes almost perfectly flat. Alan proved that as long as inflation continues long enough to make our observable Universe, it makes space flat enough to last until the present day without a Big Crunch or Big Chill.

In fact, inflation typically continues a lot longer than that, ensuring that space remains essentially perfectly flat until the present day.

In other words, inflation theory made a testable prediction back in the eighties: our space should be flat. As we saw in the last two chapters, we've now performed this test to better than 1% precision, and inflation passed the test with flying colors!

Who Paid for the Ultimate Free Lunch?

Inflation is like a great magic show—my gut reaction is: *This can't possibly obey the laws of physics!* Yet under close enough scrutiny, it does.

First of all, how can one gram of inflating matter turn into two grams when it expands? Surely, mass can't just be created from nothing? Interestingly, Einstein offered us a loophole through his special relativity theory, which says that energy E and mass m are related according to the famous formula $E = mc^2$. Here $c = 299,792,458$ meters per second is the speed of light, and because it's such a large number, a tiny amount of mass corresponds to a huge amount of energy: less than a kilogram of mass released the energy of the Hiroshima nuclear blast. This means that you can increase the mass of something by adding energy to it. For example, you can make a rubber band very slightly heavier by stretching it: you need to apply energy to stretch it, and this energy goes into the rubber band and increases its mass.

A rubber band has *negative pressure* because you need to work to expand it. For a substance with positive pressure, like air, it's the other way around: you need to do work to compress it. In summary, the inflating substance has to have negative pressure in order to obey the laws of physics, and this negative pressure has to be so huge that the energy required to expand it to twice its volume is exactly enough to double its mass.

Another puzzling feature of inflation is that it causes accelerated expansion. In high school, I was taught that gravity is an attractive force, so if I have a bunch of expanding stuff, then shouldn't gravity instead *decelerate* the expansion, trying to ultimately reverse the motion and pull things back together? Again Einstein comes to the rescue with a loophole, this time from his general relativity theory, which says that gravity is caused not only by mass, but also by pressure. Since mass can't be negative, the gravity from mass is always attractive. But positive pressure also causes attractive gravity, which means that negative pressure causes repulsive gravity! We just saw that an inflating substance has huge negative pressure. Alan Guth calculated that the repulsive gravitational force caused by its negative pressure is three times stronger than the attractive gravitational force caused by its mass, so the gravity of an inflating substance will blow it apart!

In summary, an inflating substance produces an antigravity force that blows it apart, and the energy that this antigravity force expends to expand the substance creates enough new mass for the substance to retain constant density. This process is self-sustaining, and the inflating substance keeps doubling its size over and over again. In this way, inflation creates everything we can observe with our telescopes from almost nothing. This prompted Alan Guth to refer to our Universe as "the ultimate free lunch": inflation predicts that its total energy is very close to zero!

But according to the Nobel Prize–winning economist Milton Friedman, "there's no such thing as a free lunch," so who paid the energy bill for all that galactic grandeur that we observe around us in our Universe? The answer is that gravity did, because the gravitational force injected energy into the inflating matter by stretching it out. But if the total energy of everything can't change and heavy objects have loads of positive energy according to Einstein's $E = mc^2$ formula, then this means that gravity must have gotten stuck with a corresponding amount of negative energy! That's in fact exactly what's happened. The gravitational field, which is responsible for all gravitational forces, has negative energy. And it gets more negative energy every time gravity accelerates something. Consider, for example, a distant asteroid. If it's moving only slowly, it has very little motion energy. If it's far from Earth's gravitational pull, it also has very little gravitational energy (so-called potential energy). If it gradually falls toward Earth, it will pick up great speed and motion energy—perhaps enough to create a huge crater on impact. Since the gravitational field started with almost no energy and then released all this positive energy, it now has negative energy left.

We've now tackled another question from our list at the beginning of Chapter 2: ***Doesn't creation of the matter around us from almost nothing by inflation violate energy conservation?*** We've seen that the answer is no: all the required energy was borrowed from the gravitational field.

I have to confess that, although this process doesn't violate the laws of physics, it makes me nervous. I just can't shake the uneasy feeling that I'm living in a Ponzi scheme of cosmic proportions. If you'd visited Bernie Madoff before his 2008 arrest for embezzling $65 billion, you'd have thought that he was surrounded by real wealth that he actually owned. Yet on closer scrutiny, it turned out that he'd effectively purchased it with borrowed money. Over the years, he doubled the scale of his operation over and over again by cleverly leveraging what he had to borrow even more from naive investors. An inflating universe does exactly the same thing: it doubles its size over and over again by leveraging the energy that it already has to borrow even more energy from the gravitational field. Just like Madoff, the inflating universe exploits an inherent instability in the system to create apparent grandeur out of nothing. I just hope that our Universe proves less unstable than Madoff's. . . .

The Gift That Keeps on Giving

Inflation Encore

Like many successful scientific theories, inflation got off to a rough start. Its first firm prediction, that space was flat, seemed inconsistent with mounting observational evidence. As we saw in the last chapter, Einstein's gravity theory says that space can only be flat if the cosmic density equals a particular critical value. We use the symbol Ω_{total} (or just Ω or "Omega" for short) to denote how many times denser our Universe is than this critical density, so inflation predicted that $\Omega = 1$. While I was a grad student, however, our measurements of the cosmic density from galaxy surveys and other data kept getting better, suggesting the much lower value $\Omega \approx 0.25$, and it became increasingly embarrassing for Alan Guth to travel from conference to conference stubbornly insisting that $\Omega = 1$ despite what his experimental colleagues told him. But Alan stuck to his guns, and history proved him right. As we saw in the last chapter, the discovery of dark energy revealed that we'd only been counting about a quarter of the density, and when we counted dark energy, too, we measured $\Omega = 1$ to better than 1% precision (see Table 4.1).

The discovery of dark energy gave a huge credibility boost to inflation also for another reason: Now you could no longer dismiss the assumption of a nondiluting substance as nutty and unphysical, because dark energy is precisely such a substance! So the epoch of inflation that created our Big Bang ended 14 billion years ago, but a new epoch of inflation has begun. This new phase of inflation driven by dark energy is just like the old one but in slow motion, doubling the size of our Universe not every split second but every 8 billion years. So the interesting debate is no longer about whether inflation happened or not, but about whether it happened once or twice.

Sowing the Seed Fluctuations

The hallmark of a successful scientific theory is that you get more out of it than you put into it. Alan Guth showed that, with one single assumption (a tiny speck of a hard-to-dilute substance), you could solve three separate cosmological conundrums: the Bang problem, the horizon problem and the flatness problem. Above we saw how inflation did more: it predicted $\Omega = 1$, which was accurately confirmed about two decades later. However, that wasn't all.

We ended the last chapter by asking where the galaxies and the large-scale cosmic structure ultimately came from, and much to everybody's surprise, inflation answered this question, too! And what an answer it gave! The idea was first proposed by two Russian physicists, Gennady Chibisov and Viatcheslav Mukhanov, and when I first heard it, I thought it sounded absurd. Now I think it's a leading candidate for the most radical and beautiful synthesis of ideas in scientific history.

In short, the answer is that the cosmic seed fluctuations came from quantum mechanics, the theory of the microworld that we'll explore in Chapters 7 and 8. But I learned in college that quantum effects are important only for the very smallest things we study, such as atoms, so how can they possibly have any relevance to the very largest things we study, such as galaxies? Well, one of the beauties of inflation is that it connects the smallest and largest scales: during the early stages of inflation, the region of space that now contains our Milky Way Galaxy was much smaller than an atom, so quantum effects could have been important. And indeed they were: as we'll see in Chapter 7, the so-called Heisenberg uncertainty principle of quantum mechanics prevents any substance, including the inflating material, from being completely uniform. If you try to make it uniform, quantum effects force it to start wiggling around, spoiling the uniformity. When inflation stretched a subatomic region into what became our entire observable Universe, the density fluctuations that quantum mechanics had imprinted were stretched as well, to sizes of galaxies and beyond. As we saw in the last chapter, gravitational instability took care of the rest, amplifying these fluctuations from the tiny 0.002%-level amplitudes with which quantum mechanics had endowed them into the spectacular galaxies, galaxy clusters and superclusters that now adorn our night sky.

The best part is that this isn't just qualitative blah blah, but a rigorous quantitative story where everything can be accurately calculated. The

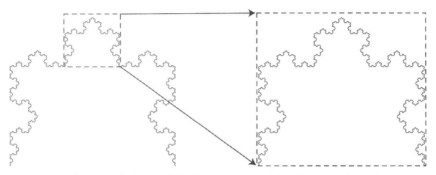

Figure 5.6: This so-called snowflake fractal, invented by the Swedish mathematician Helge von Koch, has the remarkable property that it's identical to a magnified piece of itself. Inflation predicts that our baby Universe was similarly indistinguishable from a magnified piece of itself, at least in an approximate statistical sense.

power-spectrum curve I've plotted in Figure 4.2 is a theoretical prediction for one of the very simplest inflation models, and I find it remarkable how well it matches all the measurements. Inflation models can also predict three of the measured cosmological parameters that I listed in Table 4.1. I've already mentioned one of these predictions: $\Omega = 1$. The other two involve the nature of the cosmic-clustering patterns that we explored in the last chapter. In the simplest inflation models, the amplitude of the seed clustering (called Q in the table) depends on how fast the inflating region doubles its size, and with a doubling time around 10^{-38} seconds, the prediction matches the observed value $Q \approx 0.002\%$.

Inflation also makes an interesting prediction for the seed clustering "tilt" parameter (called n in the table). To understand this, we need to look at the jagged curve in Figure 5.6, which is what mathematicians call self-similar, fractal or scale-invariant. All of these words basically mean that if I replace the image by a magnified piece of it, you can't tell the difference. Since I can repeat this zoom trick as many times as I want, it's clear that even a trillionth of the curve must look identical to the whole thing. Interestingly, inflation predicts that to a good approximation, our baby Universe was scale-invariant, too, in the sense that you couldn't tell the difference between a random cubic centimeter of it and a greatly magnified piece of it. Why? Well, during the inflation epoch, magnifying our Universe was basically equivalent to waiting a little, until everything doubled in size yet again. So if you could have time-traveled back to the inflation epoch, seeing that the statistical properties of the fluctuations were scale-invariant would have been equivalent to seeing that these properties didn't change over time.

But inflation predicts that these properties hardly change over time for a very simple reason: the local physical conditions that generate the quantum fluctuations hardly change over time either, since the inflating substance isn't noticeably changing its density or other properties.

The tilt parameter n in Table 4.1 measures how close the inflating universe was to scale-invariant. It contrasts the amount of clustering on large and small scales, and is defined so that $n = 1$ means perfectly scale-invariant (the same clustering on all scales), $n < 1$ means more clustering on large scales, and $n > 1$ means more clustering on small scales. Mukhanov and other inflation pioneers had predicted that n would be quite close to 1. When my friend Ted and I moonlighted on the magicbean computer back in Chapter 4, it was to make the most accurate measurement to date of n. Our result was $n = 1.15 \pm 0.29$, confirming that yet another prediction from inflation was looking good.

The n business gets even more interesting. Because inflation eventually has to end, the inflating substance has to gradually dilute ever so slightly during inflation—otherwise nothing would change and inflation would continue forever. In the simplest inflation models, this decrease in the density causes the amplitude of generated fluctuations to decrease as well. This means that the fluctuations generated later on have lower amplitude. But fluctuations generated later didn't get stretched as much before inflation ended, so they correspond to fluctuations on smaller scales today. The upshot of all this is the prediction that $n < 1$. To predict something more specific, you need a model for what the inflating substance is made of. The simplest such model of all, pioneered by Andrei Linde (Figure 5.1), is known in geek-speak as a "scalar field with quadratic potential" (it's basically a hypothetical cousin of a magnetic field), and it predicts that $n = 0.96$. Now take another look at Table 4.1. You'll see that the n measurement has now gotten about 60 times more accurate since those wild magicbean days, and that the latest measurement is $n = 0.96 \pm 0.005$, tantalizingly close to what was predicted!

Andrei Linde is one of inflation's pioneers, and has inspired me a lot. I'll hear someone explain something and think it's complicated. Then I'll hear Andrei's explanation of the same thing and realize that it's simple when I think about it in the right way—his way. He has a dark but warm sense of humor that undoubtedly helped him survive back in the Soviet Union, and has a mischievous glint in the eye regardless of whether he's discussing personal things or cutting-edge science.

All these measurements will keep getting more accurate in the years to come. We also have the potential to measure several additional numbers that inflation models make predictions for. For example, in addition to intensity and color, light has a property called polarization—bees can see it and use it to navigate, and although our human eyes don't notice it, our polarized sunglasses let light through only if it's polarized in a particular way. Many popular inflation models predict a rather unique signature in the polarization of the cosmic microwave–background radiation: quantum fluctuations during inflation generate what's known as *gravitational waves*, vibrations in the very fabric of spacetime, and these in turn distort the cosmic microwave–background pattern in a characteristic way.

One morning in 2014, Alan Guth sent me an email marked "CONFIDENTIAL," inviting me to a March 17 press conference at Harvard where a discovery of these gravitational waves was to be announced. Wow! The room was packed with physicists and journalists, and both Alan and Andrei were all smiles. John Kovac and his colleagues from the BICEP2 experiment reported that through three painstaking years of careful microwave measurements from the South Pole, they had detected humungous gravitational waves close to a billion light-years long. Making such strong gravitational waves requires extreme violence. For example, a cataclysmic collision of two black holes squeezing more than the Sun's mass into a volume smaller than a city can create gravitational waves that the U.S.-based LIGO experiment hopes to detect, but these waves are only about as big as the pair of objects creating them. So what could possibly have created the vast waves BICEP2 saw, given that our universe seems to contain no objects large enough to make them? In my opinion, the only compelling explanation for these waves is that inflation made them, by violently doubling the size of space in about a hundredth of a trillionth of a trillionth of a trillionth (10^{-38}) of a second and repeating it at least eighty times.

So how seriously should we take inflation? It had emerged as the most successful and popular theory for what happened early on even before BICEP2, as experiments gradually confirmed one of its predictions: that our universe should be large, expanding and approximately homogenous, isotropic and flat, with tiny fluctuations in the cosmic baby pictures that were roughly scale invariant, adiabatic and Guassian. To me and many of my cosmology colleagues, the gravitational waves discovered by BICEP2, if confirmed by other experiments, provide the smoking-gun evidence that really clinches it, because we lack any other compelling explanations for them. So, if the BICEP2 claims stand the test of time, then although it sounds crazy, it seems like inflation really happened: our entire observable universe was once much smaller than an atom.

If we take inflation seriously, then we need to start correcting people claiming that inflation happened shortly *after* our Big Bang, because it happened *before* it, creating it. It is inappropriate to define our Hot Big Bang as the beginning of time, because we don't know whether time actually had a beginning, and because the early stages of inflation were neither strikingly hot nor big nor much of a bang. I think that the early stages of inflation are better thought of as a *Cold Little Swoosh*, because at that time our universe was not that hot (getting a thousand times hotter once inflation ended), not that big (less massive than an apple and less than a billionth of the size of a proton), and not much of a bang (with expansion velocities a trillion trillion times slower than after inflation).

Eternal Inflation

Our discussion of inflation so far might sound like the typical life cycle of a successful physics idea: new theory solves old problems. Further predictions. Experimental confirmation. Widespread acceptance. Text-books rewritten. It sounds as though it's time to give inflation the tra-ditional scientific retirement speech: "Thank you, inflation theory, for your loyal service in tying up some loose ends regarding the ultimate origins of our Universe. Now please go off and retire in neatly compart-mentalized textbook chapters, and leave us alone so that we can work on other newer and more exciting problems that aren't yet solved." But like a tenacious aging professor, inflation refuses to retire! In addition to being the gift that keeps on giving within its compartmentalized subject area of early-universe cosmology, as we saw above, inflation has given us more radical surprises that were quite unexpected—and to some of my colleagues, also quite unwelcome.

Unstoppable

The first shocker is that inflation generally refuses to stop, forever pro-ducing more space. This was discovered for specific models by Andrei Linde and Paul Steinhardt. An elegant proof of the existence of this effect was given by Alex Vilenkin, a friendly soft-spoken professor at Tufts University, and the one who invited me to give that talk that put

Alan Guth to sleep. While he was a student back in his native Ukraine, he refused a request from the KGB to testify against a fellow student who was critical of the authorities, despite being warned of "consequences." Although he'd been admitted to physics grad school at the University of Kharkiv, the most prestigious physics program in the Soviet Union, the permission he required for moving there was never granted. Nor was he able to get any normal jobs. He spent a year struggling as a night watchman at a zoo before finally managing to leave the country. Whenever I get annoyed by a bureaucrat, thinking of Alex's story transforms my frustration into grateful realization of how small my problems are. Perhaps his disposition to stick with what he believes is right despite authority pressure helps explain why he persisted and discovered things that other great scientists dismissed.

Alex found that the question of where and when inflation ends is quite subtle and interesting. We know that inflation ends in at least *some* places, since 14 billion years ago, it ended in the part of space that we now inhabit. This means that there must be some physical process which can get rid of the inflating substance, causing it to decay into ordinary non-inflating matter, which then keeps expanding, clustering, and ultimately forming galaxies, stars and planets as we described in the last chapter. Radioactivity famously makes unstable substances decay into others, so let's suppose that the inflating substance is similarly unstable. This means that there's some time scale called the half-life during which half of the inflating substance will decay. As illustrated in Figure 5.7, we now have an interesting tug-of-war between the doubling caused by inflation and the halving caused by decay. For inflation to work, the former has to win so that the total inflating volume grows over time. This means that the doubling time of the inflating substance has to be shorter than its half-life. The figure illustrates such an example, where inflation triples the size of space while one third of the inflating substance decays away, over and over again. As you can see, the total volume of space that's still inflating keeps doubling forever. In parallel, non-inflating regions of space are continuously being produced by the decay of inflating space, so the amount of non-inflating volume, where inflation has ended and galaxies can form, keeps doubling, too.

This perpetual property of inflation turned out to be much more general than originally expected. Andrei Linde, who coined the term "eternal inflation," discovered that even the very simplest inflation model that he'd proposed, which we talked about above, inflated eter-

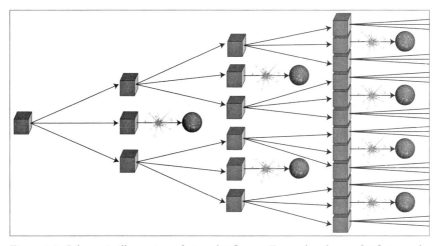

Figure 5.7: Schematic illustration of eternal inflation. For each volume of inflating substance (symbolized by a cube) that decays into a non-inflating Big Bang universe like ours, two other inflating volumes don't decay, instead tripling their volume. The result is a never-ending process where the number of Big Bang universes increases as 1, 2, 4, etc., doubling at each step. So what we call our Big Bang (one of the flashes) isn't the beginning of everything, but the end of inflation in our part of space.

nally through an elegant mechanism related to the quantum fluctuations that generated our cosmological seed fluctuations.

By now, a very large class of inflation models have been analyzed in detail by researchers around the globe, and it's been found that almost all of them lead to eternal inflation. Although most of these calculations are rather complicated, the schematic illustration of Figure 5.7 captures the essence of why inflation is generally eternal: for inflation to work in the first place, the inflating substance needs to expand faster than it decays, and this automatically makes the total amount of inflating stuff grow without limit.

The discovery of eternal inflation has radically transformed our understanding of what's out there in space on the largest scales. Now I can't help but feel that our old story sounds like a fairy tale, with its single narrative in a simple sequence: "Once upon a time, there was inflation. Inflation made our Big Bang. Our Big Bang made galaxies." Figure 5.7 illustrates why this story is too naive: it yet again repeats our human mistake of assuming that all we know of so far is all that exists. We see that even our Big Bang is just a small part of something much grander, a treelike structure that's still growing. In other words, what we've called our Big Bang wasn't the ultimate *beginning*, but rather the *end*—of inflation in our part of space.

How to Make an Infinite Space in a Finite Volume

That kindergartner in Chapter 2 asked whether space goes on forever. Eternal inflation gives a clear answer: *space isn't just huge—it's infinite.* With infinitely many galaxies, stars and planets.

Let's explore this notion more carefully. Although the schematic nature of Figure 5.7 doesn't make this clear, we're still talking about just one single connected space. Right now (we'll return to what "right now" means below), some parts of this space are expanding very fast because they contain inflating matter, other parts are expanding more slowly because inflation has ended there, and yet other parts, like the region that's inside our Galaxy, are no longer expanding at all. So does inflation end? The detailed inflation research we mentioned above shows that the answer is: yes and no. It ends and it doesn't end, in the following sense:

1. In almost all parts of space, inflation will eventually end in a Big Bang like ours.
2. There will nonetheless be some points in space where inflation never ends.
3. The total inflating volume increases forever, doubling at regular intervals.
4. The total post-inflationary volume containing galaxies also increases forever, doubling at regular intervals.

But does this really mean that space is infinite already? This brings us back to another one of our questions from Chapter 2: ***How could an infinite space get created in a finite time?*** It sounds impossible. But as I mentioned, inflation is like a magic show where seemingly impossible tricks happen through creative use of the laws of physics. Indeed, inflation can do something even better, which I think is its most amazing trick of all: *it can create an infinite volume inside a finite volume!* Specifically, it can start with something smaller than an atom and create an infinite space inside of it, containing infinitely many galaxies, without affecting the exterior space.

Figure 5.8 illustrates how inflation does this trick. It shows a slice through space and time, where the left and right edges correspond to two points where inflation never ends, and the bottom edge corresponds to a time when the entire region between these two points is inflating. It's hard to draw an expanding three-dimensional space, so I've ignored both the

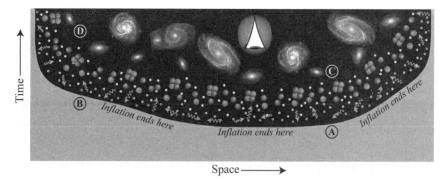

Figure 5.8: As described in the text, inflation can create an infinite universe inside of what looks like a subatomic volume from the outside. An observer inside will view A as simultaneous with B, C as simultaneous with D, the infinite U-shaped surface where inflation ends as her *time zero*, the infinite U-shaped surface where atoms form as her *time 400,000 years*, etc. For simplicity, this cartoon ignores both the expansion of space and two of the three space dimensions.

expansion and two of the three space dimensions in the picture, because neither of these two complications affect the basic argument. Eventually, inflation will end everywhere except at the left and right edges; the curved boundary shows the exact time when it ends at different places. Once inflation ends in a given region, the traditional Big Bang story from the last two chapters starts unfolding there, with a hot cosmic fusion reactor eventually cooling to form atoms, galaxies, and perhaps observers like us.

Here's the key part of the trick: according to Einstein's theory of general relativity, an observer living in one of these galaxies will perceive space and time differently than I've defined them with my axes in the drawing. Our physical space doesn't come with centimeter marks built in the way a ruler does, nor does our Universe come with a bunch of clocks pre-installed. Instead, any observer needs to define her own measurement rods and clocks, which in turn define her notion of space and time. This idea can lead to one of Einstein's core insights, immortalized by the slogan "It's all relative": that different observers can perceive space and time in different ways. In particular, simultaneity can be relative. Suppose you email an astronaut friend on Mars:

Hey, how are things over there?

Ten minutes later, she gets your message, which was transmitted at the speed of light using radio waves. While you're waiting, you receive an email from Nigeria, offering cheap Rolex watches. Another ten minutes later, you get her reply:

Good, but I miss Earth!

Now which event happened first, you receiving the spam or your astronaut friend sending her message? Amazingly, Einstein discovered that this simple question has no simple answer. Instead, the correct answer depends on the velocity of the person answering it! For example, if I'm zooming past Earth toward Mars in a spaceship, intercept all three emails, and analyze the situation, I'll determine that according to my onboard clock, your friend on Mars sent the message before you got the spam. If I'm flying in the opposite direction, I'll determine that you got the spam first. Confusing? That's what most of Einstein's colleagues thought as well when he presented his relativity theory, but countless experiments have since confirmed that this is how time works. The only circumstance when we can definitely say that an event on Mars happened before an event on Earth is if we can send a message from Mars after the Mars event that reaches Earth before the Earth event.

Now let's apply this to the situation in Figure 5.8. For someone outside of this region, it might make sense to define space and time as the horizontal and vertical directions, respectively, just as I've drawn the figure, so that the four events I've circled happened in the order A, B, C, D. Moreover, B definitely happened before D because you could imagine sending a message from B to D, and similarly, A definitely happened before C. But can we really be sure that A happened before B, given that the two events are too far apart for light to have time to reach one from the other? Einstein's answer is no. Indeed, for an observer living in one of these galaxies, it makes more sense to define inflation as having ended at a particular fixed time, since the end of inflation corresponds to her Big Bang, so according to her, the events A and B are simultaneous! As you can see, the "Inflation ends" surface is *not* horizontal. In fact, it's infinite, since it bends up like the letter U toward the left and right edges of the plot where we agreed that inflation never ends. This means that as far as she's concerned, her Big Bang occurred at a single instant in a truly infinite space! Where did the infinity come sneaking in from? You can see that it snuck in via the infinite future time available, by her space direction being curved progressively more upward.

She'll similarly conclude that her space is infinite at later times. For example, if she builds a cosmic microwave–background experiment to take baby pictures of her 400,000-year-old universe, the plasma surface she's imaging corresponds to the surface in the picture where protons and electrons combine into transparent (invisible) hydrogen atoms. Since you can see that this is also an infinite U-shaped surface, she'll perceive her 400,000-year-old universe as having been infinite. She'll also consider events C and D simultaneous, since they lie on the U-shaped surface where the first galaxies form, and so on. Because you can stack an infinite number of these U-shapes inside each other, she'll feel that her universe is infinite in both space and future time—even though it all neatly fits into an initially subatomic region according to the outside observer. The fact that space expands inside doesn't necessarily increase the amount of room it all takes as seen from outside: remember that Einstein allows space to stretch and produce more volume from nothing, without taking it from someplace else. In practice, this infinite universe might look something like a subatomic black hole from the outside. In fact, Alan Guth and collaborators even explored the speculative possibility of doing this trick yourself for real: creating in your laboratory something that looks like a small black hole from the outside and that looks like an infinite universe from the inside—as to whether this is really possible, the jury is still out. If you're harboring demiurgic urges, I highly recommend Brian Greene's instructions for "aspiring universe creators" in his book *The Hidden Reality*.

We began our exploration of inflation earlier in this chapter by lamenting the unsatisfactory answers that Friedmann's classic Big Bang theory gave to some basic questions, so let's conclude our exploration by reviewing how inflation answers them:

Q: What caused our Big Bang?
A: *The repeated doubling in size of an explosive subatomic speck of inflating material.*
Q: Did our Big Bang happen at a single point?
A: *Almost: it began in a region of space much smaller than an atom.*
Q: Where in space did our Big Bang explosion happen?
A: *In that tiny region—but inflation stretched it out to about the size of a grapefruit growing so fast that the subsequent expansion made it larger than all the space that we see today.*
Q: How could our Big Bang create an infinite space in a finite time?
A: *Inflation produces an infinite number of galaxies by continuing forever. According to general relativity, an observer in one of these galaxies will view space and time differently, perceiving space as having been infinite already when inflation ended.*

In summary, inflation has radically transformed our understanding of our cosmic origins, replacing the awkward unanswered questions of Friedmann's Big Bang model by a simple mechanism that creates our Big Bang from almost nothing. It has also given us more than we asked for: a space that isn't just huge but truly infinite, with infinite numbers of galaxies, stars and planets. And as we'll see in the next chapter, that's just the tip of the iceberg.

THE BOTTOM LINE

- There are serious problems with the earliest stages of Friedmann's Big Bang model.
- Inflation theory solves them all, and explains the mechanism that caused the Big Bang.
- Inflation explains why space is so flat, which we've measured to about 1% accuracy.
- It explains why, on average, our distant Universe looks the same in all directions, with only 0.002% fluctuations from place to place.
- It explains the origins of these 0.002% fluctuations as quantum fluctuations stretched by inflation from microscopic to macroscopic scales, then amplified by gravity into today's galaxies and cosmic large-scale structure.
- It explains the origin of the enormous gravitational waves discovered in 2014.
- Inflation even explains cosmic acceleration, which nabbed a 2011 Nobel Prize, as inflation restarting, in slow motion, doubling the size of our Universe not every split second but every 8 billion years.
- Inflation theory says that our Universe grew much like a human baby: an accelerating growth phase, in which the size doubled at regular intervals, was followed by a more leisurely decelerating growth phase.
- Inflation created our Hot Big Bang, and inflation's early stages are better thought of as a *Cold Little Swoosh*, because it was neither strikingly hot nor big nor much of a bang.
- What we call our Big Bang wasn't the beginning but the end—of inflation in our part of space—and inflation typically continues forever in other places.
- Inflation generically predicts that our space isn't just huge, but infinite, filled with infinite galaxies, stars and planets, with initial conditions generated randomly by quantum fluctuations.

6

Welcome to the Multiverse

*If the doors of perception were cleansed every thing would appear to man
 as it is, Infinite.
For man has closed himself up, till he sees all things thro' narrow chinks
 of his cavern.*
                    —William Blake, *The Marriage of Heaven and Hell*

*Two things are infinite: the universe and human stupidity; and I'm not
sure about the universe.*
                    —attributed to Albert Einstein

Are you ready for controversy? The science we've explored so far in this
book has by now become mostly mainstream and well accepted. We
now enter the controversial, which many of my physics colleagues will
argue passionately either for or against.

The Level I Multiverse

Is there another copy of you reading this book, deciding to put it aside
without finishing this sentence, while you're reading on? A person living
on a planet called Earth, with misty mountains, fertile fields and sprawl-
ing cities, in a solar system with seven other planets? The life of this
person has been identical to yours in every respect—until now, that is,
when your decision to read on signals that your two lives are diverging.

You probably find this idea strange and implausible, and I must confess that this is my gut reaction, too. Yet it looks like we might just have
to live with it, since the simplest and most popular cosmological model
today predicts that this person actually exists in a galaxy about $10^{10^{29}}$
meters from here. This proposition doesn't even assume speculative
modern physics, but merely that space is infinite and rather uniformly
filled with matter. Your alter ego is simply a prediction of eternal inflation, which, as we've seen in the last chapter, agrees with all current

observational evidence and is implicitly used as the basis for most calculations and simulations presented at cosmology conferences.

What's a Universe?

Before we start talking in earnest about other universes, it's crucial that we're clear on what we mean by our own Universe. This is the terminology we'll use in this book:

Term	Definition
Physical reality	*Everything that exists*
Our Universe	*The part of physical reality we can in principle observe*

If we ignore the quantum complications of the next chapter, the following universe definition is equivalent.

> **Our Universe:** *The spherical region of space from which light has had time to reach us during the 14 billion years since our Big Bang—basically this:*

In the past chapters, we also referred to this region as *our observable Universe*. Geekier-sounding synonyms that are popular with astronomers are our *horizon volume*, or *the region within our particle horizon*. Astronomers also like to talk about our *Hubble volume*, whose size is in the same ballpark, defined as the region within which galaxies are receding slower than light.

Given that other universes may exist, I find it a bit arrogant referring to our own as *the* universe, so I try to avoid using that term altogether. But this is clearly a matter of taste, since New Yorkers refer to their town as "the city," and Americans and Canadians refer to their joint baseball championship as "the World Series."

Although you might find these definitions reasonable, please beware that some people use these words differently, which can cause confusion. In particular, some people use the phrase I eschew, "the universe," to mean everything that exists, in which case, by definition, there can't be any parallel universes.

Now that we've defined our Universe, how big is it? As we discussed, our Universe is a spherical region with Earth at the center. The stuff near the edges of our Universe, from which light has only now reached us after a 14-billion-year space journey, is currently about 5×10^{26} meters away from us.* As far as we currently know, our Universe contains about 10^{11} galaxies, 10^{23} stars, 10^{80} protons and 10^{89} photons (particles of light).

This is certainly a lot of stuff, but could there exist even more, farther away in space? As we saw, inflation predicts that there is. Your doppelgänger's universe (page 126), if it exists, is a sphere of the same size centered over there, none of which we can see or have any contact with yet, because light or other information from there hasn't had time to reach us. This is the simplest (but far from the only) example of parallel universes. I like to call this kind, a distant region of space the size of our Universe, a *Level I* parallel universe. All the Level I parallel universes together form the *Level I multiverse*. Table 6.1 defines all the different types of multiverses we explore in this book and how they're interrelated.

By our very definition of *universe*, one might expect the notion that our observable Universe is merely a small part of a larger multiverse to be forever in the domain of metaphysics. Yet the epistemological borderline between physics and metaphysics is defined by whether a theory is experimentally testable, not by whether it's weird or involves unobservable entities. Technology-powered experimental breakthroughs have therefore expanded the frontiers of physics to incorporate ever more abstract (and at the time counterintuitive) concepts such as a round rotating Earth, an electromagnetic field, time slowdown at high speeds, quantum superpositions, curved space and black holes. As we'll see below, it's becoming increasingly clear that theories grounded in modern physics can in fact be empirically testable, predictive and falsifiable even if they involve a multiverse. Indeed, in the rest of this book, we'll be exploring as many as four distinct levels of parallel universes, so that, to me, the most interesting question isn't whether there's a multiverse (since Level I isn't that controversial), but rather how many levels it has.

What Are Level I Parallel Universes Like?

Suppose inflation really happened and made our space infinite. Then there are infinitely many Level I parallel universes. Moreover, as Fig-

---

* As we saw in Chapter 3, this is more than 14 billion light-years because light gets helped along by the expansion of space.

ure 5.8 illustrates, all of the infinite space was created full of matter which, much like here in our own Universe, gradually formed atoms, galaxies, stars and planets. This means that most of the Level I parallel universes shared our own cosmic history in broad brushstrokes. However, most of them differ from our Universe in the details, because they started out slightly differently. The reason they did is that, as we saw in the previous chapter, the seed fluctuations responsible for all cosmic structure were generated by quantum fluctuations which are for all practical purposes random (see page 107).

Our physics description of the world is traditionally split into two parts: how things start out and how things change. In other words, we have initial conditions and we have laws of physics specifying how the initial conditions evolve over time. Observers living in parallel universes at Level I observe the exact same laws of physics as we do, but with different initial conditions than those in our Universe. For example, the particles start out in slightly different places, moving with slightly different speeds. It's these slight differences that ultimately determine what happens in their universes: which regions turn into galaxies, which regions become intergalactic voids, which stars get planets, which planets get dinosaurs, which planets get their dinosaurs killed by an asteroid collision, and so on. In other words, the quantum-induced differences between parallel universes get amplified over time into very different histories. In summary, students in Level I parallel universes would learn the same thing in physics class but different things in history class.

But would those students exist in the first place? It feels extremely unlikely that your life turned out exactly as it did, since it required so many things to happen: Earth had to form, life had to evolve, the dinosaurs had to go extinct, your parents had to meet, you had to get the idea to read this book, etc. But the probability of all these outcomes happening clearly isn't zero, since it in fact happened right here in our Universe. And if you roll the dice enough times, even the most unlikely things are guaranteed to happen. With infinitely many Level I parallel universes created by inflation, quantum fluctuations effectively rolled the dice infinitely many times, guaranteeing with 100% certainty that your life would occur in one of them. Indeed, in infinitely many of them, since even a tiny fraction of an infinite number is still an infinite number.

And an infinite space doesn't contain only exact copies of you. It contains many more people who are almost like you, yet slightly different. So if you were able to go meet the closest person out there in space who looked like your spitting image, this person would probably speak an alien language you couldn't understand and would have experienced a life quite different from yours. But out of all your infinitely many look-alikes out there on other planets, there's also one who speaks English, lives on a planet identical to Earth, and has experienced a life completely indistinguishable from yours in all ways. This person subjectively feels exactly like you feel. Yet there may be some very minor difference in how the particles move in your alter ego's brain that's too subtle to make a perceptible difference now, but which in a few seconds will make your alter ego put this book aside while you read on, causing your two lives to start diverging.

This raises an interesting philosophical point that will come back and haunt us in Chapter 11: if there are indeed many copies of "you," with identical past lives and memories, this kills the traditional notion of determinism: you can't predict your own future—even if you have complete knowledge of the entire past and future history of the cosmos! The reason you can't is that there's no way for you to determine which of these copies is "you" (they all feel that they are). Yet their lives will typically begin to differ eventually, so the best you can do is predict probabilities for what you'll experience from now on.

In summary, in an infinite space created by inflation, everything that can happen according to the laws of physics does happen. And it happens an infinite number of times. This means that there are parallel universes where you never get a parking ticket, where you have a different name, where you've won a million-dollar lottery, where Germany won World War II, where dinosaurs still roam Earth, and where Earth never formed in the first place. Although each of these outcomes occur in an infinite number of universes, some occur in a larger fraction than others, and making sense of this raises a host of intriguing issues that we'll tackle in Chapter 11.

Are Parallel Universes Unscientific?

Hold on!!! Did I just go bananas??? I mean, so far in this book, I've mostly written about stuff that I hope you found pretty reasonable. Sure, some of the scientific discoveries I wrote about were controversial at the time, but at least they're accepted by the scientific mainstream today. But then things started going kind of crazy in this chapter. And this last business about infinite copies of us doing everything we can imagine—this just sounds nuts. Totally nuts. So before going any farther down this rabbit hole, we need to pause for a sanity check. First of all, is it really science to talk about such crazy things that we can't even observe, or have I crossed the line into pure philosophical speculation?

Let's be more specific. The influential Austro-British philosopher Karl Popper popularized the now widely accepted adage "If it's not falsifiable, then it's not scientific." Physics is all about testing mathematical theories against observation: if a theory can't be tested even in principle, then it's logically impossible to ever falsify it, which, by Popper's definition, means that it's unscientific. It follows then that the only thing that can have any hope of being scientific is a *theory*. Which brings us to a very important point:

> *Parallel universes are not a theory, but a prediction of certain theories.*

Of theories such as inflation. Parallel universes (if they exist) are *things*, and things can't be scientific, so a parallel universe can't be scientific any more than a banana can.

Therefore, we must reformulate our question about philosophical speculation in terms of theories, which leads to the following crucial question:

Are theories predicting the existence of unobservable entities unfalsifiable and therefore unscientific? This is where I think it gets really interesting, because this question has a clear answer: *For a theory to be falsifiable, we need not be able to observe and test all its predictions, merely at least one of them.* Consider the following analogy:

Theory	Prediction
General relativity	Black-hole interiors
Inflation (Chapter 5)	Level I parallel universes
Inflation + landscape (Chapter 6)	Level II parallel universes
Collapse-free quantum mechanics (Chapter 8)	Level III parallel universes
External reality hypothesis (Chapter 10)	Level IV parallel universes

Because Einstein's theory of general relativity has successfully predicted many things that we *can* observe, such as the detailed motion of Mercury around the Sun, the bending of light by gravity, and the gravitational slowing of clocks, we consider it a successful scientific theory and take seriously also its predictions for things we *can't* observe—for example, that space continues inside black-hole event horizons* and that (contrary to early misconceptions) nothing funny happens right at the horizon. Analogously, the successful predictions of inflation that we've described in the last two chapters make inflation a scientific theory, which makes it reasonable to take seriously its other predictions as well—both testable predictions such as what future cosmic microwave-background experiments should measure and seemingly untestable predictions such as the existence of parallel universes. The last three examples in the table on page 124 involve theories we'll describe later in the book that predict additional types of parallel universes.

Another important thing about physics theories is that if you like one, you have to buy the whole package. You're not allowed to say: "Well, I like how general relativity explains Mercury's orbit, but I don't like black holes, so I'm going to opt out of that feature." You can't buy general relativity with the black holes removed the way you can buy coffee with the caffeine removed. General relativity is a rigid mathematical theory with no adjustments possible; you have to either accept *all* its predictions, or you have to start over from scratch and invent a different mathematical theory that agrees with all of general relativity's successful predictions while simultaneously predicting that black holes can't exist. This turns out to be extremely difficult, and so far, all such attempts have failed.

In the same way, parallel universes aren't optional in eternal inflation. They come as part of the package, and if you don't like them, then you have to find a different mathematical theory that solves the bang problem, the horizon problem and the flatness problem, that generates the cosmic seed fluctuations—and doesn't predict parallel universes. This, too, has proven difficult, which is why more and more of my colleagues are—often grudgingly—beginning to take parallel universes seriously.

Evidence for Level I Parallel Universes

Okay, so we've settled one thing: we don't need to feel guilty for talking about parallel universes in this book, even though it's supposed to be a scientific book. But just because something is scientific, it doesn't

---

* Although you can, in principle, enter a black hole and observe what happens inside (if its tidal forces don't "spaghettify" you first), you won't be able to publish your findings in a scientific journal, since you effectively went there with a one-way ticket.

have to be correct, so let's take a closer look at the evidence for parallel universes.

Earlier in this chapter, we saw that the Level I multiverse, including your doppelgängers, is a logical consequence of eternal inflation. We've also seen that inflation is currently the most popular early-universe theory in the scientific community, and that inflation is typically eternal, thus producing the Level I multiverse. In other words, the best evidence for the Level I multiverse is the evidence we have for inflation. Does this prove that your doppelgängers exist? Certainly not! At this point, we can't be 100% certain that inflation is eternal, or even that it happened at all. Fortunately, inflation research is now a very active field both theoretically and experimentally, so we're likely to gain more evidence for or against eternal inflation (and consequently for or against the Level I multiverse) in the years ahead.

So far, our entire discussion in this chapter has been in the context of inflation. But does the Level I multiverse stand and fall with inflation? No! For there to be no Level I parallel universes at all, there must be no space whatsoever beyond the region we can see. I don't have a single science colleague who's argued for such a small space, and someone arguing for it could be likened to an ostrich with its head in the sand, claiming that only what it can see can exist. We all accept the existence of things that we can't see but could see if we moved or waited, like ships beyond the horizon. Objects beyond our cosmic horizon have similar status, since our observable Universe grows by roughly a light-year every year as light from farther away has time to reach us.*

What about evidence for our doppelgängers? If we tease apart our arguments above, we see that the "everything that can happen does happen" property of the Level I multiverse follows from two logically distinct assumptions, both of which could conceivably be correct even without inflation:

1. **Infinite space and matter:** *Early on, there was an infinite space filled with hot expanding plasma.*
2. **Random seeds:** *Early on, a mechanism operated such that any region could receive any possible seed fluctuations, seemingly at random.*

---

* If the cosmic expansion continues to accelerate (currently an open question), the observable Universe will eventually stop growing: all galaxies beyond a certain distance will eventually recede faster than light and be forever invisible to us.

Let's explore these two assumptions in turn. I think the second one is a pretty reasonable assumption, regardless of inflation. We've observed that these random-looking seed fluctuations exist, so we know that *some* mechanism made them. We've measured their statistical properties carefully using cosmic microwave–background and galaxy maps, and their random properties are consistent with what's known to statisticians as a "Gaussian random field," which satisfies assumption 2. Moreover, if inflation didn't happen and distant spatial regions were never able to communicate with each other (Figure 5.2), then this mechanism would be guaranteed to roll the dice independently in each region.

What about the assumption of infinite space and matter? Well, an infinite space rather uniformly filled with matter used to be the standard assumption in mainstream cosmology even long before inflation was invented, and is now part of what's known as the cosmological standard model. Yet this assumption and its Level I multiverse implications used to be controversial; indeed, an assertion along these lines was one of the heresies for which the Vatican had Giordano Bruno burned at the stake in 1600. Those of us who have published on this topic more recently, including George Ellis, Geoff Brundrit, Jaume Garriga and Alex Vilenkin, have thus far avoided the stake, but let's nonetheless take a critical look at the infinite space and infinite matter assumption.

We saw in Chapter 2 that although the simplest model of space (dating back to Euclid) is infinite, Einstein's general relativity allows various elegant ways in which space can be finite. If space curves back on itself like a hypersphere (Figure 2.7), then the total volume of this hypersphere must be at least a hundred times larger than the part of it that we can observe (our Universe) in order to explain why our visible part of space is so flat that cosmic microwave–background experiments haven't detected any curvature. In other words, even if we live in a finite space of the hypersphere kind, then there are at least a hundred Level I parallel universes.

What about a finite space of the torus (bagel) kind that we explored in Chapter 2, where space is flat but you nonetheless return to your starting point if you travel some distance? Such a space is like that of one of those computer games where you can fly off the screen and instantly reenter on the other side, so if you could see far enough in front of you, you'd see the back of your own head—and infinitely many regularly spaced copies of you in all directions, much like if you were standing in a mirror-covered room. If our space has this property, what's the small-

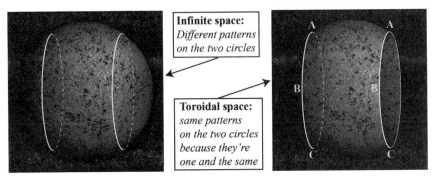

Figure 6.1: If you fly to the right, past the circle in the toroidal universe, you immediately reenter at the corresponding point on the left circle—leave at A, reenter at A, etc.; the two As are actually the same physical point. This means that the cosmic microwave–background patterns along the two circles should look similar to us, since they're actually one and the same.

est size it could have? It clearly has to be much larger than our Galaxy, since our telescopes don't show infinite copies of the Milky Way lined up in tidy rows. But if the size were, say, 10 billion light-years, this test would fail: we wouldn't see the nearest copy of our Galaxy because it didn't exist 10 billion light-years ago. Fortunately, there's an even more sensitive test: we can look for a recognizable object such as a bright galaxy 5 billion light-years away, and then look for the same object 5 billion light-years in the opposite direction. Such searches have also come up empty-handed. The most sensitive test of all is to use the most distant thing we can see, the cosmic microwave background, and look for matching patterns in opposite directions as in Figure 6.1—many research teams, Angélica and I included, tried this and found nothing. Also, if space has a finite volume, only certain perturbation frequencies are allowed, just as the air in a flute can only vibrate at certain special frequencies. This distorts the microwave-background power spectrum in a particular way that Angélica and others have looked for without finding anything. In summary, it's still possible that space is finite, but finite space models have been severely constrained by observations in recent years, so the only spaces still allowed have a volume that's comparable to or greater than our Universe. This makes it really tough to avoid at least a handful of parallel universes. Moreover, having exactly one universe right now would require a strange unexplained "Why now?" coincidence, since there would have been more than one universe when light had only had time to reach us from a smaller fraction of space.

Enough about infinite space. What about the infinite-matter part of the assumption? Before inflation, it was often justified by appealing to the so-called *Copernican principle*, that we humans don't occupy any special place in the cosmos: if there are galaxies around here, there should be galaxies everywhere.

What do recent observations have to say about it? Specifically, how uniform is the matter distribution on large scales? In an "island universe" model where space is infinite but all the matter is confined to a finite region, almost all members of the Level I multiverse would be dead, consisting of nothing but empty space. Such models have been popular historically, originally with the island being Earth and the celestial objects visible to the naked eye, and in the early twentieth century, with the island being the known part of the Milky Way Galaxy. The island-universe model has been demolished by recent observations. The 3-D galaxy maps from the last chapter have shown that the spectacular large-scale structure observed (galaxy groups, clusters, superclusters, walls) gives way to dull uniformity on large scales, with no coherent structures larger than about a billion light-years.

The larger the scale we observe, the more uniformly filled with matter our Universe looks (Figure 4.6). Barring conspiracy theories where our Universe is designed to fool us, the observations thus speak loud and clear: space as we know it appears to continue far beyond the edge of our Universe, teeming with galaxies, stars and planets.

Where Are the Level I Parallel Universes?

We've seen that if they exist, then Level I parallel universes are simply universe-sized parts of our space that are so far away that light from them hasn't yet had time to reach us. Does the fact that we're at the center of our Universe mean that we're somehow in a special place in space? Well, if you're walking on a large field when fog has cut the visibility to 50 meters, you'll feel like you're at the center of a fog sphere, beyond which (akin to the edge of our Universe) you can't see anything. But that doesn't mean that you're in any sort of special place, at the center of anything fundamental, because everyone else on that field will find themselves at the center of their own fog spheres. In the same way, any observers anywhere in space will find themselves at the centers of their universes. Also, there are no physical boundaries between neighboring universes, just as there's no special boundary 50 meters into the fog—the field and the fog have the same properties over there as where you are. Moreover, universes can overlap just as fog spheres can: just as someone 30 meters away on the field can see both you and regions that you can't see, the universe of someone in a galaxy 5 billion light-years from us would contain both Earth and regions of space that lie outside of our Universe.

If eternal inflation or something else created an infinite number of such parallel universes, then how far away is the nearest identical copy of our own? According to classical physics, a universe can be arranged in infinitely many different ways, so there's no guarantee that you'd ever find an exactly identical one. Classically, there are infinitely many options even for the distance between two particles, since it requires infinitely many decimal places to specify. However, there's clearly only a finite number of universe possibilities that our collective human civilization can ever distinguish between in practice, since our brains and computers can store only a finite amount of information. Moreover, we can only measure things with finite accuracy—our current record in physics is measuring a quantity to about sixteen decimal places.

Quantum mechanics limits the variety even at a fundamental level. As we'll explore in the next two chapters, quantum mechanics adds a sort of intrinsic fuzziness to nature that makes it meaningless to talk about where things are beyond a certain level of precision. The result of this limitation is that the total number of ways in which our Universe can be arranged is finite. A conservative estimate, erring on the high side, is that there are at most $10^{10^{118}}$ possible ways in which a universe the size of ours can be arranged.* An even more conservative bound, known as the holographic principle, says that a volume the size of our Universe can be arranged in, at most, $10^{10^{124}}$ ways.† Otherwise, you'd have to pack so much stuff into it that it would form a black hole larger than itself.

These are huuuuuuge numbers, even larger than the famed googol-

---

* This is an extremely conservative estimate, simply counting all possible quantum states that a universe (horizon volume) can have that are no hotter than 10^8 degrees. Although the actual calculation requires quantum-mechanical details, the number 10^{118} can be roughly understood as the number of protons that the so-called Pauli exclusion principle would allow you to pack into a universe at this temperature (our own Universe contains only about 10^{80} protons). If each of these 10^{118} slots can be either occupied or unoccupied, there are $2^{10^{118}} \sim 10^{10^{118}}$ possibilities.
† That's two to the power of the surface area of our Universe measured in so-called Planck units. The books by Lenny Susskind and Brian Greene in the "Suggestions for Further Reading" section describe the holographic principle in detail and how it was developed from ideas of Gerard t'Hooft, Lenny Susskind, Charles Thorn, Raphael Bousso, Jacob Bekenstein, Stephen Hawking, Juan Maldacena and others.

plex. Little boys tend to obsess about big things, and I've overheard my sons and their friends try to outdo each other by naming ever bigger numbers. After trillions, octillions and so on, someone inevitably drops the G-bomb: googolplex. After which, a moment of awed silence ensues. As you may know, a googolplex is one followed by a googol zeros, where a googol is one followed by a hundred zeros. So it's $10^{10^{100}}$, which isn't one followed by a hundred zeros, but one followed by 10, 000,000,000,000,000,000,000,000,000,000,000,000,000,000, 000,000,000,000,000,000,000,000,000,000,000,000,000,000, 000 zeros! This number is so large that you couldn't write it out even in principle, since it contains more digits than there are atoms in our Universe. I always suspected that Google was an ambitious company. When I visited them for a conference, I discovered that they call their corporate campus the Googleplex.

Although $10^{10^{118}}$ is huge beyond astronomical, it's still nothing compared with infinity. This means that if eternal inflation made a space containing infinitely many Level I parallel universes, then you'll find it containing all possibilities. Specifically, you'll have to check on average about $10^{10^{118}}$ universes until you find a copy of any particular kind of universe, as illustrated in Figure 6.2. So if you could travel in a straight line until you reached the closest identical copy of our own Universe, you'd need to journey about $10^{10^{118}}$ universe diameters. If you're willing

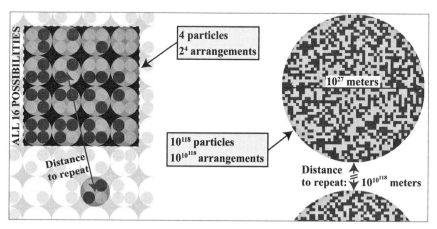

Figure 6.2: In a toy universe where four different locations can each hold one of two kinds of particles, there are only 2^4 possible arrangements (top left). This means that in a Level I multiverse of such universes, on average you have to check 16 universes until you find a repeat of a particular universe. If our Universe can similarly contain $10^{10^{118}}$ particles arranged in $10^{10^{118}}$ different ways, then you'll have to travel past about $10^{10^{118}}$ parallel universes before reaching an identical copy.

to look in all directions to find our closest copy, the distance to the closest one comes out be about the same, which is also about the same as $10^{10^{118}}$ meters, given the funny mathematical behavior of double exponents (powers of powers).*

Closer by, about ~ $10^{10^{91}}$ meters away, there should be a sphere of radius 100 light-years identical to the one centered here, so all perceptions that we have during the next century will be identical to those of our counterparts over there. About ~ $10^{10^{29}}$ meters away, there should be an identical copy of you. Indeed, there are probably copies of you much closer than that, since the planet formation and evolutionary processes that have tipped the odds in your favor are at work everywhere. There are probably at least 10^{20} habitable planets in our own Universe volume alone.

The Level II Multiverse

Earlier, I called inflation the gift that keeps on giving, because every time you think it can't possibly predict something more radical than it already has, it does. If you felt that the Level I multiverse was large and hard to stomach, try imagining an infinite set of distinct ones, some perhaps with apparently different laws of physics. Andrei Linde, Alex Vilenkin, Alan Guth and their colleagues have shown that this is what inflation typically predicts, and we'll refer to it as the Level II multiverse.

Many Universes in One Space

How can physics possibly allow such craziness? Well, we saw in Figure 5.8 how inflation could create an infinite volume inside of a finite volume. As Figure 6.3 illustrates, there's no reason why inflation can't do this in several adjacent volumes, ending up with several infinite regions (Level I multiverses), as long as inflation is eternal and never ends at the boundaries between them. This means that if you live in one of these Level I multiverses, it's impossible for you to visit a neighboring one: inflation keeps creating intervening space faster than you can

---

* If you're a math buff, note that $10^{10^{118}}$ universe diameters ≈ $10^{10^{118}} \times 10^{27}$ m = $10^{10^{118}+27}$ m ≈ $10^{10^{118}}$ m. If you're willing to look in all directions to find our closest copy, then you need to explore a spherical volume around us containing about $10^{10^{118}}$ universes, whose radius exceeds that of our Universe by a factor $(10^{10^{118}})^{1/3} = 10^{10^{118}/3} \approx 10^{10^{117.53}} \approx 10^{10^{118}}$.

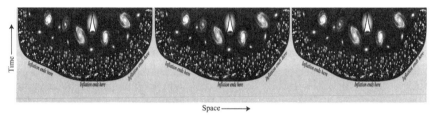

Figure 6.3: If eternal inflation creates three infinite regions using the mechanism from Figure 5.8, then travel between them is impossible because inflation keeps creating new space between you and your destination faster than you can travel through it.

travel through it. I imagine trying this with my kids in the backseat of my rocket:

"Dad, are we there yet?"

"We have one light-year left to go."

"Dad, are we there yet?"

"We have two light-years left to go."

In other words, although these other parts of the Level II multiverse are in the same space as we are, they're more than infinitely far away in the sense that we'd never reach them even if we traveled at the speed of light forever. In contrast, you can in principle travel to an arbitrarily distant part of our Level I multiverse if you're patient enough and the cosmic expansion decelerates.*

I've simplified things in Figure 6.3 by ignoring the fact that space is expanding. The eternally inflating regions in the figure, which I've drawn as thin vertical bars separating the U-shaped Level I multiverses, will in fact expand rapidly, and eventually, parts within them will stop inflating, giving rise to additional U-shaped regions. This makes things even more interesting, giving the Level II multiverse a treelike structure as illustrated in Figure 6.4. Any inflating region keeps expanding rapidly, but inflation eventually ends in various parts of it, forming U-shaped regions that each constitute an infinite Level I multiverse. This tree continues growing forever, creating an infinite number of such U-shaped regions—all of them together form the Level II multiverse. Within each such region, the end of inflation transforms the inflating substance into particles that eventually cluster into atoms, stars

---

* If the dark energy sticks around so that our cosmic acceleration continues, then even most Level I parallel universes will remain forever separate, with the intervening space stretching faster than light can travel through it. We don't yet understand dark energy well enough to know whether this will happen.

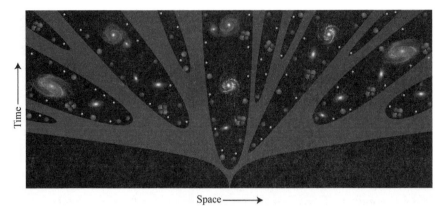

Figure 6.4: The expansion of space and the fact that inflation keeps ending in certain places gives the Level II multiverse a treelike structure. Inflation continues in the tree-shaped gray part of space and time, and each U-shaped region where inflation has ended is an infinite Level I multiverse.

and galaxies. Alan Guth likes to call each Level I multiverse a "pocket universe," because it conveniently fits into a small part of the tree.

Diversity!

Earlier in this chapter, I mentioned that the Level II multiverse can contain infinite regions with apparently different laws of physics. But this sounds absurd: how can the laws of physics allow different laws of physics? As we'll now see, the key idea is that *fundamental laws* of physics, which by definition hold anywhere and anytime, can give rise to a complicated physical state of affairs where the *effective laws* of physics inferred by self-aware observers vary from place to place.

If you were a fish who'd lived your entire life in the ocean, you might make the mistake of thinking of water not as a substance, but as empty space. What a human would think of as a property of water, say, the friction when swimming through it, you might misinterpret as a fundamental law of physics: "a fish in uniform motion ends up at rest—unless flapping its fins." You'd probably have no idea that water can exist in three different phases—solid, liquid and gaseous—and that your "empty space" was simply the liquid phase, a particular solution to the equations describing water.

This example may sound silly, and if a real fish were to think this, we might be tempted to laugh at it. But could it be that what we humans

Figure 6.5: Can space freeze? A fish might think of water as empty space, because it's the only medium it knows. But if a clever fish figured out the physical laws governing water molecules, it could realize that they have three different solutions, "phases," corresponding to the liquid water it knows and also to steam and ice, which it's never seen. In the same way, what we've thought of as empty space may be a medium with 10^{500} or more different phases, of which we've experienced only one.

think of as empty space is also some form of medium? Then the last laugh would be on us! As a matter of fact, there's mounting evidence that this is exactly how things are. Not only does our "empty space" seem to be a sort of medium, but it appears to have way more than three phases—perhaps about 10^{500}, and perhaps even infinitely many, which opens up the possibility that, in addition to curving, stretching and vibrating, our space may even be able to do something analogous to freezing and evaporating!

How did physicists reach this conclusion? Well, if a fish were sufficiently intelligent, it could build experiments and determine that its "space" is made of water molecules obeying certain mathematical equations. By studying these equations, it could, as illustrated in Figure 6.5, determine that they have three different solutions corresponding to the three phases of solid ice, liquid water and gaseous steam, even if it had never seen either an iceberg or a steam-producing underwater volcanic vent. In exactly the same way, we physicists are searching for equations describing our own space and its contents. We haven't yet found the final answer, but the approximate answers we've found so far tend to share a key feature: they have more than one solution (phase) that describes a uniform space. String theory, which is a leading candidate for a final answer, has been found to have perhaps 10^{500} or more solutions, and there's no indication that competing theories such as loop quantum gravity have a single unique solution either. Physicists like to

call the collection of all possible solutions the *landscape* of the theory.*
These solutions, whose properties constitute effective laws of physics,
all correspond to different possibilities allowed by the same fundamental laws of physics.

What does this have to do with inflation? Remarkably, eternal inflation has the property that it creates all possible kinds of space! It realizes
the entire landscape. In fact, for each phase that space can have, it creates infinitely many Level I multiverses full of that phase. This means
that we observers are easily tricked into making the same mistake as the
fish: because we observe space to have the same properties everywhere
in our Universe, we're tempted to mistakenly conclude that space is like
that everywhere else as well.

How does inflation do this? Well, it requires lots of energy to change
the phase of space, so the everyday processes that we can observe are
unable to do it. But back during inflation, there was an enormous amount
of energy in each small volume, enough for the previously mentioned
quantum fluctuations to occasionally cause a phase change in some tiny
region, which would then inflate to become an enormous region containing only that same phase. Moreover, a given region of space has
to be in a definite phase in order to stop inflating. This ensures that
boundary regions between two phases keep inflating forever, so that
each phase fills an entire infinite Level I multiverse.

What are these different phases of space like? Imagine that you get
a car as a birthday present, with the key in the ignition, but you have
never heard of cars before and have absolutely no information about
how they work. Being an inquisitive person, you get inside and start
messing with the various buttons, knobs and levers. Eventually, you figure out how to use it and get quite good at driving. But unbeknownst
to you, somebody has removed the letter *R* by the gearshift and messed
with the transmission so that you need to apply a crazy amount of force
to shift into Reverse. This means that unless someone tells you, you'll
probably never figure out that the car can drive backwards as well. If

---

* For detailed step-by-step accounts of how the Level II multiverse was discovered and
developed by Andrei Linde, Alex Vilenkin, Alan Guth, Sidney Coleman, Frank de Luc-
cia, Raphael Bousso, Joe Polchinski, Lenny Susskind, Shamit Kachru, Renata Kallosh,
Sandip Trivedi and others, I recommend the recent books by Brian Greene, Lenny Susskind and Alexander Vilenkin in the "Further Reading" section at the end of this book.
The Greene and Susskind books provide good introductions to string theory by two of
its pioneers.

asked to describe how the car worked, you'd incorrectly assert that, without exception, as long as the engine is running, the harder you push on the accelerator pedal, the faster the car moves forward. If in a parallel universe, the car had instead required huge force to shift into forward drive mode, you'd have concluded that this strange machine worked differently and only moved backwards.

Our Universe is very much like this car. As illustrated in Figure 6.6, it has a bunch of "knobs" that control how it works: the laws according to which things move when you do various things to them and so forth—what we're told in school are the laws of physics, including so-called constants of nature. Each setting of the knobs corresponds to one of the phases of space, so if there are 500 knobs with 10 possible settings each, there are 10^{500} different phases.

When I was in high school, I was incorrectly taught that these laws and constants were always valid, and never changed either from place to place or from time to time. Why this mistake? Because an enormous amount of energy—much more than we have at our disposal—is required to change the settings of these knobs, just as the gearshift on that car, so we didn't realize that the settings could be changed. Nor that there even were any settings to change: unlike gearshifts, nature's knobs are well hidden. They come in the form of so-called high-mass fields and other obscure entities, and huge

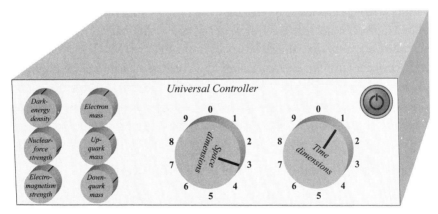

Figure 6.6: The very fabric of space and time seems to have various built-in knobs that can be dialed to different settings in different parts of the Level II multiverse. Our actual Universe seems to have thirty-two knobs that can be dialed continuously, as we'll see in Chapter 10, as well as additional ones with a discrete number of settings that control what kinds of particles that can exist.

energy is required not only to alter them, but even to detect that they exist in the first place.

So how then have physicists figured out that these knobs probably do exist, and that we could actually make our Universe work differently if we had enough energy? In the same way that you could, if you were really inquisitive, figure out that your car could in principle drive backwards: by examining in detail how its parts work! You could figure it out by carefully examining the transmission gearbox. In the same way, detailed study of the smallest building blocks of nature suggests to us that, with enough energy, they could be rearranged in a way such that our Universe would operate differently—we'll explore the workings of these building blocks in the next chapter. Eternal inflation would have provided enough energy for the quantum fluctuations to actually make all such possible rearrangements in different Level I multiverses. It acted like an extremely strong gorilla that randomly messed with all the knobs and gearshifts in a whole parking lot full of cars: by the time it was done, some fraction of them would be in Reverse.

In summary, the Level II multiverse fundamentally changes our notion of physical laws. Many of the regularities that we used to view as *fundamental laws*, which by definition hold anywhere and anytime, have turned out to be merely *effective laws*, local bylaws that can vary from place to place, corresponding to different knob settings defining space in different phases. Table 6.1 summarizes these notions and how they're related to parallel universes. This change continues an old trend: whereas Copernicus thought that it was a fundamental law that planets orbit in perfect circles, we now know that more general orbits are allowed, and that the level of non-circularity (which astronomers call "eccentricity") of an orbit is effectively a knob that can be changed only slowly and with difficulty once a solar system has formed. The Level II multiverse takes this concept to a new level by downgrading many more fundamental laws to effective laws, as we'll explore next.

Fine-Tuning as Evidence for the Level II Multiverse

So does the Level II multiverse really exist? As we've seen, evidence for eternal inflation (of which there's plenty) is evidence for the Level II multiverse, because the former predicts the latter. We also saw that if there are laws or constants of nature that can in principle vary from place to place, then eternal inflation will make them do so across the

Multiverse Terminology That We Use in This Book	
Physical reality	Everything that exists; Chapter 12 argues that this equals the Level IV multiverse
Space	The part of physical reality that's continuously connected to what we can observe; with eternal inflation, this equals the Level II multiverse
Our Universe	The part of physical reality we can in principle observe; quantum complications aside, this is the spherical region of space from which light has had time to reach us during the 14 billion years since our Big Bang
Parallel universe	A part of physical reality that can in principle be observed from somewhere else but not from here—parallel universes are not a theory, but a prediction of certain theories
Multiverse	A collection of universes
Level I multiverse	Distant regions of space that are currently but not forever unobservable; they have the same effective laws of physics but may have different histories
Level II multiverse	Distant regions of space that are forever unobservable because space between here and there keeps inflating; they obey the same fundamental laws of physics, but their effective laws of physics may differ
Level III multiverse	Different parts of quantum Hilbert space (Chapter 8); same diversity as Level II
Level IV multiverse	All mathematical structures (Chapter 12), corresponding to different fundamental laws of physics
Fundamental laws	The mathematical equations that govern physics
Effective laws	Particular solution to the mathematical equations that describe physics; can be mistaken for fundamental laws if the same solution is implemented throughout universe
Fine-tuning	Physical constants in the effective laws having values in a very narrow range allowing life; observed fine-tuning is arguable evidence for the Level II multiverse

Table 6.1: Summary of key multiverse concepts and how they're interrelated

Level II multiverse. But is there any more direct evidence that doesn't hinge so crucially on theoretical arguments?

I'm going to argue that there is: the fact that our Universe appears highly fine-tuned for life. Basically, we've discovered that many of those knobs that we discussed appear tuned to very special values, and if we could change them even by quite small amounts, then life as we know it would become impossible. Tweak the dark-energy knob and galaxies never form, tweak another knob and atoms become unstable, and so on.

Lacking pilot training, I'd feel terrified to mess with any of the knobs in an airplane cockpit, but if I could randomly mess with the knobs of our Universe, my survival odds would be even worse.

I've seen three main reactions to this observed fine-tuning:

1. **Fluke:** It's just a fluke coincidence and there's nothing more to it.
2. **Design:** It's evidence that our Universe was designed by some entity (perhaps a deity or an advanced universe-simulating life form) with the knobs deliberately fine-tuned to allow life.
3. **Multiverse:** It's evidence for the Level II multiverse, since if the knobs have all settings somewhere, it's natural that we'll exist and find ourselves in a habitable region.

We'll explore the fluke and multiverse interpretations below and the simulation interpretation in Chapter 12. But first, let's explore the fine-tuning evidence to see what all the fuss is about.

Fine-Tuned Dark Energy

As we saw in Chapter 4, our cosmic history has been a gravitational tug-of-war between dark matter trying to pull things together and dark energy trying to push them apart. Because galaxy formation is all about pulling things together, I think of dark matter as our friend and dark energy as our enemy. The cosmic density used to be dominated by dark matter, and its friendly gravitational attraction helped assemble galaxies such as our own. However, because the cosmic expansion diluted the dark matter but not the dark energy, the unkind gravitational repulsion of dark energy eventually gained the upper hand, sabotaging further galaxy formation. This means that if the dark energy had had significantly larger density, it would have started gaining the upper hand much sooner, before any galaxies had had time to form. The result would be a stillborn universe, remaining forever dark and lifeless, containing nothing more complex or interesting than nearly uniform gas. If, on the other hand, the dark-energy density were reduced enough to be significantly negative (which is allowed by Einstein's gravity theory), then our Universe would have stopped expanding, recollapsing in a cataclysmic Big Crunch before any life had had time to evolve. In summary, if you actually figure out how to change the dark-energy density by turning the dark-energy knob in Figure 6.6, then please don't turn it too far in either direction, because this would be just as bad for life as pressing the Off button.

How far could you rotate the dark-energy knob before the "Oops!" moment? The current setting of the knob, corresponding to the dark-energy density we've actually measured, is about 10^{-27} kilograms per cubic meter, which is almost ridiculously close to zero compared to the available range: the natural maximum value for the dial is a dark-energy density around 10^{97} kilograms per cubic meter, which is when the quantum fluctuations fill space with tiny black holes, and the minimum value is the same with a minus sign in front. If rotating the dark-energy knob in Figure 6.6 by a full turn would vary the density across the full range, then the actual knob setting for our Universe is about 10^{-123} of a turn away from the halfway point. That means that if you want to tune the knob to allow galaxies to form, you have to get the angle by which you rotate it right to over 120 decimal places! Although this sounds like an impossible fine-tuning task, some mechanism appears to have done precisely this for our Universe.

Fine-Tuned Particles

In the next chapter, we'll explore the microworld of elementary particles. There are many knobs there, too, determining particles' masses and how strongly particles interact with each other, and the science community has gradually come to realize that many of these knobs are fine-tuned as well.

For instance, if the electromagnetic force were weakened by a mere 4%, then the Sun would immediately explode as its hydrogen fused into so-called diprotons, an otherwise nonexistent kind of neutron-free helium. If it were significantly strengthened, previously stable atoms such as carbon and oxygen would radioactively decay away.

If the so-called weak nuclear force were substantially weaker, there would be no hydrogen around, since it would all have been converted to helium shortly after our Big Bang. If it were either much stronger or much weaker, the neutrinos from a supernova explosion would fail to blow away the outer parts of the star, and it's doubtful whether life-supporting heavier elements such as iron would ever be able to leave the stars where they were produced and end up in planets such as Earth.

If electrons were much lighter, there could be no stable stars, and if they were much heavier, there could be no ordered structures such as crystals and DNA molecules. If protons were 0.2% heavier, they'd decay into neutrons unable to hold on to electrons, so there would be no atoms. If they were instead much lighter, then neutrons inside of atoms would decay into protons, so there would be no stable atoms except for hydrogen. Indeed, the proton mass depends on another knob that has a very wide range of variation and needs to be fine-tuned to thirty-three decimal places to get any stable atoms other than hydrogen.

Fine-Tuned Cosmology

Many of these fine-tuning examples were discovered in the seventies and eighties by Paul Davies, Brandon Carter, Bernard Carr, Martin Rees, John Barrow, Frank Tipler, Steven Weinberg and other physicists. And more examples just kept turning up. My first foray into this was with Martin Rees, a white-haired astronomer with impeccable British manners who's one of my science heroes. I haven't seen anybody else look as happy and excited when they give a talk, and it's as if his eyes beam out enthusiasm. He was the first member of the scientific establishment to encourage me to follow my heart and pursue non-mainstream topics. In the last chapter, we saw that the cosmic seed–fluctuation amplitude was about 0.002%. Martin and I calculated that if it were much smaller, galaxies wouldn't have formed, and if it were much larger, frequent asteroid impacts and other difficulties would ensue.

This is what I was talking about when I put Alan Guth to sleep. My talk host, Alex Vilenkin, stayed awake, however, and we later teamed up to study neutrinos, ghostlike particles that our Big Bang created in abundance. We found that they, too, appeared somewhat fine-tuned, in that making them significantly heavier would sabotage galaxy formation. My MIT colleague Frank Wilczek had an idea for how the dark-matter density could vary from universe to universe, and together with Martin Rees and my friend Anthony Aguirre, we calculated that turning the dark-matter knob far from its observed value is also bad for our health.

The Fluke Explanation

So what are we to make of this fine-tuning? First of all, why can't we just dismiss it all as a bunch of fluke coincidences? Because the scientific method doesn't tolerate unexplained coincidences: saying, "My theory requires an unexplained coincidence to agree with observation" is equivalent to saying, "My theory is ruled out." For example, we've seen how inflation predicts that space is flat and the spots in the cosmic microwave background should have an average size around a degree, and that the experiments described in Chapter 4 confirmed this. Suppose that the Planck team had observed a much smaller average spot size, prompting them to announce that they'd ruled out inflation with 99.999% confidence. This would mean that random fluctuations in a flat universe *could* have caused spots to appear as unusually small as they measured, tricking them into an incorrect conclusion, but that with 99.999% probability, this wouldn't happen. In other words, inflation would require a 1-in-100,000 unexplained coincidence in order to agree with the measurement. If Alan Guth and Andrei Linde now held a joint press conference, insisting that there was no evidence against inflation because they had a gut feeling that the Planck measurements were just a fluke coincidence that should be dismissed, they'd be violating the scientific method.

In other words, random fluctuations mean that we can never be 100% sure of anything in science—there's always a tiny probability that you got really unlucky with random measurement noise, that your detector malfunctioned, or even that the whole experiment was just a hallucination. In practice, however, a theory that's ruled out at 99.999% confidence is normally considered dead as a doornail by the scientific community. Yet the theory that the dark-energy fine-tuning is a fluke requires us to believe in a much more unlikely unexplained coincidence, and is therefore ruled out at about 99.999999 . . . percent confidence, where there are about 120 nines after the decimal point.

The *A* Word

What about the Level II multiverse explanation of fine-tuning? A theory where the knobs of nature take essentially all possible values somewhere will predict with 100% certainty that a habitable universe like ours exists, and since we can only live in a habitable universe, we shouldn't be surprised to find ourselves in one.

Although this explanation is logical, it's quite controversial. After all the silly historical attempts to keep Earth as the center of our Universe, the opposite viewpoint has gotten deeply entrenched. Known as the Copernican principle, it holds that there's nothing special about our place in space and time. Brandon Carter proposed a direct competitor that he called the *weak anthropic principle*: "We must be prepared to take account of the fact that our location in the universe is necessarily privileged to the extent of being compatible with our existence as observers." Some of my colleagues view this as an objectionable step backwards, reminiscent of geocentrism. When taking fine-tuning into account, the Level II–multiverse picture does indeed violate the Copernican principle big time, as illustrated in Figure 6.7: the vast majority of all universes are stone dead, and our own is extremely atypical: it contains way less dark energy than most other ones, and also has highly unusual settings of many other "knobs."

Explaining things we can observe by introducing parallel universes that we can't observe also rubs some of my colleagues the wrong way. I remember a 1998 talk at Fermilab, home of the famous particle accelerator outside Chicago, where the audience erupted in an audible hiss when a speaker mentioned the "*A* word," *anthropic*. Indeed, to sneak under the radar and past the referee, Martin Rees and I went out of our way to avoid using the *A* word anywhere in the abstract of that first anthropic paper we wrote together. . . .

Personally, my only objection to Carter's anthropic principle is that it contains the word *principle*, suggesting that it's somehow optional. But no, the use of correct logic when confronting a theory with observation isn't optional. If most of space is uninhabitable, then we should clearly expect to find ourselves in a place that's special in the sense of being habitable. Indeed, most of space seems rather uninhabitable even if we limit ourselves to our own Universe: good luck surviving in an intergalactic void or inside a star! For example, only a thousandth of a trillionth

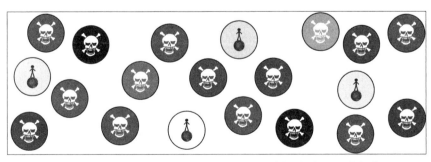

Figure 6.7: If the dark-energy density (here represented by darkness of shading) varies from universe to universe, then galaxies, planets and life will only emerge in those universes where it's the lowest. In this illustration, the habitable least-dark fraction is 20% of the universes, but the real fraction may be closer to 10^{-120}.

of a trillionth of a trillionth of our Universe lies within a kilometer of a planetary surface, so that's quite a special place, yet that's where we find ourselves and it's hardly surprising.

As a more interesting example, consider M, the mass of our Sun. M affects the luminosity of the Sun, and using basic physics, one can compute that life as we know it on Earth is only possible if M is in the narrow range between 1.6×10^{30}kg and 2.4×10^{30}kg—otherwise Earth's climate would be colder than on Mars or hotter than on Venus. The measured value is $M \sim 2.0 \times 10^{30}$kg. This apparently unexplained coincidence of the habitable and observed M values may appear disturbing given that calculations show that stars in the much broader mass range from $M \sim 10^{29}$kg to 10^{32}kg can exist: the mass of our Sun appears fine-tuned for life. However, we can explain this apparent coincidence because there's an ensemble of many such systems with different "knob settings": we now know that there are many solar systems with a range of sizes of the central star and the planetary orbits, and we should obviously expect to find ourselves living in one of the inhabitable solar systems.

The interesting point here is that we could have used this fine-tuning of our Solar System to argue that different solar systems should exist even before any were discovered. Using the exact same logic, we can use the observed fine-tunings of our Universe to argue for the existence of different universes. The only difference is whether the other predicted entities are observable or not, but this difference doesn't weaken the argument, since it never enters into the logic.

What Can We Ever Hope to Predict?

We physicists like measuring numbers. Such as these, for example:

Parameter	Observed Value
Mass of Earth	5.9742×10^{24}kg
Mass of electron	$9.10938188 \times 10^{-31}$kg
Radius of Earth's orbit in Solar System	$149,597,870,691 \times 10^{24}$m
Radius of electron's orbit in hydrogen atom	$5.29177211 \times 10^{-11}$m

We also like trying to predict such numbers from first principles. But will we ever succeed, or is this merely wishful thinking? Before making his famous discovery that planetary orbits are ellipses, Johannes Kepler had an elegant theory related to the third number in the table above: he proposed that the orbits of Mercury, Venus, Earth, Mars, Jupiter and

Saturn had exactly the same proportions to one another as six spheres nested like Russian Dolls that had between them an octahedron, icosahedron, dodecahedron, tetrahedron and cube, respectively (see Figure 7.2, page 159). Aside from the fact that his theory was soon ruled out by better measurements, its entire premise seems silly now that we know that there are other solar systems: the particular orbits we measure in our Solar System don't tell us anything fundamental about our Universe, merely something about our location in it, in this case which particular solar system we live in. In this sense, we can think of these digits as just part of our address in space, as part of our cosmic postal code. For example, to explain to an extraterrestrial mailman which solar system in our Galactic neighborhood we wanted our package delivered to, we could tell him to come to the one with eight planets whose orbits were 1.84, 2.51, 4.33, 12.7, 24.7, 51.1 and 76.5 times larger than that of the innermost planet, and he might say: "Oh, I know which solar system you're talking about!" In the same vein, we've permanently given up on predicting Earth's mass or radius from first principles because we know that many planets with different sizes exist.

But what about the mass and orbital size for an electron? These numbers are the same for all electrons in our Universe that we've checked, so we've gotten our hopes up that they may be truly fundamental properties of our physical world that we'll one day be able to compute from theory alone, in the spirit of Kepler's orbit model. Indeed, as recently as 1997, the famous string theorist Ed Witten told me that he thought string theory would one day predict how many times lighter an electron is than a proton. Yet when I last saw him at Andrei Linde's sixtieth birthday party, he confessed after some wine that he'd given up on ever predicting all constants of nature.

Why this new pessimism? Because history is repeating itself. The Level II multiverse does to the electron's mass what other planets did to Earth's mass, demoting it from being a fundamental property of nature to being merely part of our cosmic address. For any number that varies across the Level II multiverse, measuring its value simply narrows down the options for what particular universe we happen to be in.

As we'll see in Chapter 10, we've so far discovered thirty-two independent numbers built into our Universe that we're trying to measure to as many decimal places as possible. Do they all vary across the Level II multiverse, or can any of them be computed from first principles (or from some other shorter list of numbers)? We still lack a successful fun-

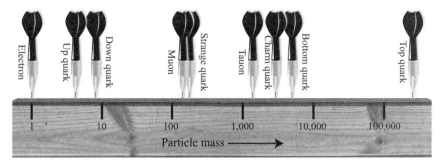

Figure 6.8: The nine masses that we've managed to measure for so-called fermion particles look rather random, as some multiverse models predict, suggesting that we'll never manage to calculate them from first principles. The scale shows how many times heavier than an electron each particle is.

damental theory of physics that can answer this question, so until we do, it's interesting to look at the measurements for some hints. Numbers that vary across the multiverse should look random to us if we're living in a random universe. Do the measured numbers look random? Well, you can judge for yourself in Figure 6.8, where I've plotted the masses of the nine fundamental particles called *fermions* in particle physics. Aside from the funny scale I've used, where the mass increases tenfold for every few centimeters you go to the right, it looks to me like nine randomly thrown darts. In fact, these nine numbers have passed some stringent statistical randomness tests with flying colors, consistent with being randomly generated from what statisticians call a uniform distribution with a slope below 10%.

All Isn't Lost

If we're living in a random *habitable* universe, the numbers should still look random, but with a probability distribution that favors habitability. By combining predictions about how the numbers vary across the multiverse with the relevant physics of galaxy formation and so on, we can make statistical predictions for what we should actually observe, and such predictions have so far agreed fairly well with data for dark energy, dark matter and neutrinos (Figure 6.9). Indeed, Steven Weinberg's first prediction of a non-zero dark-energy density was made this way.

I've had fun going through the full list of known "knobs" of our "universal controller," pondering what would happen if they were set differently. For example, please don't rotate the Figure 6.6 knobs for the

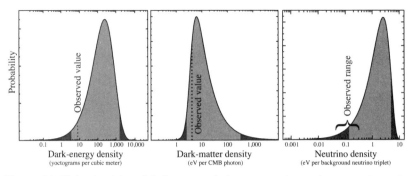

Figure 6.9: If the densities of dark energy, dark matter and neutrinos vary dramatically across a Level II multiverse, then most universes will be devoid of galaxies and lifeless, and a random observer should expect to measure values in a fairly narrow range quantified by the probability distributions shown. We should expect the measured values to fall in the central gray region with 90% probability, and indeed we do.

number of space and time dimensions, since it would be lethal. If you increase the number of space dimensions beyond three, there can be neither stable solar systems nor stable atoms. For instance, going to a four-dimensional space changes Newton's inverse-square law for the gravitational force to an inverse-cube law, for which there are no stable orbits whatsoever. I got quite excited when I figured this out, and then realized that I'd just broken my personal scooping record: the Austrian physicist Paul Ehrenfest had discovered this already back in 1917. . . . Spaces with less than three dimensions don't allow solar systems because gravity ceases to be attractive, and they're probably too simple to contain observers also for other reasons—for example, two neurons can't cross. Changing the number of time dimensions isn't as absurd as you might think, and Einstein's theory of general relativity can handle this just fine. However, I once wrote a paper showing that doing that would eliminate the key mathematical property of physics that allows us to make predictions, thus making it pointless to evolve a brain. As Figure 6.10 illustrates, this leaves three space dimensions and one time dimension as the only viable option. In other words, an infinitely intelligent baby could in principle, before making any observations at all, calculate from first principles that there's a Level II multiverse with different combinations of space and time dimensions, and that 3 + 1 is the only option supporting life. Paraphrasing Descartes, it could then think, *Cogito, ergo three space dimensions and one time dimension*, before opening its eyes for the very first time and verifying its predictions.

The entire Level II multiverse exists in a single space, so how can

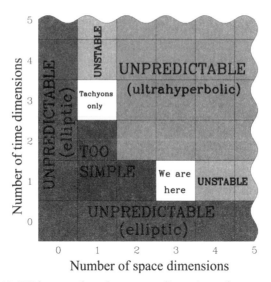

Figure 6.10: With more than three space dimensions, there are no stable atoms or solar systems. With fewer, there's no gravitational attraction. With more or less than one time dimension, physics loses all predictive power, and there would be no point in evolving a brain. In a Level II multiverse where the number of space and time dimensions varies from universe to universe, we should therefore expect to find ourselves in a universe with three dimensions of space and one of time, since all other universes are probably uninhabited.

the dimensionality vary within it? Well, according to the most popular string-theory models, it's only the *apparent* dimensionality that varies: the true space always has nine dimensions, but we don't notice six of them because they're microscopically curled up in the spirit of the cylinder from Figure 2.7: if you travel a tiny distance along one of these six hidden dimensions, you find yourself back where you started. Supposedly, all nine dimensions started out curled up, and then in our patch of space, inflation stretched three of them out to astronomical size while leaving six of them tiny and invisible. Elsewhere in the Level II multiverse, inflation stretched out different numbers of dimensions, creating worlds that seem anywhere from zero-dimensional to nine-dimensional.

Mathematicians have identified many different ways in which these extra dimensions can be curled up and filled with energy (for example, generalized magnetic fields can loop around inside the hidden dimensions), and in string theory, these many options correspond to the changeable knobs that we explored earlier. Different options can correspond not only to different physical constants in the dimensions that aren't curled up, but also to different rules for what elementary particles can exist and the effective equations that describe them. There might be Level II parallel universes where there are, say, ten rather than six kinds of quarks.

In summary, this means that although the *fundamental* equations of physics (those of string theory, perhaps) remain valid throughout the Level II multiverse, the *apparent* laws of physics that observers will uncover can change from one Level I multiverse to another. In other words, these apparent laws are *universal* not in the dictionary sense of "always applicable," but only in the literal sense of "applicable in our Universe." They're *multiversal* only at Level I, not at Level II. The fundamental equations, however, are multiversal even at Level II—they won't change until we get to Chapter 12 and the Level IV multiverse. . . .

Multiverse Halftime Roundup

We've explored lots of crazy-sounding ideas in this chapter, so let's end it by taking a step back and looking at the big picture. I think of inflation as the explanation that doesn't stop—inflating or explaining. Just as cell division didn't make merely one baby and stop, but a huge and diverse population of humans, it looks like inflation didn't make merely one universe and stop, but a huge and diverse population of parallel universes, perhaps realizing all possible options for what we used to think of as physical constants. Which would explain yet another mystery: the fact that our Universe is so fine-tuned for life. Even though most of the parallel universes created by inflation are stillborn, there will be some where conditions are just right for life, and it's not surprising that this is where we find ourselves.

My colleague Eddie Farhi likes to call Alan Guth "The Enabler," because eternal inflation enables everything that can happen to actually happen: inflation produces space for it to take place and creates initial conditions allowing the story to play itself out. In other words, inflation is a process converting potentiality into reality.

If you feel uncomfortable talking about our Level II multiverse, just say "space" instead, remembering that all of our Level I and Level II parallel universes are simply distant regions of one and the same infinite space. It's just that the structure of this space is much richer than Euclid imagined: it's expanding so that we can only see the small part of it that we call our Universe, and its faraway properties are more diverse than what we see in our telescopes. The Chapter 3 notion that our Universe is homogeneous and looks the same everywhere is just an interlude, valid only on intermediate scales: gravity makes things clumpy and interesting on smaller scales, and inflation makes things diverse and interesting on larger scales.

If you're still struggling to make inner peace with parallel universes, here's another way of thinking about them that might help. Alan Guth mentioned it in a recent MIT talk, but it has nothing to do with inflation. When we discover an object in nature, the scientific thing to do is look for a mechanism that created it. Cars are created by car factories, rabbits are created by rabbit parents and solar systems are created from gravitational collapse in giant molecular clouds. So it's quite reasonable to assume that our Universe was created by some sort of universe-creation mechanism (perhaps inflation, perhaps something totally different). Now here's the thing: all the other mechanisms we mentioned naturally produce *many* copies of whatever they create; a cosmos containing only one car, one rabbit, and one solar system would seem quite contrived. In the same vein, it's arguably more natural for the correct universe-creation mechanism, whatever it is, to create many universes rather than just the one we inhabit.

If we apply this same argument to whatever mechanism *started* inflation and ultimately produced our Level II multiverse, we conclude that it probably produced many separate Level II multiverses that are completely disconnected. However, this variant appears to be untestable, since it would neither add any qualitatively different worlds nor alter the probability distribution for their properties—all possible Level I multiverses are already realized within each of these Level II multiverses.

Inflation aside, there might be other mechanisms that create universes. An idea proposed by Richard Tolman and John Wheeler and recently elaborated on by Paul Steinhardt and Neil Turok is that our cosmic history is cyclic, going through an infinite series of Big Bangs. If it exists, the ensemble of such incarnations would also form a multiverse, perhaps with a diversity similar to that of Level II. However, the cyclic models are ruled out by the gravitational wave observations of BICEP2.

Another universe-creation mechanism, proposed by Lee Smolin, involves mutating and sprouting new universes through black holes rather than through inflation. This would produce a Level II multiverse as well, with natural selection favoring universes with maximal black-hole production. My friend Andrew Hamilton from Chapter 4 may have uncovered such a universe-creation mechanism: he's investigated an instability that occurs inside black holes shortly after they form, and it may be violent enough to trigger inflation that would create a Level I multiverse—which would be entirely contained inside the original black hole, but its inhabitants would probably neither know nor care about this fact.

In so-called braneworld scenarios, another three-dimensional world could be quite literally parallel to ours, a short distance away in a higher dimension. However, I don't think that such a world (*brane*) deserves to be called a parallel universe separate from our own, since it can interact with it gravitationally much as we do with dark matter.

Parallel universes remain highly controversial. However, there's been a striking shift in the scientific community during the past decade, where multiverses have gone from having lunatic-fringe status to being discussed openly at physics conferences and in peer-reviewed papers. I think that the success of precision cosmology and inflation has played a major role in this shift, as has the discovery of dark energy and the failure to explain its fine-tuning by other means. Even those of my colleagues who dislike the multiverse idea now tend to grudgingly acknowledge that the basic arguments for it are reasonable. The main critique has shifted from "This makes no sense and I hate it" to "I hate it."

In my opinion, our job as scientists isn't to tell our Universe how to work in order to conform to our human prejudice, but to look at it with open minds and try to figure out how it actually works.

We humans have a well-documented tendency toward hubris, arrogantly imagining ourselves at center stage, with everything revolving around us. We've gradually learned that it's instead we who are revolving around the Sun, which is itself revolving around one galaxy among countless others. Thanks to breakthroughs in physics, we may be gaining still deeper insights into the very nature of reality—indeed, in this book, we're still only two multiverse levels down, with two to go, and will start exploring the Level III multiverse in the next chapter. The price we have to pay is becoming more humble—which will probably do us good—but in return we may find ourselves inhabiting a reality grander than our ancestors imagined in their wildest dreams.

THE BOTTOM LINE

- Parallel universes are not a theory, but a prediction of certain theories.
- Eternal inflation predicts that our Universe (the spherical region of space from which light has had time to reach us during the 14 billion years since our Big Bang) is just one of infinitely many universes in a Level I multiverse where everything that can happen does happen somewhere.
- For a theory to be scientific, we need not be able to observe and test all its predictions, merely at least one of them. Inflation is the leading theory for our cosmic origins because it's passed observational tests, and parallel universes seem to be a non-optional part of the package.
- Inflation converts potentiality into reality: if the mathematical equations governing uniform space have multiple solutions, then eternal inflation will create infinite regions of space instantiating each of those solutions—this is the Level II multiverse.
- Many physical laws and constants that are unchanged across a Level I multiverse may vary across the Level II multiverse, so students in Level I parallel universes learn the same things in physics class but different things in history class, while students in Level II parallel universes could learn different things in physics class as well.
- This could explain why many constants in our own Universe are so fine-tuned for life that if they differed by small amounts, life as we know it would be impossible.
- This would also give many numbers we've measured in physics a new meaning: they're not telling us something fundamental about physical reality, but merely something about our location in it, forming part of our cosmic postal code.
- Although these parallel universes remain controversial, the main critique has shifted from "This makes no sense and I hate it" to "I hate it."

Part Two

Zooming In

7

Cosmic Legos

Everything we call real is made of things that cannot be regarded as real.
—Niels Bohr

No, this just doesn't make sense! There's got to be a mistake somewhere! I'm alone in my girlfriend's dorm room in Stockholm, studying for my first college quantum-mechanics exam. The textbook says that small things such as atoms can be in several places at once, whereas big things such as people can't. *No way!* I tell myself. *We people are made of atoms, so if they can be in several places at once, surely we can, too!* It also says that every time a person observes where an atom is, it randomly jumps to just one of the places where it previously was. But I can't find any equation defining what exactly is supposed to count as an observation. *Would a robot count as an observer? How about a single atom? And the book just said that every quantum system changes deterministically according to the so-called Schrödinger equation. Isn't that logically inconsistent with this random-jumping business?*

Flustered, I muster up the courage to knock on the door of our great expert, a physics professor on the Nobel Committee. Twenty minutes later, I emerge from his office feeling stupid, convinced that I've somehow misunderstood the whole thing. This marks the beginning of a long personal journey of mine that still continues, and leads to quantum parallel universes. It's not until a couple of years later, when I move to Berkeley to do my Ph.D., that I realize that it wasn't I who had misunderstood. I eventually learn that many famous physicists had been vexed by these problems with quantum mechanics, and I end up having lots of fun writing my own papers on the subject.

However, before telling you how I now think this all fits together (in Chapter 8), I want to take you back in time to really appreciate the craziness of quantum mechanics, and what all the fuss is about.

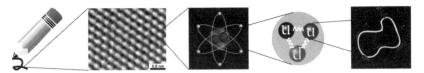

Figure 7.1: A pencil lead is made of graphite, which is made of layers of carbon atoms (this is a real image from a scanning tunneling microscope), which are made of protons, neutrons and electrons. The protons and neutrons are made of up and down quarks, which may in turn be vibrating strings. The refill pencil leads I bought to work on this book contain about 2×10^{21} atoms, so you could cut them in half at most 71 times.

Atomic Legos

Last time I asked my son Alexander what he wanted for his birthday, he said: "Surprise me! Anything is okay as long as it's Legos. . . ." I, too, love Legos, and I feel that our Universe does as well: everything is made of the same basic building blocks, as illustrated in Figure 7.1. I find it quite remarkable that the same cosmic Lego set consisting of the eighty stable atoms from the periodic table* can be used to build everything from rocks to rabbits, from stars to stereos—the only difference being how many Legos of each kind are used, and how they're arranged.

The basic Lego idea of indivisible building blocks of course has a venerable history, with our owing the term *atom* to the ancient Greek word for "indivisible." Indeed, Plato argued in his dialogue *Timaeus* that the four basic elements postulated at the time (earth, water, air and fire) consisted of four kinds of atoms, and that these atoms were invisibly small mathematical objects: cubes, icosahedra, octahedra and tetrahedra, respectively, i.e., four of Plato's five eponymous solids (Figure 7.2). For example, he argued that the sharp corners of the tetrahedron explained why fire was painful, that the ball-like shape of the icosahedron explained water's ability to flow, and that the unique ability of cubes to be compactly stacked explained Earth's solidity. Although this cute theory was eventually demolished by observational facts, some aspects of it survive, such as his suggestions that each fundamental ele-

---

* There are 80 kinds of stable atoms, containing all numbers of protons from 1 (hydrogen) through 82 (lead), except for 43 (technetium) and 61 (promethium), which are radioactive and unstable. Many of these atoms have more than one stable version corresponding to different numbers of neutrons (so-called isotopes); the total number of stable atomic isotopes is 257. There are about 338 isotopes found naturally here on Earth, if we also count about 30 isotopes with half lives longer than 80 million years and about 50 more short-lived ones.

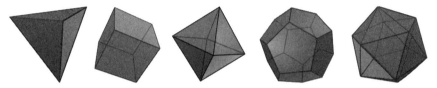

Figure 7.2: The five Platonic solids: tetrahedron, cube, octahedron, dodecahedron and icosahedron. Only the dodecahedron was excluded from Plato's atomic theory; sometimes viewed as a cult object of mysterious mythical significance, it figures in art from ancient times through Salvador Dalí's *Sacrament of the Last Supper*.

ment consists of a specific kind of atom, and that properties of a substance are determined by properties of its atoms. Moreover, I'll argue in Chapter 10 that the ultimate building blocks of our Universe are indeed mathematical in a different way than Plato suggested: not that our Universe is made of mathematical objects, but that it's a part of a single mathematical object.

It took another two millennia for the modern theory of atoms to really catch on, and the famous Austrian physicist Ernest Mach refused to believe in the reality of atoms even in the early 1900s. He'd undoubtedly have been impressed by our current ability to image individual atoms (Figure 7.1) and even manipulate them.

Nuclear Legos

The very success of the atomic hypothesis naturally led to the question of whether *atom* was a misnomer: if macroscopic objects are all made of the smaller Legos that we call atoms, might they in turn be divisible into some form of smaller Legos that could be rearranged?

I find it extremely elegant that all the atoms in our periodic table are in fact made up of merely *three* kinds of smaller Lego blocks, even fewer than the four in Plato's theory. We encountered them briefly in Chapter 3, and Figure 7.1 illustrates how these three—protons, neutrons and electrons—are arranged much like a miniature solar system with electrons orbiting the compact ball of protons and neutrons that we call the atomic nucleus. Whereas the Earth is kept in its orbit around the Sun by the attractive gravitational force between them, the electrons are kept in the atoms by the electrical force that attracts them to the protons (electrons have negative charge, protons have positive charge, and opposite charges attract). Since electrons also feel attracted to the protons in other atoms, they can help bind different atoms together into the larger structures we know as molecules. If the atomic nuclei and the electrons get shuffled around without changing the number of each kind, then we call that a chemical reaction, regardless of whether it's fast like a forest fire (which is mostly carbon and hydrogen atoms in wood and leaves combining with oxygen from the air to form carbon dioxide and water molecules) or slow like a growing tree (which is mostly the reverse reaction, powered by sunlight).

Over centuries, alchemists tried in vain to convert certain kinds of atoms into others, typically cheaper ones like lead into more expensive ones like gold. Why did they all fail? An atom is simply named according to the number of protons it contains (1 = hydrogen, 79 = gold, etc.), so what the alchemists failed to do was clearly to play Legos with the protons and move them from one atom to another. Why couldn't they do it? We now know that they failed not because they tried something impossible, but merely because they didn't use enough energy! Since the electric force causes equal charges to repel each other, the protons in atomic nuclei would fly apart unless some more powerful force held them together. The aptly named strong nuclear force does just this, and acts like a sort of nuclear Velcro that holds both protons and neutrons together as long as they get sufficiently close. It's so strong that you need extreme violence to overpower it: whereas slamming two hydrogen molecules (each consisting of a pair of hydrogen atoms) together at 50 kilometers per second can break them apart so that their atoms get separated, you'd need to crash two helium nuclei (each consisting of two protons and two neutrons) together at the dizzying speed of 36,000 kilometers per second to stand a chance of breaking them apart into separate neutrons and protons—that's about 12% of the speed of light, and fast enough to get you from New York to San Francisco in a tenth of a second.

In nature, such violent collisions happen when it gets extremely hot—millions of degrees. There were no atoms around in our early Universe except hydrogen plasma (single protons), since it was so hot that any protons or neutrons stuck together as heavier atoms were smashed apart. As our Universe gradually expanded and cooled, there was a brief period of a few minutes when collisions were still strong enough to overcome the electric repulsion between protons, but no longer strong enough to overpower the strong "Velcro" force that made them and neutrons stick together as helium: this was the period of Gamow's Big

Bang nucleosynthesis that we explored in Chapter 3. In the core of our Sun, the temperature is similarly in that magic range where hydrogen atoms can fuse into helium atoms.

The laws of economics tell us that atoms are expensive if they're rare, and the laws of physics tell us that they're rare if they require unusually high temperatures to make. Putting this together tells us that if atoms could talk, the priciest ones would tell the best stories. Garden-variety atoms such as carbon, nitrogen and oxygen (which together with hydrogen make up 96% of your body weight) are so cheap because garden-variety stars such as our Sun can produce them in their death throes, after which they can form new solar systems in a cosmic recycling event. Gold, on the other hand, is produced when a star dies in a supernova explosion so violent and rare that it, during a fraction of a second, releases about as much energy as all the other stars in our observable Universe combined. No wonder making gold eluded the alchemists.

Particle-Physics Legos

If everyday stuff is made of atoms and atoms are made of smaller pieces (neutrons, protons and electrons), then are these in turn made of some form of still smaller Legos? History has taught us the way to tackle this question experimentally: collide these smallest known building blocks together really hard and check if they break apart. This procedure has been tried with ever-larger particle colliders, but electrons still show no sign of being made of anything smaller despite having been smashed at 99.999999999% of the speed of light at the CERN Laboratory outside Geneva. Colliding protons, on the other hand, has revealed that both they and neutrons are made of smaller particles known as *up quarks* and *down quarks*. Two ups and a down make a proton (Figure 7.1), while two downs and an up make a neutron. Moreover, a slew of previously unknown particles have been produced in particle collisions (see Figure 7.3).

All of these new particles, with exotic names such as *pions, kaons, Sigmas, Omegas, muons, tauons, W-bosons* and *Z-bosons*, are unstable and decay into more familiar stuff in a split second, and clever detective work has revealed that all except the last four are made of quarks—not just ups and downs, but also four new unstable kinds known as *strange, charm,*

STANDARD MODEL

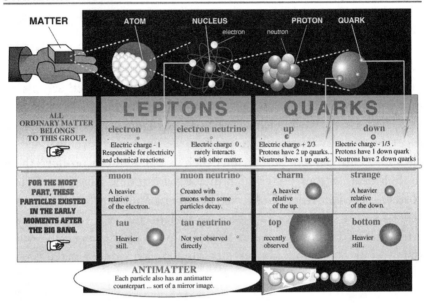

Figure 7.3: The current standard model of particle physics *(Image credit: CERN)*

bottom and *top*. The W- and the Z-bosons have been found responsible for transmitting the so-called weak force that's involved in radioactivity, and are big fat cousins of the boson we know as the photon, the particle that light is made of and which transmits the electromagnetic force. Additional boson family members known as *gluons* have been discovered to glue the quarks together into larger particles, and the recently discovered *Higgs boson* endows other particles with mass. In addition, stable ghostlike particles known as *electron neutrinos, muon neutrinos* and *tau neutrinos* have been discovered—we encountered them in the last chapter, and they are so shy that they barely interact with other particles at all: if a neutrino crashes into the ground, it typically passes right through Earth and emerges unscathed on the other side, and continues into space. Finally, almost all of these particles have an evil twin called its *antiparticle*, which has the property that, if the two collide, they can annihilate each other in a burst of pure energy. Table 7.2 summarizes the key particles and related concepts that we discuss in this book.

So far, no evidence has been found that any of these bosons, quarks, leptons (that's the family name for the electron, the muon, the tauon and the neutrinos) or their antiparticles are made of any smaller or more fundamental parts. However, since quarks are building blocks a full three levels down in the Lego hierarchy (Figure 7.1), you don't need to be Sherlock Holmes to start wondering whether there are even more levels that we're failing to discover simply because we don't have enough energy in our particle accelerators. Indeed, as we hinted at in Chapter 6, string theory suggests precisely this: that if we could slam them together with vastly (perhaps ten trillion times) more energy than today, we'd discover that everything is made of tiny vibrating strings, and that different types of vibrations of the same basic type of strings would correspond to different types of particles a bit like different types of vibrations of a guitar string correspond to different musical notes. The rival theory known as loop quantum gravity suggests that everything is made not of strings, but of a so-called spin network of quantized loops of excited gravitational fields—that's quite a mouthful there, and if you don't fully understand what that means, don't worry, since not even the most devoted practitioners of string theory and loop quantum gravity claim to fully understand their theories yet. . . . So what's everything ultimately made of? Based on the current state-of-the-art experimental evidence, the answer is clear: we simply don't know yet, but there's good reason to suspect that everything we know of so far—including the very fabric of spacetime itself—is ultimately made up of some more fundamental building blocks.

Mathematical Legos

Even though we still don't know the ultimate answer to the question of what everything is made of, we've discovered one more fascinating hint that I have to tell you about. To me, it feels pretty crazy that colliding two protons at the CERN Large Hadron Collider can produce a Z-boson weighing ninety-seven times as much as a proton. I used to think that mass was conserved: surely you can't produce a cruise ship by colliding two Ferraris, since it would weigh more than both cars combined? However, if you think that it feels as fraudulent as a Ponzi scheme to create new particles like this, remember that Einstein taught us that energy E can be converted into mass m by the formula $E = mc^2$, where c is the speed of light. So if you have loads of motion energy at your disposal in a particle collision, then part of that energy is indeed allowed to take the form of new particles. In other words, the total *energy* is conserved (stays the same), but a particle collision repackages this available energy in new ways, which may include putting some into new particles that weren't there to start with. The exact same thing happens with *momentum:** the total amount is conserved, but it gets redistributed during the collision just as in pool when the cue ball slows down while sending a stationary ball flying into a pocket. One of the most important discoveries in physics has been that there are additional quantities which, just as energy and momentum, appear to always be conserved: *electric charge* is the most familiar example, but there are also other kinds of conserved quantities known, with names such as *isospin* and *color*. There are also quantities that are conserved in many important circumstances, notably lepton number (the number of leptons minus the number of antileptons) and baryon number (the number of quarks minus the number of antiquarks, all divided by three so that neutrons and protons count as +1). Table 7.1 lists the amounts (called *quantum numbers*) of these quantities that various particles have. You'll notice that many of them are whole numbers or simple fractions, and that three of the masses aren't well measured.

Particle name	Mass in MeV	Charge	Spin	Isospin	Baryon number	Lepton number
Proton	938.3	1	1/2	1/2	1	0
Neutron	939.6	0	1/2	1/2	1	0
Electron	0.511	−1	1/2	−1/2	0	1
Up quark	1.5–4	2/3	1/2	1/2	1/3	0
Down quark	4–8	−1/3	1/2	−1/3	1/3	0
Electron neutrino	$< 10^{-6}$	0	1/2	1/2	0	1
Photon	0	0	1	0	0	0

Table 7.1: All known elementary particles are described by their own unique sets of *quantum numbers*, and this table shows a sample. The particles are purely mathematical objects in the sense that they have no properties at all beyond their quantum numbers. The mass shown corresponds to how much energy you'd need to create the particle at rest. The funny unit MeV is the amount of motion energy an electron picks up if you use a million volts to accelerate it.

I remember this old Cold War joke about how, in the West, everything that wasn't forbidden was allowed, while in the East, everything that wasn't allowed was forbidden. Intriguingly, particle physics seems

---

* The *momentum* of something measures the punch it packs if it crashes into something; or, more rigorously, the amount of time it would take you to stop it times the average force with which you'd need to push it. The momentum p of something with mass m moving with velocity v is simply given by $p = mv$ as long as v is far below the speed of light.

to prefer the former: every reaction that isn't forbidden (for violating some conservation law) appears to actually occur in nature. This means that we can think of the fundamental Legos of particle physics as being not the particles themselves, but the conserved quantities! So particle physics is simply rearranging energy, momentum, charge and other conserved quantities in new ways. For example, Table 7.1 shows that the cookbook recipe for making an up quark is to combine 2/3 units of charge, 1/2 unit of spin, 1/2 unit of isospin, 1/3 unit of baryon number, and top it all off with a few MeV of energy.

So what are quantum numbers like energy and charge made of? Nothing—they're just numbers! A cat has energy and charge, too, but it also has many other properties besides these numbers such as its name, smell and personality—so it would sound crazy to say that the cat is a purely mathematical object completely described by those two numbers. Our elementary-particle friends, on the other hand, are completely described by their quantum numbers, and appear to have no intrinsic properties at all besides these numbers! In this sense, we've now come full circle back to Plato's idea: the fundamental Legos out of which everything is made appear to be purely mathematical in nature, having no properties except mathematical properties. We'll return to this idea in more detail in Chapter 10, and see that it's just the tip of a mathematical iceberg.

At a more technical level, some particle physicists like to glibly answer the question "What's a particle?" by saying, "It's an element of an irreducible representation of the symmetry group of the Lagrangian." That's quite a mouthful, and enough to stop most budding conversations dead in their tracks, but it's a completely mathematical thing, just a bit more general than the concept of a set of numbers. And yes, sure, string theory or a competitor may deepen our understanding of what particles really are, but all the leading theories out there simply replace one mathematical entity with another. For example, if the quantum numbers from Table 7.1 turn out to correspond to different types of superstring vibrations, then you shouldn't think of these strings as fuzzy little objects with intrinsic properties like being made out of braided golden-brown cat hairs, but rather as purely mathematical constructs that physicists have dubbed "strings" simply to emphasize their one-dimensional nature and to make an analogy with something that feels less mathematical and more familiar.

In summary, nature has a hierarchical Lego structure. If my son Alexander plays normally with his birthday present, all he can rearrange are the factory-made Lego pieces. If he'd play atom Lego by setting them on fire, immersing them in acid, or using some alternative method to rearrange their atoms, he'd be doing chemistry. If he'd play nucleon Legos by rearranging their neutrons and protons into different kinds of atoms, he'd be doing nuclear physics. If he'd smash his pieces together near the speed of light to rearrange the energy, momentum, charge, etc., of their neutrons, protons and electrons into new particles, he'd be doing particle physics. The Legos at the deepest level appear to be purely mathematical objects.

Particle-Physics Cheat Sheet	
Momentum	The punch something packs if it crashes into something or, more rigorously, the amount of time it would take you to stop it times the average force with which you'd need to push it
Angular momentum	How much something spins or, more rigorously, the amount of time it would take you to make it stop spinning times the average torque (twisting force) you'd need to use
Spin	The angular momentum of a single particle spinning around its center
Conserved quantity	Quantity that remains constant over time and can neither be created nor destroyed. Examples: energy, momentum, angular momentum, electric charge
Atom	Electrons orbiting around a nucleus of protons and neutrons; the number of protons in an atom determines its name (1 = hydrogen, 2 = helium, etc.)
Electron	Negatively charged particle that electric currents are made of
Proton	Positively charged particle found in atomic nuclei, made of two up quarks and a down quark
Neutron	Particle without electric charge that's found in atomic nuclei, made of two down quarks and an up quark
Photon	Particle of light
Gluon	Particle that help glue quark triplets together into protons and neutrons
Neutrino	Particle that's so stealthy that it can usually pass right through Earth without interacting with anything
Fermion	Particle that can't be in the same place and state as an identical particle. Examples: electrons, quarks, neutrinos
Boson	Particle that likes to be in the same place and state as an identical particle. Examples: photons, gluons, Higgs particle

Table 7.2: Summary of key physics terms for understanding the microworld

Photon Legos

It's not only "stuff" that's made of Lego-like building blocks. As we mentioned in Part I of this book, so is light, being composed of particles called *photons*, inferred by Einstein in 1905.

Four decades earlier, James Clerk Maxwell had discovered that light is an electromagnetic wave, a type of electrical disturbance. If you could carefully measure the voltage between two points in a beam of light, you'd find that it oscillates over time; the frequency f of this oscillation (how many times per second it oscillates) determines the color of the light, and the strength of the oscillation (the maximum number of volts you measure) determines the intensity of the light. Our Omniscope, from back in Chapter 4, measures such voltages. We humans give these electromagnetic waves different names, depending on their frequency (by increasing frequency, we call them radio waves, microwaves, infrared, red, orange, yellow, green, blue, violet, ultra-violet, x-rays, gamma rays), but they're all forms of light and they're all made of photons. The more photons an object emits each second, the brighter it looks.

Einstein realized that the amount of energy E in a photon was given by its frequency f through the simple formula $E = hf$, where h is the constant of nature known as Planck's constant. The constant h is tiny, so a typical photon has very little energy in it. If I lie on the beach for just a second, I get warmed by about a sextillion (10^{21}) photons, which is why it feels like a continuous flow of light. However, if my friends have sunglasses blocking 90% of the light, and I put on twenty-one pairs at once, then only about one of the original photons would get through each second, which a sensitive photon detector could confirm.

Einstein got the Nobel Prize because he used this idea to explain the so-called photoelectric effect, whereby the ability of light to knock electrons out of metal had been found to depend only on the frequency of the light (the energy of the photons), not on the intensity (the number of photons). Lower-frequency photons just don't have enough energy for the task, just as you can't break a glass window by throwing tennis balls with low energy no matter how many you throw. The photoelectric effect is related to the processes used in present-day solar cells and the image sensors in digital cameras.

My namesake Max Planck won the 1918 Nobel Prize for showing that the photon idea also solved another outstanding mystery: why the previously calculated heat radiation of a glowing hot object didn't come out right. The rainbow (Figure 2.5) reveals the spectrum of sunlight, that is, how much light there is at different frequencies. People knew that the temperature T of something is a measure of how rapidly its particles are moving around, and that the typical motion energy E of a particle was given by the formula $E = kT$, where k is a number known as Boltzmann's constant. When particles in the Sun collide, roughly a quantity kT of motion energy can be converted into light energy. Unfortunately, the detailed prediction for the rainbow was the so-called ultraviolet catastrophe: that the intensity of light would increase forever toward the right in Figure 2.5 (toward higher frequencies), so that you'd get blinded by gamma rays when you looked at any warm object, say your best friend. You're saved by the fact that light is made of particles: the Sun can radiate light energy only one photon at a time, and the typical energy kT available for making a photon falls far short of the amount of energy hf required to make even a single gamma ray.

Above the Law?

So if everything is made of particles, then what are the laws of physics that govern them? Specifically, if we know what all the particles in our Universe are doing right now, then what equation lets us calculate what they'll be doing in the future? If there is such an equation, then you might hope that it will allow us to—at least in principle—predict all aspects of the future from the present, from the future trajectory of a just-hit baseball to the winners of the 2048 Olympic Games: just figure out what all the particles will do, and there's your answer.

The good news is that there does seem to be just such an equation, called the Schrödinger equation (Figure 7.4). The bad news is that it doesn't predict exactly what the particles will do, and that almost a century after the Austrian physicist Erwin Schrödinger wrote it down, physicists still argue about what to make of it.

What everybody does agree on is that microscopic particles don't obey the classical laws of physics that we're taught in school. Since an

Figure 7.4: The Schrödinger equation lives on. Since I took this photo in 1996, the inscription font has mysteriously changed. Will quantum weirdness never end?

atom is reminiscent of a miniature solar system (Figure 7.1), it would seem quite natural to assume that its electrons orbit the nucleus according to Newton's laws just as the planets orbit the Sun. Indeed, when you do the math, things look promising at first. You can spin a yo-yo in a circle around your head by pulling with a force on its string; if the string snapped, the yo-yo would move in a straight line with constant speed, so the force with which you pull on it is required to deflect it from this straight-line motion to go in a circle. In our Solar System, it's not a string but the Sun's gravity that provides this force, and in an atom, the electric attraction of the nucleus provides the force. If you do the calculation for an orbit the size of a hydrogen atom, you'll predict that the electron orbits just about as fast as we measure in the lab—quite a theoretical triumph! However, to be more accurate, we need to include one more effect in the math: an electron that's accelerating (changing its speed or its direction of motion) will radiate away energy—your mobile phone exploits this by jiggling electrons around in its antenna so that radio waves get transmitted. Since energy is conserved, this radiated energy has to come from somewhere. In your phone, it gets taken from the battery, but in a hydrogen atom, it gets taken from the motion energy of the electron, causing it to fall farther and farther "down" toward the atomic nucleus, just as upper atmosphere air resistance causes satellites in low-Earth orbit to lose motion energy and eventually fall down. This means that the electron orbit isn't a circle, but a death spiral (Figure 7.5): after about 100,000 orbits, the electron has crashed into the proton and the hydrogen atom has collapsed, at the ripe old age of about 0.02 nanoseconds.*

This is bad. Really bad. Here, we're not talking about some minor 1% discrepancy between theory and experiment, but about a prediction that all hydrogen atoms (as well as all other atoms) in our Universe will collapse in a billionth of the time it took you to read the last word in this sentence. Indeed, since most hydrogen atoms have been around for about 14 billion years, they've lasted more than twenty-eight orders of magnitude longer than classical physics predicts—this held the dubious record as the worst-ever quantitative failure of physics until it was overtaken by the 123-order-of-magnitude mismatch between prediction and measurement for the dark-energy density that we mentioned in Chapter 3.

When physicists assumed that elementary particles obeyed the classical laws of physics, they ran into other problems as well. For example, it was found that the amount of energy needed to heat very cool objects was smaller than predicted. There were also further problems, but we don't need to flog a dead horse, since the message from nature is crystal clear: microscopic particles violate the laws of classical physics.

So are microscopic particles above the law? No, they obey a different law: Schrödinger's.

Quanta and Rainbows

To explain how atoms worked, the Danish physicist Niels Bohr introduced a radical idea in 1913: perhaps it wasn't just matter and light that was quantized (that came in Lego-like discrete chunks), but aspects of *motion* as well. What if motion isn't continuous but jumpy as in the computer game PAC-MAN or in an old Chaplin movie where the frame rate is too slow? Figure 7.5 shows Bohr's atom model: circular orbits are allowed only if the circles have certain magical sizes. There's a small-

---

* The electron makes about $1/8\pi\alpha^3 \sim 10^5$ orbits before crashing into the proton, where $\alpha \approx 1/137.03599968$ is the dimensionless strength of the electromagnetic force, a.k.a. the fine-structure constant. You'll find a nice derivation of the death spiral here: http://www.physics.princeton.edu/~mcdonald/examples/orbitdecay.pdf.

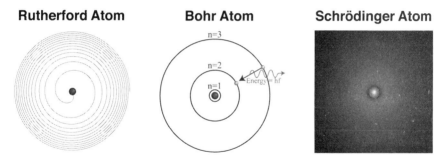

Rutherford Atom   **Bohr Atom**   **Schrödinger Atom**

Figure 7.5: Our evolving understanding of the hydrogen atom. The classic "solar sys-tem" model of Ernest Rutherford was unfortunately unstable, with the electron spiral-ing into the proton in the center (I'm showing what it would look like if the electric force were twenty times stronger; otherwise it would spiral around about a hundred thousand times and make my plot illegible). The Bohr model confines the electron to discrete orbits numbered $n = 1$, 2, 3, etc., between which it can jump when emitting or absorbing photons, but fails for all atoms except hydrogen. The Schrödinger model has a single electron in many places at once, in an "electron cloud" whose shape is given by a so-called wavefunction ψ.

est allowed orbit labeled $n = 1$, and then there are larger orbits labeled $n = 2$, etc., the radii of which are n^2 times as large as the smallest one.*

The first and most obvious success is that Bohr's atom can't collapse like the classical one to its left in Figure 7.5; when the electron is in the innermost orbit, there's simply no smaller orbit for it to jump to. But Bohr's model explains much more. The higher orbits have more energy than the lower ones, and total energy is conserved, so whenever the electron makes a PAC-MAN-like jump down to a lower orbit, the extra energy must get emitted from the atom in the form of a photon (see Figure 7.5), and in order to jump back up to a higher orbit, the electron must be able to pay the energy cost by absorbing an incom-ing photon with the required energy. Since there's only a discrete set of orbit energies, this means that the atom can only emit or absorb photons with certain magical energies. In other words, an atom can only emit or absorb light of certain special frequencies. This solved a long-standing mystery: the rainbow of sunlight (Figure 2.5) was known to have dark lines at certain mysterious frequencies (certain colors were

---

* Actually, what Bohr did, which was mathematically equivalent, was to assume that the *angular momentum* of the electron orbit was quantized, and was only allowed to equal some multiple n of what's called the *reduced Planck constant* $\hbar$, defined as $h/2\pi$. Analo-gously to momentum, you can think of the angular momentum of a spinning object as a measure of the amount of time it would take you to make it stop spinning times the average torque (twisting force) you'd need to use. Something orbiting in a circle of radius r with momentum p has angular momentum rp.

missing), and by studying hot glowing gases in the laboratory, it had been observed that each type of atom had its unique spectral fingerprint in the form of the frequencies of light that it could emit and absorb. Bohr's atom model didn't just explain the existence of these spectral lines, but also their exact frequencies for hydrogen.*

That was the good news, for which Bohr won a Nobel Prize (as did most of the others I mention in this chapter). The bad news was that Bohr's model didn't work for any atoms other than hydrogen, except if all but one of their electrons were removed.

Making Waves

Despite these early successes, physicists still didn't know what to make of these strange and seemingly ad hoc quantum rules. What did they really mean? *Why* is angular momentum quantized? Is there a deeper explanation for this?

Louis de Broglie proposed one: that the electron (and indeed all particles) has wavelike properties the way photons do. In a flute, standing sound waves can vibrate only at certain special frequencies, so could something analogous be determining the frequencies with which electrons could orbit in atoms?

Two waves can pass through each other unaffected, like the circular waves in the water tank in Figure 7.6 (left); at any time, their effects simply add together. In some places, we see peaks of the two waves adding up to an even higher peak (so-called constructive interference), in others we see a peak from one wave canceling the trough from the other to leave the water completely undisturbed (so-called destructive interference). On the surface of the Sun (Figure 7.6, center), sound waves in the hot gas/plasma have been observed. If such a wave propagates all the way around the Sun (right), then it will cancel itself out with destructive interference unless it performs exactly a whole number of oscillations as it goes around, thereby staying in sync with itself. This means that, just as a flute, the Sun vibrates only with certain special frequencies.†

---

* The energies of the orbits are E_1/n^2 where E_1 is the known energy of the lowest orbit, so by jumping between two orbits n_1 and n_2, the electron can emit photons of all energies of the form $\left(\dfrac{1}{n_2^2} - \dfrac{1}{n_1^2}\right) E_1$.

† The same phenomenon has been observed in car tires at very high speeds, and the resulting sound waves traveling around the tube in resonance can damage your budget.

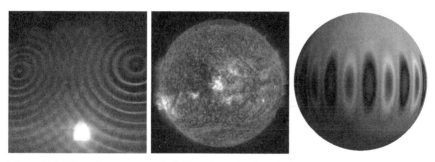

Figure 7.6: Waves in a water tank (left) and on the Sun (right)

In his 1924 Ph.D. thesis, de Broglie applied this reasoning to waves going around the hydrogen atom instead of the Sun, and obtained the exact same frequencies and energies as the Bohr model had predicted. A more direct demonstration of particles behaving as waves is given by the double-slit experiment illustrated in Figure 7.7.

This wave picture also gives a more intuitive picture of why atoms don't collapse as classical physics predicted: if you try to confine a wave to a very small space, it immediately starts spreading out. For example, if a raindrop lands in a water tank, it will initially disturb the water

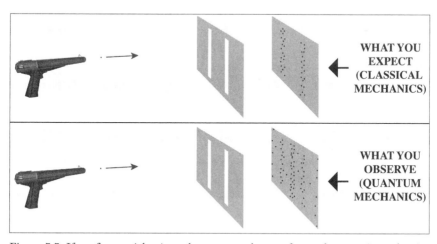

Figure 7.7: If we fire particles (say, electrons or photons from a laser gun) at a barrier with two vertical slits, classical physics predicts that they'll hit our detector in two vertical strips behind the slits. In contrast, quantum mechanics predicts that each particle will act like a wave and pass through *both* slits in a quantum superposition, interfere with itself, and form an interference pattern akin to Figure 7.6. Performing this famous double-slit experiment shows that quantum mechanics is correct: one detects particles along a whole series of vertical strips.

only in the small area where it landed, but this disturbance soon starts spreading outward in all directions as a series of circular waves, like the ripples in Figure 7.6 (left). This is the essence of the *Heisenberg uncertainty principle:* Werner Heisenberg showed that if you confine something to a small region of space, then it will have lots of random momentum, which tends to make it spread out and become less confined. In other words, an object can't simultaneously have an exact position and an exact velocity!* This means that if a hydrogen atom tries to collapse as in Figure 7.5 (left) by sucking the electron into the proton, then the increasingly confined electron will get enough momentum and speed to come flying back out to a higher orbit again.

De Broglie's thesis made waves, and in November 1925, Erwin Schrödinger gave a seminar about it in Zurich. When he was finished, Peter Debye said in effect: "You speak about waves, but where is the wave equation?" Schrödinger went on to produce and publish his famous wave equation (Figure 7.4), the master key for so much of modern physics. An equivalent formulation involving tables of numbers called matrices was provided by Max Born, Pasqual Jordan and Werner Heisenberg around the same time. With this new powerful mathematical underpinning, quantum theory made explosive progress. Within a few years, a host of hitherto unexplained measurements had been successfully explained, including spectra of more complicated atoms and various numbers describing properties of chemical reactions. Eventually this quantum physics gave us the laser, the transistor, the integrated circuit, computers and smartphones. Further successes of quantum mechanics involve its extension, quantum field theory, which underpins present-day frontier research such as the search for dark-matter particles.

What's the hallmark of good science? There are several science definitions that I like, and one of them is *data compression*, explaining a lot with a little. With a good scientific theory, you get more out of it than you put into it. I just applied standard data-compression software to the text file containing this chapter draft, and it compressed it threefold, using regularities and patterns that it found in my prose. Let's compare this to quantum mechanics. I just downloaded a list of over 20,000 spec-

---

* Specifically, if a particle has position uncertainty Δx and momentum uncertainty Δp, then the Heisenberg uncertainty principle states that $\Delta x \, \Delta p \geq \hbar/2$ where $\hbar$ is the reduced Planck constant $h/2\pi$ as before. Mathematically, the uncertainty for each quantity is defined as the standard deviation of its probability distribution.

tral lines from http://physics.nist.gov/cgi-bin/ASD/lines1.pl that have had their frequency painstakingly measured in laboratories around the world, and by capturing the patterns and regularities in these numbers, the Schrödinger equation can data-compress them down to just three numbers: the so-called fine-structure constant $\alpha \approx 1/137.036$, which gives the strength of electromagnetism, the number 1836.15, which is how many times heavier the proton is than the electron, and the orbital frequency of hydrogen.* That's the equivalent of data compressing this whole book down to a single sentence!

Erwin Schrödinger is one of my physics superheros. When I was a postdoc at the Max Planck Institute for Physics in Munich, the copying machine in the library used to take eons to warm up, and I'd pass the time by pulling classic books from the shelves. Once I pulled out *Annalen der Physik* from 1926, and was amazed to see that essentially everything we'd covered in my graduate quantum classes had been worked out in four of his 1926 papers. I admire him because he wasn't just brilliant, but also a freethinker: he questioned authority, thought for himself and did what he felt was right. After getting Max Planck's job as professor in Berlin, one of the most prestigious posts in the world, he gave it up because he wouldn't tolerate Nazi persecution against his Jewish colleagues. He then turned down a job offer from Princeton because they wouldn't accept his unorthodox family decisions (he lived with two women and had a child with the one he wasn't married to). Indeed, when I made a pilgrimage to his grave during a 1996 ski vacation in Austria, I discovered that his freethinking didn't go down well in his home village either: you'll see in the photo I took (Figure 7.4) that the small town of Alpbach has buried their most famous citizen ever in a quite modest grave right at the edge of the cemetery. . . .

Quantum Weirdness

But what did it all mean? What were these waves that Schrödinger's equation described? This central puzzle of quantum mechanics remains a potent and controversial issue to this day.

---

* In fact, the last one arguably shouldn't count because we could redefine our unit of time so that it equals 1. If you want still more accuracy to get all the measured decimal places right, you need only toss in a few more numbers to better model the exact masses of the different atomic nuclei (neutrons weigh about 0.1% more than protons, and so on).

When we physicists describe something mathematically, we usually need to describe two separate things:

1. Its state at a given time.
2. The equation describing how this state will change over time.

For example, to describe the orbit of Mercury around the Sun, Newton described the state of Mercury by six numbers: three for the position of its center (say, the x-, y- and z-coordinates) and three for the components of the velocity in these directions.* For the equation of motion, he used Newton's law: that the acceleration is given by the gravitational pull toward the Sun, which depends on the inverse square of the distance to the Sun.

In his solar-system atom model (Figure 7.5, middle), Niels Bohr changed the second part of the description by introducing quantum jumps between special orbits, but he kept the first part. Schrödinger was even more radical, and changed the first part too: he abandoned the very idea that a particle *has* a well-defined position and velocity! Instead, he described the state of a particle by a new mathematical beast called a *wavefunction*, written ψ, which describes the extent to which the particle is in different places. Figure 7.5 (right) shows the square† of the wavefunction, $|\psi|^2$, for the electron in a hydrogen atom in an $n = 3$ orbit, and you can see that rather than being in a particular place, it seems to be on all sides of the proton equally, while preferring certain radii over others. How intense the "electron cloud" of Figure 7.5 (right) is in different places corresponds to the extent to which the electron is in these places. Specifically, if you experimentally go looking for the electron, you find that the square of the wavefunction gives the probability that you'll find it in different places, so some physicists like to think of the wavefunction as describing a *probability cloud* or *probability wave*. In particular, you'll never find a particle in places where its wavefunction equals zero. If you want to stir up a cocktail party by sounding like a

---

* If vector calculus floats your boat, then think of the state as simply the position vector **r** and its time derivative $\dot{\mathbf{r}}$ (the velocity vector).

† If you're a math aficionado and like complex numbers, you'll be pleased to know that the wavefunction for a particle specifies a complex number ψ (**r**) for each place **r** in space. What I'm casually calling the "square" of the wavefunction throughout this book for brevity is actually $|\psi|^2$, the square of the absolute value $|\psi|$ of the wavefunction, which is defined as the real part squared plus the imaginary part squared. If you're not a math aficionado, then don't worry, since you can understand the key arguments in this book anyway.

Figure 7.8: The wavefunction ψ on the verge of collapse

quantum physicist, another buzzword you'll need to drop is *superposition*: a particle that's both here and there at once is said to be in a superposition of here and there, and its wavefunction describes all there is to know about this superposition.

These quantum waves are strikingly different from the classical waves from Figure 7.6: a classical wave that you're surfing on is made of water and the thing which has a wavy shape is the water surface, but the thing which is wavy or cloudlike in a hydrogen atom isn't water or any kind of substance at all: there's only a single electron there, and what's wavy is its wavefunction, the extent to which it is in different places.

The Collapse of Consensus

In summary, Schrödinger altered the classical description of the world in two ways:

1. The *state* is described not by positions and velocities of the particles, but by a wavefunction.
2. The *change* of this state over time is described not by Newton's or Einstein's laws, but by the Schrödinger equation.

These discoveries by Schrödinger have been universally celebrated as among the most important achievements of the twentieth century, and they created a revolution in both physics and chemistry. But they also left people tearing their hair out in confusion: if things could be in several places at once, why did we never observe that (while sober)? This puzzle became known as the *measurement problem* (in physics, *measurement* and *observation* are synonyms).

After much debate and discussion, Bohr and Heisenberg came up with a remarkably radical remedy that became known as the *Copenhagen interpretation*, which to this day is taught and advocated in most quantum-mechanics textbooks. A key part of it is to add a loophole to the second item mentioned above, postulating that change is only governed by the Schrödinger equation *part of the time*, depending on whether an *observation* is taking place. Specifically, if something is *not* being observed, then its wavefunction changes according to the Schrödinger equation, but if it *is* being observed, then its wavefunction *collapses* so that you find the object only in one place. This collapse process is both abrupt and fundamentally random, and the probability that you find the particle in any particular place is given by the square of the wavefunction. The wavefunction collapse thus conveniently gets rid of schizophrenic superpositions and explains our familiar classical world where we see things in only one place at a time. Table 7.3 summarizes the key quantum concepts that we've explored so far, and how they're interrelated.

There are other elements to the Copenhagen interpretation as well, but the part above is what's most agreed on. I've gradually discovered that those of my colleagues who hail Copenhagen as their favorite interpretation of quantum mechanics usually disagree with each other about some of those other elements, making it more appropriate to talk about the "Copenhagen interpretations." The relativity pioneer Roger Penrose quipped: "There are probably more different attitudes to quantum mechanics than there are quantum physicists. This is not inconsistent because certain quantum physicists hold different views at the same time." Indeed, even Bohr and Heisenberg held slightly different views about what it implied about the nature of reality. However, all physicists back then agreed that the Copenhagen interpretation worked great for simply getting on with business as usual in the lab.

Not everyone was thrilled, however. If wavefunction collapse really happened, then this would mean that a fundamental randomness was built into the laws of nature. Einstein was deeply unhappy about this interpretation, and expressed his preference for a deterministic universe with the oft-quoted remark "I can't believe that God plays dice." After all, the very essence of physics had been to predict the future from the present, and now this was supposedly impossible not just in practice, but even in principle. Even if you were infinitely wise and knew the wavefunction of the entire Universe, you couldn't calculate what the

Quantum-Mechanics Cheat Sheet	
Wavefunction	Mathematical entity describing the quantum state of an object. The wavefunction of a particle describes the extent to which it's in different places
Superposition	Quantum-mechanical situation where something is in more than one state at once, for example in two different places
Schrödinger equation	Equation that lets us predict how the wavefunction will change in the future
Hilbert space	Abstract mathematical space where the wavefunction lives
Wavefunction collapse	Hypothesized random process whereby the wavefunction changes abruptly in violation of the Schrödinger equation, giving a measurement a definite outcome. Lack of wavefunction collapse implies Hugh Everett's Level III multiverse
Measurement problem	The controversial question of what happens to the wavefunction during a quantum measurement: does it collapse or not?
Copenhagen interpretation	A set of assumptions including that the wavefunction collapses during measurements
Everett interpretation	The assumption that the wavefunction never collapses—implies the Level III multiverse (Chapter 8)
Decoherence	A censorship effect derivable from the Schrödinger equation, whereby superpositions become unobservable unless they're kept secret from the rest of the world—makes the wavefunction appear to collapse during measurements even if it actually doesn't (Chapter 8)
Quantum immortality	The idea that we're subjectively eternal if the Level III multiverse exists. I suspect that there's no quantum immortality because the continuum is an illusion (Chapter 11).

Table 7.3: Summary of key quantum-mechanics concepts (Hilbert space and the last three concepts will be introduced in the next chapter)

wavefunction would be in the future, because as soon as someone in our Universe made an observation, the wavefunction changed randomly.

Another aspect of collapse that caused consternation was that observation was upgraded to such a central concept. When Bohr exclaimed, "No reality without observation!" it seemed to put humans back on center stage. After Copernicus, Darwin and others had gradually deflated our human hubris and warned against our egocentric tendencies to assume that everything revolved around us, the Copenhagen interpretation made it seem as if we humans in some sense created reality by just looking at it.

Finally, some physicists were irked by the lack of mathematical rigor.

Whereas traditional physical processes would be described by mathematical equations, the Copenhagen interpretation had no equation specifying what constituted an observation, that is, exactly when the wavefunction would collapse. Did it really require a human observer, or was consciousness in some broader sense sufficient to collapse the wavefunction? As Einstein put it: "Does the Moon exist because a mouse looks at it?" Can a robot collapse the wavefunction? What about a webcam?

The Weirdness Can't Be Confined

Loosely speaking, the Copenhagen interpretation of quantum mechanics suggests that small things act weird but big things don't. Specifically, things as small as atoms are usually in several places at once, but things as big as people aren't. The above-mentioned gripes aside, this is a tenable view as long as the weirdness stays confined to the microworld and doesn't somehow leak into the macroworld, like an evil genie being confined to a bottle, unable to get big and wreak havoc. But does it really stay confined?

One of the things that bothered me back in that Stockholm dorm room at the beginning of this chapter was that big things are made of atoms, and so since atoms can be in several places at once, they can be, too. But just because they *can* doesn't mean that they *will*: you might hope that there are no physical processes that amplify microscopic weirdness into macroscopic weirdness. Schrödinger himself shattered such hopes with a diabolical thought experiment: Schrödinger's cat is trapped in a box with a cyanide canister that's opened if a single radioactive atom decays. After a while, the atom will be in a superposition of decayed and not decayed, which causes the entire cat to be in a superposition of dead and alive. In other words, a seemingly innocent microsuperposition involving a single atom is amplified over time into a macrosuperposition where a cat containing octillions of particles is in two states at once. Moreover, such weirdness amplification happens all the time, even without sadistic contraptions. You may have heard about chaos theory: how the laws of classical physics can exponentially amplify tiny differences, such as a Beijing butterfly perturbing the air and ultimately causing a Stockholm storm. An even simpler example is a pencil balanced on its tip, where a microscopic nudge of the initial tilt can determine the direction in which it will ultimately come crashing down. Whenever such chaotic dynamics are at play, the initial position of a single atom can make all the difference, so if that atom is in two places at once, you'll end up with macroscopic things in two places at once.

Such weirdness amplification clearly happens whenever we make quantum measurements: if you measure the position of a single atom that's in two places at once* and write the result on a piece of paper, then the particle position will determine the motion of your hand, and your pencil will therefore end up in two places at once.

Last but not least, such weirdness amplification happens regularly even within your brain. Whether a given neuron fires at a given time depends on whether the sum of all its input signals exceeds a certain threshold, and this can make neural networks highly unstable, much like the weather and the balanced pencil. This was exactly what was happening on the opening page of this book, when I was biking to school and decided whether to look right. Suppose my snap decision was such a close call that it came down to whether a single calcium atom would enter a particular synaptic junction in my prefrontal cortex, causing a particular neuron to fire an electrical signal that would trigger a whole cascade of activity by other neurons in your brain which collectively encode *Let's look!* So if that calcium atom started in two slightly different places at once, then half a second later, my pupils would have been pointing in two opposite directions at once, and before long, my entire body will be in two different places at once, one of them being the morgue, making this my own version of the Schrödinger's cat experiment—with me in the role of the cat. . . .

Quantum Confusion

So there I was in my girlfriend's dorm room in Stockholm, deeply frustrated and confused. Now you know why. My first quantum exam was coming up, and the more I thought about the Copenhagen interpretation that my textbook presented as obvious and absolute truth, the more disturbed I felt. Quantum weirdness clearly couldn't be confined to the microworld. Schrödinger's cat was out of the bag. I didn't mind

---

* One classic experiment that does this involves sending a single silver atom through a so-called Stern-Gerlach apparatus, which will put it in two different places depending on its spin.

the weirdness per se, but here's what really bothered me back then: suppose that you personally perform Schrödinger's cat experiment. If that textbook is right, then the cat's wavefunction collapses and it becomes definitely dead or definitely alive at the instant when you personally look at it. But what if I'm standing outside your lab and consider the wavefunction describing all the particles that make up the cat, you and everything else in your lab? Surely all those particles should obey the Schrödinger equation regardless of whether they're part of a living being or not? And in that case, the book implies that the cat's wavefunction collapses only when I myself enter the lab and observe what's going on, not at the earlier time when you took a look. And in that case, before I looked, you yourself would have been in a superposition of feeling guilty about killing the cat and relieved that it made it. In other words, at best the Copenhagen interpretation was incomplete, refusing to answer the question of when precisely the wavefunction collapsed. At worst, it was inconsistent, since the wavefunction of our whole Universe would never collapse from the viewpoint of someone in a parallel universe who could never observe us.

Please join me in the next chapter to explore what quantum mechanics is really telling us about the nature of reality. Perhaps we Swedes have a genetic predisposition toward badmouthing our southwestern neighbors, but when I think about the Copenhagen interpretation, I just can't get this *Hamlet* quote out of my mind: "Something is rotten in the state of Denmark."

THE BOTTOM LINE

- Everything, even light and people, seems to be made of particles.
- These particles are purely mathematical objects in the sense that their only intrinsic properties are mathematical properties—numbers with names like *charge, spin* and *lepton number*.
- These particles don't obey the classical laws of physics.
- Mathematically, the state of these particles (which should perhaps be called "wavicles") can't be described by six numbers (representing their position and velocity), but by a wavefunction, describing the extent to which they are in different places.
- This gives them properties both of traditional particles (they're either here or there) and of waves (they can be in several places at once in a so-called superposition).
- Particles aren't allowed to be in only one place (the Heisenberg uncertainty principle), which prevents atoms from collapsing.
- The future behavior of particles is described not by Newton's laws, but by the Schrödinger equation.
- This equation shows that innocent microscopic superpositions can get amplified into crazy macroscopic superpositions such as Schrödinger's cat, and you personally being in two places at once.
- The textbook formulation postulates that the wavefunction sometimes "collapses," violating the Schrödinger equation and introducing fundamental randomness into nature.
- Physicists argue passionately about what this all means.
- The textbook formulation of quantum mechanics is either incomplete or inconsistent.

8

The Level III Multiverse

When you come to a fork in the road, take it.

—Yogi Berra

"Wow—it's beautiful down there!" The San Francisco Bay glistened in the evening sun, and I felt even more excited than when my parents gave me my first-ever magic set. I was glued to my window, trying to make out all the famous landmarks I was seeing for the very first time. Since saving up enough money as a cheese salesman to take the train to Spain at age seventeen, I'd fallen more and more in love with travel. Since reading Feynman in college, I'd fallen more and more in love with physics. Now, after twenty-three years of living with ice and snow, I'd get to spend four years doing both! In what seemed to me one of the coolest places on Earth, and the perfect place to have crazy ideas.

Through a wild stroke of luck, I'd been admitted to Berkeley for physics grad school, and even though my expectations were perhaps unreasonably high, those four years ended up superseding them on all counts. I found Berkeley to be every bit as inspiring, wild and crazy a place as I'd hoped. I ended up with an Australian girlfriend the very day after I arrived. I found it convenient to hail from an obscure country that most people couldn't find on the map: my nationality allowed me to be as crazy as I wanted, quickly earning the nickname "Mad Max," and get away with it—people would give me the benefit of the doubt and assume that this was normal behavior in Sweden. Not that I needed to make excuses. A student who ended up living across the street from me would only attend class naked, and made national news when he got expelled. A physics classmate with whom I did homework problems doubled as a porn actor to help finance his studies. The guy across the hall from me in International House got arrested with a gun and a list of names of "People to Destroy."* So if your most crazy traits were being Swedish and having strange physics ideas, you blended right in.

Back in high school, my friend Magnus Bodin had inspired me with his contrarian philosophy. Since everyone else sent their letters in rectangular envelopes, he made triangular ones. Ever since, when I see the majority do things one way, I instinctively look for alternatives. For example, all my classmates spent ages on electromagnetism homework during our first year, so I talked our professor into letting me skip this in return for an oral exam at the end of the course. Instead, I spent endless hours in the library feeding my curiosity, learning all sorts of amazing physics that wasn't in the textbooks—and which has kept helping me to this day. It also freed me up to pursue research on the side.

For the first time in my life, I made friends who shared my obsession with crazy physics questions, and it felt amazing to sit up late at night with these kindred souls speculating about the ultimate nature of reality. Justin Bendich, whose scruffy appearance reminded me of Shaggy from *Scooby-Doo*, was a gold mine of information and would give thoughtful answers to even my wackiest questions. Bill Poirier was obsessed with information theory, and together we came up with a cool information theory–based improvement of the Heisenberg uncertainty principle that had us extremely excited until I found an article about it in the library. I felt like the luckiest guy on the planet: I'd figured out what I really, really wanted to do, and I was doing it.

The Level III Multiverse

My new teachers were inspiring, too. I learned quantum mechanics much more thoroughly from Eugene Commins, whose dry humor livened up the blackboards full of equations. I once raised my hand and asked, "Isn't that like adding apples and pears?"—a common Swedish expression. "No," he replied. "It's like adding apples and oranges."

Although his one-year course taught me many useful technical tools, it never answered my burning quantum questions. In fact, it didn't even

* The student newspaper *The Daily Cal* published a quote from me, followed by "according to a Swedish student who lived across the hall and requested to remain anonymous," and for days afterward, my friends would come up and say: "Hey, Max, you look so anonymous today!"

ask them, leaving me stuck struggling with them on my own. Was quantum mechanics inconsistent? Did the wavefunction really collapse? If so, when? If not, then why didn't we see things in two places at once, and where did the randomness and probabilities of quantum mechanics come from?

I'd heard that, back in 1957, Princeton grad student Hugh Everett III had proposed a truly radical answer involving parallel universes, and I was curious to learn the details. However, this idea was generally ignored and rarely taught. Although I met a few people who had heard of it, none of them had actually read his Ph.D. thesis describing it, which was buried in an out-of-print book. All our library had was a radically abbreviated version where the parallel-universe business was never explicitly mentioned. But in November 1990, my searches paid off, and I finally found that elusive book. Rather appropriately, I found it in a Berkeley store specializing in radical publications, where they also carried titles such as *The Anarchist's Cookbook*.

Everett's Ph.D. thesis totally blew me away. I felt like scales had fallen from my eyes. Suddenly it all made sense to me! Everett had been bothered by exactly the same things that bothered me, but rather than just leaving it at that, he'd pushed ahead, explored possible solutions, and discovered something remarkable. When you have a radical idea, it's so easy to say to yourself, "Of course that can't work," and drop it. But if you hold the thought just a little longer and ask yourself, "Well, why exactly can't it work?" and find that you're struggling to come up with a logically watertight answer, then you might be onto something big.

So what was Everett's radical idea? It's amazingly simple to state:

> *The wavefunction never collapses. Ever.*

In other words, the wavefunction that fully describes our Universe just changes deterministically at all times, always governed by the Schrödinger equation, regardless of whether there are observations taking place or not. So the Schrödinger equation rules supreme, without ifs, ands or buts. This means that you can think of Everett's theory as "Quantum Mechanics Lite": take the usual textbook version of the theory and simply drop the postulate that talks about wavefunction collapse and probabilities.

This surprised me, because the rumors I'd heard suggested that Everett postulated crazy-sounding stuff like parallel universes and that our Universe would split into parallel universes whenever you made an observation. Indeed, even today, many of my physics colleagues still think that this is what Everett assumed. Reading Everett's book taught me a lesson not only in physics but also in sociology: I learned the importance of going back and checking the source material for yourself rather than relying on secondhand information. It's not only in politics that people get misquoted, misinterpreted and misrepresented, and Everett's Ph.D. thesis is a great example of something that, to first approximation, everyone in physics has an opinion about and almost nobody has read.*

I just couldn't put his book down. His logic was beautiful: he didn't assume any of that crazy-sounding stuff, but it all followed as consequences of his assumption! At first it seemed so simple that it couldn't possibly work. After all, Niels Bohr and his collaborators were smart people and had invented wavefunction collapse for a reason, to explain why experiments seemed to have definite outcomes. But Everett realized something amazing: even if experiments didn't have definite outcomes, it would still *seem* as though they did!

Figure 8.1 shows an example of how I think about this. In this thought experiment that I'll call "Quantum Cards," you take a card with a perfectly sharp bottom edge, balance it on a table, and bet $100 that it's going to land face-up when it falls. You keep your eyes closed until you hear that the card has fallen, then look to see whether you've won or lost your bet. According to classical physics, it will in principle stay balanced forever.† According to the Schrödinger equation, it will fall down in a few seconds even if you do the best possible job balancing it, because the Heisenberg uncertainty principle states that it can't be in only one position (straight up) without moving. Yet since the initial state was left-right symmetric, the final state must be so as well. This implies that it falls down in both directions at once, in superposition.

When you open your eyes and look at the card, you're making an

---

* His thesis finally went online in 2008, and you can read it here: http://www.pbs.org/wgbh/nova/manyworlds/pdf/dissertation.pdf. The notion that at certain magic instances, reality undergoes some sort of metaphysical split into two branches that subsequently never interact isn't only a misrepresentation of Everett's thesis, but also inconsistent with Everett's postulate that the wavefunction never collapses, since the subsequent developments could in principle make the branches interfere with each other. According to Everett, there is, was and always will be only one wavefunction, and only decoherence calculations (which I'll explain later in this chapter), not postulates, can tell us when it's a good approximation to treat two branches as non-interacting.

† In practice, this unstable card will of course get toppled in no time by a tiny air current, so it would be better to take a normal card with a thick bottom edge and use a quantum device such as Schrödinger's radioactive-atom trigger to nudge it one way or the other.

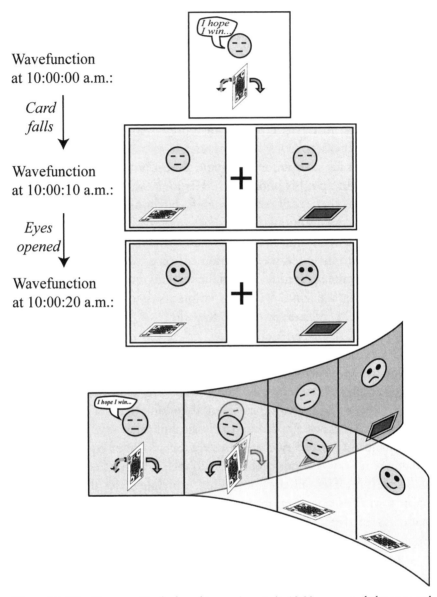

Figure 8.1: The Quantum Cards thought experiment: At 10:00 a.m., you balance a card on its edge, bet $100 on it falling face-up, and close your eyes. Ten seconds later, the card has fallen down both to the left and to the right in quantum superposition, so the wavefunction describes the card being in two places at once. Another ten seconds later, you've opened your eyes and looked at the card, so the wavefunction describes your being happy and sad at once. Although there's still only one wavefunction and one quantum reality (within which particles making up both the card and you are in two places at once), Everett realized that this is in practice as if our Universe has split into two parallel universes (bottom), with a definite outcome in each of them.

observation. So according to the Copenhagen interpretation, the wave-function would collapse and you'd see the card either face-up or face-down, with 50% probability for each outcome. You'd be either smiling about easy profits or cursing yourself for being tricked into wasting a hundred bucks on a silly physics experiment—and the laws of physics wouldn't predict which, since it would have been caused by inherent randomness in nature. And according to Everett? Well, to him, there was nothing magical about observation: it was just a physical process like any other, but one characterized by transfer of information—in this case from the card to your brain. If the wavefunction had described the card as only face-up, you'd have gotten happy, and vice versa. Combining these facts with the Schrödinger equation, he could easily calculate exactly what would happen to the wavefunction: it would change to describe a superposition of two different configurations of the particles that made up you and the card: one where the card was face-up and you were cheerful, and one where it was face-down and you were disappointed. There are three key insights here:

1. The experiment puts your mind into two states at once. It's basically a nonlethal version of Schrödinger's cat experiment, with you in the role of the cat.
2. These two mind states are completely unaware of each other.
3. The state of your mind becomes linked with the state of the card, in such a way that everything is consistent. (The wavefunction doesn't describe any particle configuration where you perceive the card face-up when it's face-down.)

It's easy to prove that the Schrödinger equation always keeps things consistent like this. For example, if your broke friend enters the room and asks you what's up, the state of all the particles (making up the card, you and your friend) evolves into a quantum superposition of "card down/you sad/friend empathizes" and "card up/you happy/friend asks you for loan."

Putting all this together, as illustrated in Figure 8.1, Everett realized that even though there's still only one wavefunction and one quantum reality (within which many of the particles making up our Universe are in two places at once), this is in practice as if our Universe has split into two parallel universes! At the end of this experiment, there will be two different versions of you, each subjectively feeling just as real as the other, but completely unaware of each other's existence.

This was when my head really started to spin, because the Quantum Cards experiment is just one particular example of how microscopic quantum weirdness gets amplified into macroscopic quantum weirdness. As we discussed in the last chapter, such amplification of small differences into big differences happens virtually all the time, like when a cosmic ray–particle hit does/doesn't give someone a cancerous mutation, when today's atmospheric conditions do/don't evolve into a Category 4 hurricane next year, or when you use your neurons to make decisions. In other words, parallel-universe splitting is happening constantly, making the number of quantum parallel universes truly dizzying. Since such splitting has been going on ever since our Big Bang, pretty much any version of history that you can imagine has actually played out in a quantum parallel universe, as long as it doesn't violate any physical laws. This makes vastly more parallel universes than there are grains of sand in our Universe. In summary, Everett showed that if the wavefunction never collapses, then the familiar reality that we perceive is merely the tip of an ontological iceberg, constituting a minuscule part of the true quantum reality.

As you remember, we encountered parallel universes in Chapter 6 as well, but of a different kind. To avoid confusing ourselves with an overdose of parallel universes, let's review the terminology we agreed on in Chapter 6. By *our Universe*, we mean the spherical region of space from which light has had time to reach us during the 14 billion years since our Big Bang, with its classical observed properties (which galaxies are where, what the history books say, etc.). In Chapter 6, we called other such spherical regions far away in our large or infinite space *Level I parallel universes* or *Level II parallel universes*, depending on whether they had our effective laws of physics or not. Let's call the quantum parallel universes that Everett discovered *Level III parallel universes*, and the collection of all of them the *Level III multiverse*. Where are these parallel universes? Whereas the Level I and Level II kinds are far away in our good old three-dimensional space, the Level III ones can be right here as far as these three dimensions are concerned, but separated from us in what mathematicians call *Hilbert space*, an abstract space with infinitely many dimensions where the wavefunction lives.*

After being dismissed and almost completely ignored for a decade,

---

* The wavefunction corresponds to a single point in this infinite-dimensional space, and the Schrödinger equation says that this point will orbit around the center of the space at a fixed distance.

Everett's version of quantum mechanics first began to get popularized by the famous quantum-gravity theorist Bryce DeWitt, who called it the *Many Worlds interpretation*—a name that stuck. When I later met Bryce, he told me he'd at first complained to Hugh Everett, saying that he liked his math, but was really bothered by the gut feeling that he just didn't *feel* like he was constantly splitting into parallel versions of himself. He told me that Everett had responded with a question: "Do you feel like you're orbiting the Sun at thirty kilometers per second?" "Touché!" Bryce had exclaimed, and conceded defeat on the spot. Just as classical physics predicts both that we're zooming around the Sun and that we won't feel it, Everett showed that collapse-free quantum physics predicts both that we're splitting and that we won't feel it.

Sometimes it's hard to reconcile what I believe with what I feel. Fast-forward to May 1999, and I'm waiting for the stork to arrive with my first son. I feel anxious, and hope that the delivery will end well. But at the same time, my physics calculations have convinced me that it will both end well and end badly, in different parallel universes. And in that case, what do I mean by hoping? Perhaps I mean that I hope that I'll end up in one of those parallel universes where things went well? No, that's nonsense, since I'll end up in all of these parallel universes, and am jubilant in some and devastated in others. Hmmm. Perhaps I mean that I hope that the delivery will go well in most of the parallel universes? No, that's nonsense as well, since the percentage where things go well can in principle be calculated using the Schrödinger equation, and it's illogical to have hopes about something that's already predetermined. But apparently—and perhaps fortunately—my emotions aren't completely logical.

The Illusion of Randomness

I had more questions. It was well known that if you repeated a quantum experiment many times, you'd typically get different results seemingly at random: for example, you can measure the spin direction of lots of identically prepared atoms in such a way that you'll get a seemingly random sequence of results—say "clockwise," "counterclockwise," "clockwise," "clockwise," "counterclockwise," etc. Quantum mechanics won't predict the outcomes, merely the probability of different outcomes. But this probability business was all part of the collapse postulate from the

Copenhagen interpretation, so after Everett dropped it, how could he get quantum mechanics to predict anything random? There's nothing random at all about the Schrödinger equation: if you know the wavefunction of our Universe right now, it will in principle let you predict what the wavefunction will be at any time in the future.

In the fall of 1991, I signed up for an unusual course on the interpretation of quantum mechanics that was taught by a fellow grad student, Andy Elby. His dorm room used to be next to my girlfriend's, and his door would be adorned with helpful advice such as "How to procrastinate in 7 easy steps." Like me, he was very interested in what quantum mechanics really meant, and as part of his course, he let me give two lectures about Everett's work. This was an exciting rite of passage for me, since it was my first time ever giving a talk about physics, and I spent much of it on how Everett explained randomness. First of all, if you do the Quantum Cards experiment (Figure 8.1), both copies

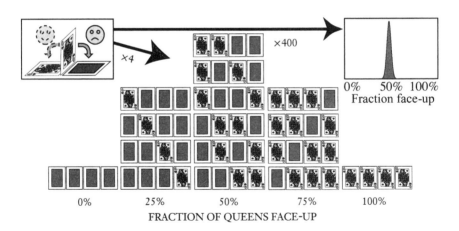

FRACTION OF QUEENS FACE-UP

Figure 8.2: The origin of quantum probabilities. According to quantum physics, a card perfectly balanced on its edge will by symmetry fall down in both directions at once, in what's known as a superposition. If you've bet money on the queen landing face-up, the state of the world will become a superposition of two outcomes: your smiling with the queen face-up and your frowning with the queen face-down. In each case, you're unaware of the other outcome and feel as if the card fell randomly. If you repeat this experiment with four cards, there will be $2 \times 2 \times 2 \times 2 = 16$ outcomes (see figure). In most of these cases, it will appear to you that queens occur randomly, with about 50% probability. Only in two of the sixteen cases will you get the same result all four times. If you repeat 400 times, most of the 2^{400} outcomes have about 50% queens (top right). According to a famous theorem, you'll observe queens 50% of the time in almost all cases in the limit where you repeat the card experiment infinitely many times. Almost all of the copies of you in the final superposition will therefore conclude that the laws of probability apply even though the underlying physics isn't random and, as Einstein put it, "God doesn't play dice."

of you afterward (each effectively in a separate parallel universe) will see a definite outcome. Both will feel that this outcome was random in the sense that there was no way to predict it: for any predicted outcome, the opposite outcome also occurred in an equally real universe. Now what about probabilities—where do they come sailing in from? Well, if you repeat this experiment with four cards, there will be $2^4 =$ 16 outcomes (Figure 8.2), and in most of these cases, it will appear to you that queens occur randomly, with roughly 50% probability. Only in two of the sixteen cases will you get the same result all four times. As you repeat the experiment more and more times, things start getting interesting. According to a 1909 theorem by the French mathematician Émile Borel, you'll observe queens 50% of the time in almost all cases (in all cases except for what mathematicians call a set of measure zero) if you repeat the card experiment infinitely many times. Almost all of the copies of you in the final superposition will therefore conclude that the laws of probability apply even though the underlying physics (the Schrödinger equation) isn't random at all.

In other words, the subjective perception of a copy of you in a *typical* parallel universe is a seemingly random sequence of wins and losses, behaving as if generated through a random process with probabilities of 50% for each outcome. This experiment can be made more rigorous if you take notes on a piece of paper, writing "1" every time you win and "0" every time you lose, and place a decimal point in front of it all. For example, if you lose, lose, win, lose, win, win, win, lose, lose and win, you'd write ".0010111001." But this is just what real numbers between zero and one look like when written out in binary, the way computers usually write them on the hard drive! If you imagine repeating the Quantum Cards experiment infinitely many times, your piece of paper would have infinitely many digits written on it, so you can match each parallel universe with a number between zero and one. Now what Borel's theorem proves is that almost all of these numbers have 50% of their decimals equal to 0 and 50% equal to 1, so this means that almost all of the parallel universes have you winning 50% of the time and losing 50% of the time.*

---

* It's interesting to note that Borel's theorem made a strong impression on many mathematicians of the time, some of whom felt that the whole concept of probability was too philosophical to qualify as rigorous mathematics. Suddenly Borel confronted them with a theorem at the heart of classical mathematics that could be reinterpreted in terms of probabilities even though the theorem itself never mentioned probabilities at all. Borel would undoubtedly have been interested to know that his work showed the emergence of probabilities "out of the blue" not only in mathematics, but in physics as well.

It's not just that the percentages come out right. The number ".010101010101 . . ." has 50% of its digits equal to 0 but clearly isn't random, since it has a simple pattern. Borel's theorem can be generalized to show that almost all numbers have random-looking digits with no patterns whatsoever. This means that in almost all Level III parallel universes, your sequence of wins and losses will also be totally random, without any pattern, so that all that can be predicted is that you'll win 50% of the time.

It gradually hit me that this illusion of randomness business really wasn't specific to quantum mechanics at all. Suppose that some future technology allows you to be cloned while you're sleeping, and that your two copies are placed in rooms numbered 0 and 1 (Figure 8.3). When they wake up, they'll both feel that the room number they read is completely unpredictable and random. If in the future, it becomes possible for you to upload your mind to a computer, then what I'm saying here will feel totally obvious and intuitive to you, since cloning yourself will be as easy as making a copy of your software. If you repeated the cloning experiment from Figure 8.3 many times and wrote down your room number each time, you'd in almost all cases find that the sequence of zeros and ones you'd written looked random, with zeros occurring about 50% of the time.

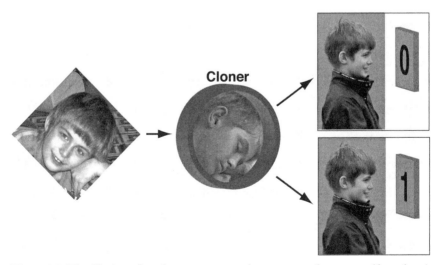

Figure 8.3: The illusion of randomness occurs whenever you clone yourself, so there's really nothing specifically quantum-mechanical about it. If some future technology allows my son Philip to be cloned while he's asleep, and his two copies are placed in rooms numbered 0 and 1, then they'll both feel that the room number they read on awakening is completely unpredictable and random.

In other words, causal physics will produce the illusion of random-ness from your subjective viewpoint in any circumstance where you're being cloned. The fundamental reason that quantum mechanics appears random even though the wavefunction evolves deterministically is that the Schrödinger equation can evolve a wavefunction with a single you into one with clones of you in parallel universes.

So how does it feel when you get cloned? It feels random! And every time something fundamentally random appears to happen to you, which couldn't have been predicted even in principle, it's a sign that you've been cloned.

Hugh Everett's work is still controversial, but I think that he was right and that the wavefunction never collapses. I also think that he'll one day be recognized as a genius on par with Newton and Einstein—at least in most parallel universes. Unfortunately, in this particular uni-verse, his work was almost completely dismissed and ignored for over a decade. He didn't get a job in physics, became rather bitter and with-drawn, smoked and drank too much, and died of an early heart attack in 1982. I've learned more about him recently because I got to meet his son, Mark, at the shooting of a TV documentary called *Parallel Worlds, Parallel Lives*. The producer wanted me to explain his dad's work to him, and I felt lucky and honored: back when I stood there in that radical Berkeley bookstore, I couldn't in my wildest dreams have imagined that I'd one day get this personal connection to one of my physics super-heroes. Mark is a rock star, and if you've seen *Shrek*, you've heard him sing. His dad's fate has really tormented his family, and you can hear it in many of his songs. He and his sister had almost no contact with

Figure 8.4: Hugh Everett's rock-star son, Mark, pondering his dad's theory with me in 2007.

their dad even though they lived together. His sister committed suicide, leaving a note saying that she was going to visit her dad in a parallel universe.

Since I believe that Hugh Everett's parallel universes are real, I can't help thinking about what they're like. In our Universe, he was rejected from the Princeton Physics Department for grad school, went to the Math Department, and transferred to Physics a year later. Because of his limited time, his quantum work was his only work. In many other universes, I think he was admitted to the Princeton Physics Department from the start and had time to make his mark with more mainstream research first, making his subsequent quantum ideas harder to ignore. This launched him on a career similar to that of Einstein, whose special theory of relativity was also met with initial suspicion (especially coming from a guy working outside academia as a patent clerk), but couldn't be ignored because Einstein had already made a name for himself with previous discoveries. Just as Einstein stayed in academia and went on to discover general relativity, Everett, too, got the stability of a professorship and made further breakthroughs as remarkable as his first—ah, how I wonder what he discovered. . . .

One event that I think Everett would have enjoyed took place in late August 2001, at Martin Rees's house in Cambridge, where he'd gathered many of the world's leading physicists for an informal meeting about parallel universes and related topics. To me, this was the first time when parallel universes started feeling scientifically respectable (albeit still controversial). I think many participants stopped feeling guilty and embarrassed about harboring such interests once they saw who else was there, and jokingly said things like, "Uh . . . what are you doing at a suspect meeting like this? . . ." During a long and intense group discussion about parallel universes, I suddenly realized that part of the discord was caused by mere misunderstandings rooted in crude language usage: different people were using the term *parallel universe* to refer to several quite different ideas! *Wait,* I thought, *there are two—no three—different kinds! No—four!* After thinking it through, I raised my hand and proposed the Level IV multiverse classification scheme that I'm using in this book.

As brilliant as it was, Everett's thesis left one important question unanswered: if a large object can really be in two places at once, why don't we ever observe that? Sure, if you measure its position, the two copies of you in the two resulting parallel universes will each find it in a definite place. But that answer turns out not to be good enough, because careful experiments show that large objects *never* act like they're in two places at once, even if you don't look at them. In particular, they never display wavelike properties that make so-called quantum interference patterns. It wasn't just Everett's thesis that lacked an answer to this puzzle—there were no answers in my textbooks either.

Quantum Censorship

Holy guacamole! It works!!! It's late November 1991 in Berkeley, it's dark outside, and I'm at home at my desk frantically scribbling math symbols on a piece of paper. I felt a surge of excitement of a kind I'd never experienced before. Wow. Can it really be that I—little inconsequential me—have just discovered something really important? I just have to find out.

I think that often, in science, the hardest part isn't finding the right answer, but finding the right question. If you hit on a really interesting and well-formulated physics question, then it can take on a life of its own, automatically telling you what calculation you need to do to answer it, and the rest is almost automatic: even if the math takes hours or days, it feels a lot like mechanically pulling in a fishing line to see what you've caught. I'd just stumbled upon one of those lucky questions.

I'd learned that the business about the wavefunction collapsing could be elegantly summarized mathematically in terms of a table of numbers called a *density matrix* in quantum-physics jargon, which encodes not only the state of something (its wavefunction), but also my perhaps incomplete knowledge of what the wavefunction is.* For example, if something can be in only two different places, my knowledge about it can be described by a two-by-two table of numbers, as in these two examples:

---

* Density matrices are generalizations of wavefunctions. For every wavefunction, there's a corresponding density matrix, and there's a corresponding Schrödinger equation for density matrices. If you're a mathematically inclined reader and think of the wavefunction ψ as a complex number ψ_i for each possible state i, then the corresponding density matrix is $\rho_{ij} = \psi_i \psi_j^*$, where the star denotes complex conjugation. If you don't know the wavefunction of an object, and know only the probability that it has certain particular wavefunctions, then you should use the density matrix that's the weighted average of the density matrices corresponding to these wavefunctions.

$$\begin{pmatrix} 0.5 & 0.5 \\ 0.5 & 0.5 \end{pmatrix} = \text{"It's here } and \text{ there at the same time."}$$

$$\begin{pmatrix} 0.5 & 0 \\ 0 & 0.5 \end{pmatrix} = \text{"It's here or there—I just don't know which."}$$

In both cases, the probability that I'll find it in either place is 0.5, and that's encoded by the two numbers on the diagonal of each matrix (the 0.5 in the upper left corner and the 0.5 in the lower right corner). The other two numbers in each table, "the off-diagonal elements of the density matrix" as we call them in geek-speak, encode the difference between quantum and classical uncertainty: when they, too, equal 0.5, we have a quantum superposition on our hands (Schrödinger's cat is dead and alive, say), but when they equal zero, we're effectively dealing with good old classical uncertainty, such as when I can't remember where my keys are. So if you manage to replace these off-diagonal numbers by zeros, then you've turned *and* into *or* and collapsed the wavefunction!

As we saw in the last chapter, the Copenhagen interpretation of quantum mechanics says that if your friend observes the object without telling you the outcome, then she'll collapse the wavefunction so that the object is either here or there, and you simply don't know which. In other words, the Copenhagen interpretation says that an observation somehow makes these off-diagonal numbers zero. I wondered if there might be some less mysterious physical process that did the same thing. If you have an isolated system that's not interacting with anything else, then it's easy to prove using the Schrödinger equation that those pesky numbers will never go away. But real systems are almost never isolated, and I asked myself what effect that would have. For example, as you read this sentence, air molecules and photons are constantly colliding with you. So if something is in two places at once, what happens to the two-by-two table of numbers describing it when something else bounces off it?

This was one of those wonderful self-answering questions, and the rest was automatic. I simply considered the object and the colliding particle together as a single isolated system, and used the Schrödinger equation to calculate what would happen. A couple of hours later, I sat there with pages full of math symbols and gasped: those off-diagonal numbers changed to very close to zero, just as if the wavefunction had collapsed! It hadn't *really* collapsed, of course, and those parallel universes were still alive and well, but here was a brand-new effect that looked like wavefunction collapse and smelled like wavefunction collapse, and just as a true collapse would have done, made it impossible to ever observe the object in two places at once. So the quantum weirdness doesn't go away, it just gets censored!

I concluded that quantum mechanics requires secrecy: an object can only be found in two places at once in a quantum superposition as long as its position is kept secret from the rest of the world. If the secret gets out, all quantum superposition effects become unobservable, and it's for all practical purposes as if it's either here or there and you simply don't know which. If a lab technician measures the position and writes it down, the information is obviously out. But even if a single photon bounces off the object, the information about its whereabouts is out: it gets encoded in the subsequent position of the photon. As illustrated in Figure 8.5, a nanosecond later, the photon will be in two quite different places depending on the position of the object, so by measuring where the photon is, you'll find out where the mirror is.

Back in the beginning of the last chapter, I was wondering whether you needed a human observer to collapse the wavefunction, or if a robot would suffice. Now I was convinced that consciousness had nothing to do with it, since even a single particle could do the trick: a single photon bouncing off of an object had the same effect as if a person observed it. I realized that quantum observation isn't about consciousness, but simply about the transfer of information. Finally I understood why we never see macroscopic objects in two places at once even if they're in two places at once: it's not because they're big, but because they're hard to isolate! A bowling ball outdoors typically gets struck by about 10^{20} photons and 10^{27} air molecules every second. It's by definition impossible for me to see something without it getting struck by photons, since I can only see it when photons (light) bounce off it, so a bowling ball that's in two places at once will have its quantum superposition ruined even before I have a chance to become consciously aware of it. In contrast, if you pump out as many air molecules as you can with a good vacuum pump, an electron can typically survive for about a second without colliding with anything, which is plenty enough time for it to demonstrate funky quantum-superposition behavior. For example, it takes only a quadrillionth as long (about 10^{-15} seconds) for an electron to orbit an atom, so there will be almost no effect on its ability to be on all sides of the atom at once.

Moreover, if an air molecule bounces off of a bowling ball and encodes information about its position in its own position (as in Figure 8.5), this molecule will soon collide with many other molecules, which will get the information, too. It's a lot like when Wikileaks posts classified information online: it gets copied, then the copies get copied, and soon the cat is so out of the bag that it's in practice impossible to make the information secret again. And if you can't make the information secret again, then the quantum superposition can't be restored. Now I finally understood why Level III parallel universes stay parallel!

I felt that I was on a roll that night. I also worked this stuff out in more quantitative detail. For example, most things can be not just in two places but in many, and I worked out this case, too, as illustrated in Figure 8.6. Basically, I discovered that a photon mostly destroys the quantum superposition, but lets a bit of it survive: a superposition only as wide as its wavelength. A photon of wavelength 0.0005 millimeter essentially acts like an observer who can only measure the position of something to an accuracy of 0.0005 millimeter. We saw in the last chapter that *all* particles act like waves and have a wavelength, and I showed that when any particle whatsoever bounces off something, quantum superpositions wider than the wavelength get destroyed.

For years now, I'd known that I loved physics and wanted to dedicate my life to it. But I'd always wondered whether I had it in me to be able to contribute to it, as opposed to just learning about it and cheering it on from the sidelines of the field. As I finally drifted off to sleep that night, for the first time in my life, I thought: *Yes I can!* Might my discov-

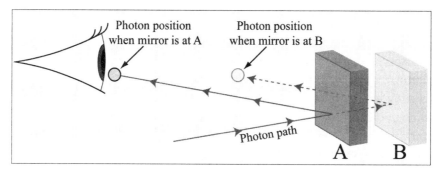

Figure 8.5: If you take a flash photo in a dark room, the photons returning to your camera have encoded information about what's in the room. The figure shows how even a single photon can "measure" things: after a photon has bounced off a mirror, it encodes information about the mirror's position in its own position. If the mirror is at both A and B in a quantum superposition, then it doesn't matter whether it's a human or just a photon that finds out where it is: in either case, the quantum superposition is effectively destroyed.

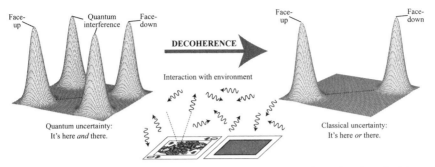

Figure 8.6: Your knowledge of the position of the fallen card can be described by a so-called density matrix, which can be represented as a bumpy surface as illustrated above. The height of the surface along the diagonal (dashed line) gives the probability that you'll find the card in various places, whereas the height of the surface elsewhere, loosely speaking, specifies the amount of quantum weirdness, the extent to which the card is in more than one place at once. The left density matrix corresponds to the card being equally in both of the two places depicted underneath, in quantum superposition, as revealed by the two peaks labeled "Quantum interference." After a photon bounces off of the card, decoherence eliminates these two peaks, giving the density matrix on the right, which corresponds to the card effectively being in only one of the two places, you simply not knowing which. The slight widths of these peaks correspond to some remaining quantum uncertainty around the face-up and face-down positions.

ery become known as the "Tegmark effect"? I knew that whatever happened, I'd never forget my excitement that evening. I felt so fortunate for all the opportunities I'd been given and for all the inspiring people who'd enabled me to join the great adventure of science. It seemed almost too good to be true. And it was. . . .

Two weeks later, I'd expanded my calculations into a first draft of a paper which I called "Apparent Wave Function Collapse Caused by Scattering," *scattering* being the technical term we use for the behavior of particles bouncing off stuff. This was the first time ever I was writing a paper for publication, and I felt like when I was a little kid on Christmas Eve. My left-handed handwriting had always been hideous (pretty much every school assignment would come back with comments such as "Work on neatness!"), and it was exciting to see my illegible scribbles transform into beautiful typeset equations. At the same time, it was funny how paranoid I was getting that someone had already discovered what I had and I'd somehow missed it. I figured that something this basic would have been mentioned in the textbooks and taught in my grad quantum class if it were known, but nonetheless, I almost trembled each time I opened a suspicious reference during my literature search. So far, so good. . . .

Anticipating my publishing debut, I even went ahead and changed my surname to something more unique, from my dad's name *Shapiro* to my mom's name *Tegmark*. I'd enjoyed having the name *Shapiro* back in Sweden, since it was so unusual: we used to be the only family in the whole country who had it. To my horror, I discovered that it was about as unique in international academia as *Andersson* had been back home. The last straw for me was when I did a database search for physics papers by "M. Shapiro" and got thousands of hits. There were even three M. Shapiros in my own Berkeley Physics Department, one of whom (Marjorie) taught me particle physics! In contrast, my mom and her relatives were the only Tegmarks on the planet, as far as I could tell. I was a bit concerned that my dad might misinterpret the name change as some sort of rejection of him, but when I asked him about this, he assured me he didn't mind with a Shakespeare quote: "What's in a name?"

The Joys of Getting Scooped

It wasn't until a month later, after I'd returned from Christmas holidays in Sweden and was just about to submit the paper, that it all came crashing down. All that time. All that enthusiasm. All that fun. All that excitement. All that hope. And—*boom!*—it took just a few minutes for it all to go up in smoke. Who lit the match? Andy Elby, actually. By telling me what a Polish physicist named Wojciech Zurek had already done. Forget the Tegmark effect—it already had a name: *decoherence*. In fact, I soon learned that the German physicist Dieter Zeh had discovered the effect already back in 1970.

At first I didn't feel much, as usual when I get bad news. Then I joked about it with my friends Wayne, Justin and Ted. Then I went home, without realizing that I was really close to the edge, and got into a stupid argument with my girlfriend about something utterly trivial: she'd made just enough rice for herself and a girlfriend, handing me some frozen rice from the freezer instead. All of a sudden I felt so sad that I wanted to cry, but didn't even manage to accomplish that.

Gradually, I've come to totally change my feelings about getting scooped. First of all, the main reason I'm doing science is that I delight in discovering things, and it's every bit as exciting to rediscover something as it is to be the first to discover it—because at the time of the discovery, you don't know which is the case. Second, since I believe that there are other more advanced civilizations out there—in parallel universes if not in our own—*everything* we come up with here on our particular planet is a rediscovery, and that fact clearly doesn't spoil the fun. Third, when you discover something for yourself, you probably understand it more deeply and you certainly appreciate it more. From studying history, I've also come to realize that a large fraction of all breakthroughs in science were repeatedly rediscovered—when the right questions are floating around and the tools to tackle them are available, many people will naturally find the same answers independently. From quantum class, I remember Eugene Commins's deadpan quip: "It's called the Klein-Gordon equation because it was discovered by Schrödinger."

I've rediscovered many other things since, and what you usually find is that you've rediscovered all the basic stuff, and that you've also worked out some interesting details others hadn't and vice versa, enabling you to still salvage a toned-down publication that acknowledges the prior work and adds something to it. This time it was almost spooky: I'd make a top-ten list of natural sources of decoherence, from obvious stuff such as air and sunlight to hard-to-shield things such as background radioactivity and neutrinos from the Sun—and then I found a beautiful paper by Zeh and his student Erich Joos from six years earlier with a virtually identical table. I still had enough new stuff in my paper (http://arxiv.org/pdf/gr-qc/9310032.pdf) to manage to get it published in a less prestigious journal, but if I'd hoped to start my publishing career with a great splash, this felt like more of a belly flop.

In hindsight, the most hilarious scooping I've ever had wasn't this first one, but in 1995, when I'd invented a technique for measuring the quantum state (wavefunction or density matrix) of a particle. I'll never forget how my jaw dropped the night I was going to submit it and stood there like an idiot, staring at a published article in the empty library: these guys hadn't just scooped me, but they'd made a really elaborate and pedagogical figure that was virtually identical to my plot, and they'd coined exactly the same obscure name as I had for the technique: *phase-space tomography*. All I could do was exclaim "HURF!"—a special word my brother Per and I have invented which really captured the moment.

I eventually got to meet many of these intimidating anonymous competitors, and discovered that they were all really nice people. Zeh and Zurek both sent me encouraging emails about my work and invited me to visit them and give talks. In 2004, I visited Wojciech Zurek in Los Alamos and discovered one of the most amazing perks of being a scientist: you get invited to visit exotic places where you spend all your time talking with fascinating people—and you get to call it work! And they even pay for your trip! Wojciech Zurek had big burly hair and a wild impish glint in his eyes, revealing his taste for adventure in both research and recreation. He once persuaded me to climb beneath a rock overhang in the cordoned-off area next to Iceland's mighty Gullfoss waterfall and go within a meter of the falling water—when the cascade suddenly shifted direction, I wondered how many parallel universes had just lost two decoherence theorists in one fell swoop. When I visited Dieter Zeh and his group in Heidelberg in 1996, I was struck by how few accolades he'd gotten for his hugely important discovery of decoherence. Indeed, his curmudgeonly colleagues in the Heidelberg Physics Department had largely dismissed his work as too philosophical, even though their department was located on "Philosopher Street." His group meetings had been moved to a church building, and I was astonished to learn that the only funding that he'd been able to get to write the first-ever book on decoherence came from the German Lutheran Church.

This really drove home to me that Hugh Everett was no exception: studying the foundations of physics isn't a recipe for glamour and fame. It's more like art: the best reason to do it is because you love it. Only a small minority of my physics colleagues choose to work on the really big questions, and when I meet them, I feel a real kinship. I imagine that a group of friends who've passed up on lucrative career options to become poets might feel a similar bond, knowing that they're all in it not for the money but for the intellectual adventure.

Whenever the person next to me on the plane asks me science questions, I'm reminded of the correct way to think about competition and getting scooped. There in the airplane seat, I'm the ambassador from Physics Land, taking great joy and pride in describing not what I've personally done, but what we physicists as a community have done. Sometimes I scoop them, more often they scoop me, but the key point is that together we can learn from each other, inspire each other, and accomplish more than any single person could in their wildest dreams. It's a wonderful community, and I feel extremely fortunate to get to be part of it.

Why Your Brain Isn't a Quantum Computer

"Sir Roger Penrose is incoherent, and Max Tegmark says he can prove it." Whoa! I was reading the first line of a news article in the February 4, 2000, issue of the journal *Science*, and felt rather taken aback. I'd never called this famous mathematical physicist incoherent, but journalists tend to like both conflict and puns, and I'd written a paper (http://arxiv .org/abs/guant-ph/9907009) arguing that one of Penrose's ideas was killed by decoherence.

In recent years, there's been a surge of interest in building so-called quantum computers, which would exploit the weirdness of quantum mechanics to solve certain problems faster. For example, if you bought this book online, your credit-card number was encrypted with a method based on the fact that multiplying two 300-digit prime numbers together is quick, but factoring the resulting 600-digit number (figuring out which two numbers it's the product of) is hard, and would take longer than the age of our Universe on today's best computers. If a large quantum computer can be built, then a hacker could use it to get the answer quite quickly and steal your money, using a quantum algorithm invented by my MIT colleague Peter Shor. As quantum-computing pioneer David Deutsch puts it, "Quantum computers share information with huge numbers of versions of themselves throughout the multiverse," and can get answers faster here in our Universe by, in a sense, getting help from these other versions. A quantum computer could also simulate the behavior of atoms and molecules quite efficiently, replacing measurements in chemistry labs in the same way that simulations on traditional computers have replaced measurements in wind tunnels. Many modern computers calculate faster by using multiple processors in parallel. A quantum computer can be thought of as the ultimate parallel computer, using the Level III multiverse as a computational resource and in a certain limited sense running different parallel calculations in parallel universes.

Before building such a machine, major engineering hurdles need to be overcome, such as isolating the quantum information well enough that decoherence doesn't ruin quantum superpositions. There's still a long way to go: whereas the computer in your cell phone probably stores billions of bits of information (zeros and ones), the state-of-the-art quantum computers in labs around the world can store only a handful. However, Penrose and others made a shocking suggestion: perhaps you already have a quantum computer—in your head! They suggested that our brains (or at least parts of them) are quantum computers, and that this is a key to understanding consciousness.

Since decoherence spoils quantum effects, I decided to use the deco-herence formulas that I'd been scooped on to check whether Penrose's idea really worked. I first did the math for neurons (Figure 8.7), the hundred billion or so nerve cells that, like wires, transmit electrical signals in your brain. Neurons are thin and long: if you laid yours out in a row, they'd wrap around Earth about four times. They transmit electrical signals by transporting sodium and potassium atoms which each have an electron missing (and therefore have a positive electric charge). If you connect a voltmeter to a resting neuron, you'll measure 0.07 volts between the inside and outside of the cell. If one end of the neuron gets triggered to lower this voltage, then voltage-sensitive gates in the cell wall open up, charged sodium atoms start gushing through, this voltage

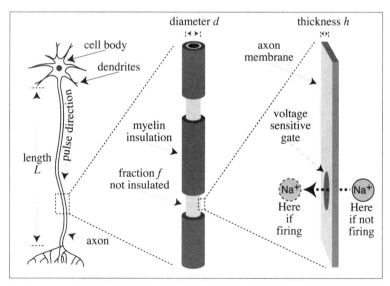

Figure 8.7: Schematic illustration of a neuron (left), a section of its long wire-shaped part called the axon (center), and a piece of its axon membrane (right). The axon is typically insulated with an insulation material called myelin that has small bare patches every half millimeter or so where voltage-sensitive sodium and potassium gates are concentrated. If the neuron is in a superposition of firing and not firing, then roughly a million sodium atoms (with chemical symbol Na) are in a superposition of being inside and outside the cell (right).

drops further and more atoms gush in. This chain reaction, called *firing*, propagates down the length of the neuron at a speed of up to 200 miles per hour, while about a million sodium atoms enter the cell. The axon quickly recovers, and fast neurons can repeat this firing process over a thousand times per second.

Now suppose that your brain really is a quantum computer, and that neuron firing is in some way involved in the computation. Then an individual neuron must be able to be in a superposition of firing and not firing, which means that about a million sodium atoms are in two places at once, both inside and outside the neuron. As we discussed above, a quantum computer only works as long as its state is kept secret from the outside world, so how long could a neuron keep secret whether it was firing or not? When I plugged in the numbers, the answer I got was "not very long at all," or to be more specific, about 10^{-20} (ten billionths of a trillionth) of a second. That's how long it would typically take before a random water molecule bumped into one of the million sodium atoms and discovered where it was, thereby destroying the quantum super-position. I also did the math for another model that Penrose had pro-posed, where the quantum computation was done not by neurons but by microtubules, parts of the scaffolding in cells, and found that they suffered decoherence after about 10^{-13} (100 quadrillionths) of a second. For my thoughts to correspond to a quantum computation, they'd need to finish before decoherence kicked in, so I'd need to be able to think fast enough to have 10,000,000,000,000 thoughts each second. Perhaps Roger Penrose can think that fast, but I sure can't. . . .

It's really not that surprising that your brain doesn't work as a quan-tum computer: my colleagues who are trying to build quantum com-puters go to great lengths to fight decoherence, typically isolating their devices in a cold, dark vacuum to keep their states secret from the rest of the world, while your brain is a warm and wet place whose parts aren't isolated. However, some people complained about my paper, and I got to experience my first scientific controversy. In particular, Stuart Hameroff, one of the quantum-consciousness pioneers, said he felt I'd "laid a stink bomb in the field" and caused problems for quantum-consciousness researchers. "Are you a hit man for scientific orthodoxy?" he asked me.

I found this rather ironic, since I'm normally on the opposite side from scientific orthodoxy, and tend to instinctively side with the under-dog who pursues contrarian ideas. Also, I hadn't made these calculations hoping for a particular result, but simply to find out what the answer was. In fact, I'd probably have been happier with the opposite conclusion, since it would have felt really cool to have my own quantum computer. With two coauthors, Hameroff went on to publish a critique of my paper which I felt was flawed,* and I couldn't help feel that sometimes scientists get attached to an idea with an almost religious fervor, so that no facts can dissuade them. I wondered if the impressive-sounding technical terminology was really just an attempt to rationalize this argument: "Consciousness is a mystery and quantum mechanics is a mystery, so they must be related."

I finally met Stuart Hameroff in 2009 and found him to be quite a jovial and friendly fellow. We had lunch together in New York and, interestingly enough, weren't able to pinpoint a single calculation or measurement that we disagreed on, so we just politely agreed to disagree on what this all meant for consciousness.

Subject, Object and Environment

I have a confession to make: my brain-decoherence calculation was just an excuse. It wasn't the real reason I wrote that paper. I had an idea that I was really excited about and really wanted to publish, but figured that it would be viewed as too philosophical to get accepted for publication. So I came up with what I called my Trojan Horse strategy: hiding the philosophical part that I wanted to sneak past the referees behind pages and pages of respectable-looking equations. Amusingly, this strategy worked in the sense that the paper got accepted, but failed in the sense that people paid attention only to the masking material: the business about the brain not being a quantum computer.

---

* They claimed that the microtubule model I'd tested wasn't from Roger Penrose's book, but in 2006, Stuart graciously acknowledged that it was. They also argued that my calculation must be flawed because the decoherence time scale that I derived decreased as you lower the temperature of the brain, whereas you might intuitively expect the opposite. The point they overlooked is that as soon as you drop the absolute temperature by about 10%, below 0 degrees Celsius, your brain freezes and the decoherence time grows dramatically. The slight decrease in decoherence time for tiny temperature reductions reflects the well-known fact that things are more likely to bump into each other as you lower the temperature, just as slow neutrons are more likely than fast ones to strike targets in a nuclear reactor. They also argued that the brain might perform quantum computations using other mechanisms, but without specifying such mechanisms with enough detail that I could test them, and that there might be other quantum effects in the brain that weren't computations, which I'd never disagreed with in the first place.

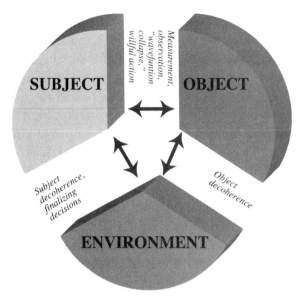

Figure 8.8: It's convenient to decompose the world into three parts: the part corresponding to your subjective perceptions (the subject), the part being studied (the object), and everything else (the environment). As indicated, the interactions between these three parts cause qualitatively very different effects, providing a unified picture including both decoherence and apparent wavefunction collapse.

So what was my hidden message? It was a unified way of thinking about quantum reality, as illustrated in Figure 8.8. Feynman had emphasized that quantum mechanics splits our Universe into two parts: the object under consideration and everything else (referred to as the *environment*). However, I felt that an important piece of the quantum puzzle was missing here: your mind. As Everett's work had shown, understanding the process of observation requires us to include a third part of our Universe as well: your mental state as an observer, labeled *subject* in Figure 8.8.*

If you're not a physicist, it might seem funny that people still talk so little about the mind in the physics community, given all the fuss about observations in quantum mechanics. After all, talking about observations without mentioning the mind feels a bit like discussing nearsightedness without mentioning the eye. I think the explanation is that, since we don't understand how consciousness works, most physi-

---

* Here I'm not referring to your entire brain, just to those aspects of it that correspond to your subjective conscious perceptions.

cists feel uncomfortable even talking about it, fearing that they'll get regarded as too philosophical. Personally, I feel that just because we don't understand something doesn't mean that we can ignore it and still expect to get correct answers.

I'll talk a lot more about our mind in the next chapter. However, to understand Figure 8.8, the details of how your mind works don't matter at all: the only assumption I'm making here is that your subjective consciousness results in some way from the remarkably complicated motions of the particles that make up your brain, and that these particles obey the Schrödinger equation just as all other particles do.

In my Trojan Horse paper, I split the Schrödinger equation into pieces: three governing the three parts of our Universe (subject, object and environment), and additional pieces governing the interactions between these parts. I then analyzed the effects of these different parts of the equation, and showed that one part gave the stuff my textbooks covered, one part gave Everett's many worlds, one part gave Zeh's decoherence, and one part gave something new. Standard textbooks have focused only on the part of the Schrödinger equation that governs the object (an atom, say), in the reductionist spirit that things should be analyzable by themselves without worrying about the greater whole that they're part of. The interaction between the subject and the object gives Everett's parallel universes (pages 186–197), spreading quantum superpositions from the object to you, the subject. The interaction between the environment and the object gives the decoherence (pages 202–205), explaining why large objects such as a queen of hearts never show signs of strange quantum behavior such as being in two places at once. It's normally hopeless to eliminate this decoherence in practice, but even in a thought experiment where you could (say, by repeating the Quantum Cards experiment in a dark, cold room with no air, with only a single photon striking the card then being seen by your eye), it wouldn't make any difference: since the card is in two places at once, so is the photon, so at least one neuron in your optical nerves would enter a superposition of firing and not firing while you looked at the card, and as we saw earlier, this superposition would decohere in about 10^{-20} seconds.

That decoherence still doesn't fully explain why you never perceive quantum weirdness, though, since your thought processes (the internal dynamics of the subject) could create weird superpositions of familiar mental states. Fortunately, here the third interaction in Figure 8.8 comes to the rescue: the interaction between the subject and the environment. The fact that neurons decohere much faster than they can process information means that if the complex neuron-firing patterns in your brain have anything to do with consciousness, then decoherence in the brain will prevent you from experiencing weird superpositions.

This subject-environment interaction also helps tie up another loose end. Wojciech Zurek had continued his decoherence research beyond what I'd rediscovered, and shown that decoherence does one more important thing for us: not only does it explain why large objects never seem to be in two places at once, but it also explains why conventional states (such as being in only one place) are so special: out of all the states that quantum mechanics allows for large objects, these conventional states are the ones that are most robust to decoherence, and therefore the ones that survive. It's a bit like why deserts tend to have cacti rather than roses: they're the most robust to the environment. In fact, a paper on this very topic, which I'd written together with my dad, was the reason Wojciech invited me to give that talk in Los Alamos.

Now, some decoherence can be reduced using clever lab equipment such as vacuum pumps and extreme refrigerators, but we can never turn off the decoherence of our neurons. We don't know how our minds work, but we do know for sure that all information that ever reaches our mind from the outside world must first pass through neurons from our sensory organs, for example the optical nerves from our eyes and the cochlear nerves from our ears, which all decohere ridiculously fast. So by the time we become subjectively aware of any observation about the outside world, things have already decohered, guaranteeing that we'll never perceive any quantum weirdness and explaining why we only perceive robust conventional states.

Among all controversies in physics, a few are so grand that they tower over the rest and last for generations. The great controversy about how to interpret quantum mechanics is clearly one of them. Another involves the *second law of thermodynamics*. It states that the entropy of an isolated system never decreases, where entropy is a quantitative measure of our *lack of information* about a system—it's essentially how many bits of information we'd need to specify its quantum state. On one hand, some scientists have elevated it to almost sacred status, and the great astrophysicist Sir Arthur Eddington had this so say: "The law that entropy always increases holds, I think, the supreme position among the laws of Nature. If someone points out to you that your pet theory of the universe is in disagreement with Maxwell's equations—then so much the worse for Maxwell's equations. If it is found to be contradicted by observation—well, these experimentalists do bungle things sometimes. But if your theory is found to be against the second law of thermodynamics I can give you no hope; there is nothing for it but to collapse in deepest humiliation." On the other hand, serious objections to the second law were made by physics titans such as Maxwell, Gibbs, Loschmidt and Poincaré, and there is still no consensus on whether they've all been satisfactorily resolved.

The way I see it, these two great controversies of quantum mechanics and thermodynamics are linked, in the sense that they can both be resolved in one fell swoop if we use the standard quantum-mechanics definition of entropy (given by John von Neumann), reject wavefunction collapse, and consider all parts of reality: subject, object and environment.

As Figure 8.8 summarizes, measurement and decoherence correspond to the object interacting with the subject and the environment, respectively. Although the processes of measurement and decoherence may appear different, entropy brings out an interesting parallelism between them involving the lack of information that we have about the object, which is a very important quantity that we call *entropy* in physics. If the object isn't interacting with anything, its entropy stays constant: you know just as much about its state a second later, since you can calculate this state from the initial state using the Schrödinger equation. If the object interacts with you, then you typically get more information about it, and its entropy decreases—after opening your eyes in Figure 8.1, there are two copies of you, each seeing a different outcome, but both of whom know how the card fell in their parallel universe and have therefore acquired one additional bit of information about the card. If the object interacts with the environment, however, you typically lose information about it, so its entropy increases: if Philip knows where his Pokémon cards are, he'll have less information about their whereabouts after Alexander messes with them. Similarly, if you know that a card is in the quantum state corresponding to being in two places at once, and then a person or a photon finds out where it is without informing you, then you've lost one bit of information about it: first you knew the quantum state, but now it's effectively in one of two quantum states and you don't know which. In summary, here's how I informally think about this: the entropy of an object decreases while you look at it and increases while you don't. Decoherence is simply a measurement that you don't know the outcome of. More rigorously, we can reformulate the second law of thermodynamics in a more nuanced way:

1. The object's entropy can't decrease unless it interacts with the subject.
2. The object's entropy can't increase unless it interacts with the environment.

The traditional formulation of the law simply corresponds to ignoring the subject. When I published a technical article about this (http://arxiv.org/pdf/1108.3080.pdf),* I included a mathematical proof of the second part (how decoherence increases entropy), but a rigorous proof of the first part (that on average, observation always reduces entropy) eluded me, even though my computer simulations strongly suggested that it was true. Then something wonderful happened, which reminded me of why I'm so fortunate to get to work at MIT: an enthusiastic twenty-year-old Armenian undergraduate student, Hrant Gharibyan, asked if I had any interesting problems he could work on. We teamed up, and he attacked my problem with great fervor, devouring math books like popcorn and mastering mathematical tools such as Schur products and spectral majorization, which aren't known to most physicists, and which I'd only learned of from my dad, who's a mathematician. And then one day when I saw Hrant, I knew from his triumphant smile that he'd solved the problem! We're hoping to publish his proof as soon as I'm done with this book.

Quantum Suicide

I used to feel that there were two kinds of physicists: the titans and the mere mortals. The titans were towering historical figures such as Newton, Einstein, Schrödinger and Feynman who possessed supernatural powers and were surrounded by legends and myths. The mere mor-

---

* If you don't mind the math, the article also explains how this result combined with inflation can explain how the entropy was so low in our early Universe, which in turn explains the so-called arrow of time (beautifully explained in the books by Sean Carroll and Dieter Zeh in the "Suggestions for Further Reading" section). It also provides a quantum-mechanical generalization of the standard procedure for updating our knowledge with new information, known as Bayes's theorem.

Figure 8.9: John Wheeler as I remember him (here, in 2004, holding a book from his ninetieth-birthday conference that I helped organize); flanked by his grad students Richard Feynman (around 1943), Hugh Everett (around 1957) and Wojciech Zurek (in 2007 by that Icelandic waterfall). *(Image credits: Pamela Bond Contractor [Ellipses Enterprises], Mark Oliver Everett, Anthony Aguirre)*

tals were the physicists I'd met who, although perhaps brilliant, were clearly just ordinary people like you and me. And then there was John Wheeler. When I saw him in January 1996, I felt overwhelmed. There he was, eighty-four years old, in the Copenhagen cafeteria where we had our conference lunch. To me, he was the "last titan." He'd worked with Niels Bohr on nuclear physics. He'd coined the term *black hole*. He'd pioneered spacetime foam. He'd had Feynman and Everett as grad students. He'd become one of my physics superheroes with his passion for wild ideas. And there he was, simply eating, like a mere mortal! I felt that I just *had* to introduce myself, or I'd never forgive myself, but I was extremely nervous as I approached his table. I'd been blown off before by people above me in the academic food chain: two different professors had turned their backs on me and walked away in mid-conversation, and yet they were mere mortals. So I was stunned by what happened. There I was, an inexperienced postdoc and a total nobody, yet Wheeler greeted me with a warm smile and invited me to join him for lunch! After hearing that I was interested in quantum mechanics, he told me about some new ideas he had about the subject of existence, and gave me a copy of some of his recent notes. He never talked down to me, and spoke to me in a way that made me feel like an equal even though clearly I wasn't. A fortnight later, I even got an email from him—an email from a titan! He wrote:

It was a great pleasure and encouragement to talk to you in Copenhagen as I believe you share my belief that under and behind quantum mechanics lies some deep and wonderful principle yet to be discovered, as Einstein's great geometric idea threw unexpected light on the power and scope of Newton's supposedly all-embracing theory. The likelihood of such a discovery is surely proportional to our belief that there is something there to be discovered.

He went on to encourage me to come to Princeton, writing, "I am eager to be able to talk to you every day." At that time, I was deciding between postdoc offers—how could I possibly turn down Princeton after that? Once I'd moved to Princeton, I started visiting him regularly, and gradually got to know him better. He and his wife came to my housewarming party. He even signed my New Jersey marriage license—in my world, this felt like having God as a witness.

In his office, he'd often get interrupted, so his favorite way to talk was while "doing orbits," walking the third-floor corridors that looped around the inner courtyards of the Princeton University physics building. His colorful stories made history come alive for me, such as when he described how it felt to see the first hydrogen bomb go off, and to meet with Klaus Fuchs, who leaked nuclear-weapon secrets to the Soviet Union. He also gave me a more personal connection with the founding fathers of my field, who for him had been mere mortals.

I showed him arguably my craziest paper ever, which explored the mathematical-universe idea that this book is leading up to, and he said he liked it. When the editor rejected it for being "too speculative" despite a positive referee report, he encouraged me to appeal the rejection, which worked. Later, we wrote an article together for *Scientific American* called "100 Years of Quantum Mysteries," in which we tried to explain both quantum parallel universes and decoherence in plain English. When I asked him whether he really believed in quantum parallel universes, he said, "I try to find time to believe in them Mondays, Wednesdays and Fridays."

I very rarely cry, but I did in 2008 when I learned that John Wheeler had died. He really touched me and inspired me, and at his memorial service, it was palpable how many others felt the same way. At the open mike during the reception afterward, when people who felt compelled spoke of him, I said a few words about how much he'd meant to me. That if I had to sum it up in only one word, it would be *inspiring*. Inspiring that someone so brilliant and famous could be so nice, "treating everyone with equal dignity," as another speaker aptly put it. And that he encouraged me to follow my heart and work on what I was really passionate about. And that the best testimony to how he had inspired people was to take a good look around the room and see how many amazing people had traveled from at least three continents to be there. The crowd felt like a veritable Who's Who of physics.

One afternoon when I was giving John a ride back home to Meadow Lakes, the retirement community where he lived, I excitedly started telling him about a totally crazy-sounding idea I'd just had, which I called "Quantum Suicide." I'd spent a lot of time wondering whether there was an experiment that could convince you that Everett's parallel universes were real, and had finally thought of one.

Surprisingly, this experiment requires only rather low-tech equipment that's readily available. However, it also requires you to be an unusually dedicated experimentalist, since it amounts to a repeated and faster version of Schrödinger's cat experiment—with you as the cat. The apparatus is a "quantum machine-gun," which fires depending on the outcome of a quantum measurement. Specifically, each time the gun is triggered, it places a particle in a superposition where it's equally in two states at once (spinning clockwise and counterclockwise, say), then measures the particle. If the particle is found to be in the first of the two states, the gun fires, otherwise it merely makes an audible click. The details of the trigger mechanism are irrelevant* as long as the time scale between the quantum measurement and the actual firing is much shorter than that characteristic of human perception, say, a hundredth of a second.

Now suppose that you start this quantum machine-gun in an automatic mode where it's triggered once every second. Regardless of whether you believe in Everett's parallel universes or not, you'll predict that you'll hear a seemingly random sequence of shots and duds such as *bang-click-bang-bang-bang-click-click-bang-click-click*. Suddenly, you do something radical: you place your head in front of the gun barrel and wait. What do you expect to perceive next? That depends on whether Everett's parallel universes are real or not! If not, then there's only one outcome of each quantum measurement, so you'll definitely be either dead or alive after the first second, with 50% probability for each. So you'd expect to perceive perhaps a click or two if you're moderately lucky, then "game over," nothing at all. The probability that you'll survive n seconds is $1/2^n$, so your chance of lasting as long as a

---

* For example, the particle could be a silver atom that has its spin measured by a so-called Stern-Gerlach apparatus, or it could be a photon that either does or doesn't pass through a half-silvered mirror.

minute is less than one in a quintillion (10^{-18}). If Everett's quantum parallel universes *are* real, on the other hand, there will be two parallel universes after the first second: one where you're alive and one where you're dead and there's blood all over the place. In other words, there's exactly one copy of you having perceptions both before and after the trigger event, and since it occurred too fast to notice, the prediction is that you'll hear *click* with 100% certainty. Wait a little longer, and you'll find this quite striking: as soon as you put your head in the firing line, the seemingly random sequence of bangs and clicks gives way to just *click-click-click-click-click-click-click*, etc. After ten clicks, you conclude that you've ruled out wavefunction collapse with 99.9% confidence, in the sense that if wavefunction collapse really happened, then the probability of being dead by now would exceed 99.9%. After a minute, you'll give it only a one-in-a-quintillion chance that Everett is wrong. To allay any concerns that the quantum machine-gun is broken, you remove your head from the firing line, and find that it, as if by magic, reverts back to firing intermittently.

If you're now convinced that Everett is right and bring a friend to witness your experiment, then there's a twist, however. Whereas you stay alive in only one parallel universe, she remains present in all of them, and typically sees you die after a few seconds. So the only thing you might succeed in convincing her of is that you were a mad scientist.

John found this interesting. I said I thought that many physicists would undoubtedly rejoice if an omniscient genie appeared at their deathbed, and as a reward for lifelong curiosity granted them the answer to a physics question of their choice. But would they be as happy if the genie forbade them from telling anybody else? Perhaps the greatest irony of quantum mechanics is that if Everett was right, then the situation is quite analogous if, once you feel ready to die, you repeatedly attempt quantum suicide: you might experimentally convince yourself that the quantum parallel universes are real,* but you can never convince anyone else!

Well, you could of course convince your friends if you made the suicide experiment a collective one, say, by connecting the quantum trigger to a nuclear bomb, so that you'd end up only with parallel universes where you and your friends were all alive or all dead. But they probably wouldn't be your friends afterward.

---

* The British philosopher Paul Almond has an interesting counterargument to this claim, which I'll tell you about in Chapter 11.

Quantum Immortality?

After I published a paper about the quantum-suicide idea, *New Scientist* and *The Guardian* ran articles about it, which generated a fair bit of attention, and it's been fun for me to see the idea subsequently appearing in various science-fiction stories. As I mentioned earlier, many people tend to have similar ideas when the time is ripe, and sure enough, I later discovered that other people had thought along similar lines previously, perhaps starting with the Austrian mathematician Hans Moravec, who mentioned the idea in his 1988 artificial intelligence book *Mind Children*. As opposed to my earlier rediscoveries, however, I felt that this one actually had some impact, by helping get the idea more widely known.

I soon got deluged with interesting email questions about quantum suicide that got me wondering more about the implications. Here's my favorite one: can you think of all potentially lethal events in nature as quantum-suicide experiments, so that you should expect subjective immortality? You can answer this question with a simple experiment: wait and see! If one day, after a long sequence of seemingly unlikely coincidences, you find yourself to be the oldest living person on Earth, then that pretty much settles it! Note that you don't expect to see *other* people get abnormally old, just as you don't expect to see other people last long if they try the quantum-suicide experiment.

So what do the laws of physics predict, assuming that Everett is right and the wavefunction never collapses? To be able to succeed, a successful quantum-suicide experiment needs to satisfy three criteria:

1. The random-number generator must be quantum, not classical (deterministic), so that you really enter a superposition of dead and alive.
2. It must kill you (or at least make you unconscious) on a time scale shorter than that on which you can become aware of the outcome of the quantum measurement—otherwise you'll have a very unhappy version of yourself for a second or more who knows s/he is about to die for sure, and the whole effect gets destroyed.
3. It must be virtually certain to really kill you, not just injure you.

Most accidents and common causes of death clearly don't satisfy all three criteria, suggesting that you won't feel immortal after all. In particular, regarding criterion 2, under normal circumstances dying isn't a binary thing where you're either dead or alive—rather, there's a whole continuum of states of progressively decreasing self-awareness. What makes the quantum suicide work is that it forces an abrupt transition. I suspect that when I get old, my brain cells will gradually give out (indeed, that's already started happening . . .), so that I keep feeling self-aware, but less and less so. This will make the final stage of death quite anticlimactic, sort of like when an amoeba croaks.

Criterion 3 places a limit on how long you can run your quantum-suicide experiment in practice before fluke events save your life. For example, my neighborhood gets a power failure on average once every few years, about once every $10^8 \approx 2^{27}$ seconds. This means that if my quantum machine-gun uses a power plug rather than battery power, I should expect to experience about twenty-seven straight clicks and then a power failure halting my experiment—because after that, there will be more parallel universes with me alive that have a disabled gun than a functioning gun. The longer I get the machine-gun to work, the crazier the flukes I should expect: for example, after to the tune of sixty-eight seconds' worth of straight clicks, I should expect my machine-gun to get struck by a meteorite. . . . In Douglas Adams's science-fiction spoof *The Hitchhiker's Guide to the Galaxy*, there's an "Infinite Improbability Drive" that makes you experience extremely unlikely events. Although such a device sounds like pure science fiction, it isn't: the quantum machine-gun effectively acts like one!

I find criterion 1 particularly interesting. Suppose your suicide device didn't rely on quantum randomness, but on something like a coin toss, where you could actually predict whether you'd get heads or tails in principle, just not in practice, because you haven't fully figured out how the coin was initially moving and done the math. Then if you started out with only one parallel universe, there would still only be one parallel universe after the first second, and you'd be either alive or dead depending on the initial position and motion of the coin, so you'd *not* feel subjectively immortal.

However, what if the Level I multiverse from Chapter 6 is real? Then there would be infinitely many parallel universes to start with that contained you in subjectively indistinguishable mental states, but with imperceptibly slight differences in the initial position and velocity of the coin. After one second, you'd be dead in half of those universes, but no matter how many times the experiment is repeated, there would always be universes where you never got shot. In other words, this sort of macabre randomized-suicide experiment can reveal the existence of not merely Level III (quantum) parallel universes, but also of parallel universes more generally.

I know. This stuff sounds seriously nuts. "Don't try this at home," as they say. Moreover, as I'll explain in Chapter 11, I've now become convinced that neither quantum suicide nor quantum immortality actually works, because they depend crucially on something that I don't think exists in nature: an infinitely divisible mathematical continuum. But who really knows? When one fateful day in the future, you think that your own life is about to end, remember this and don't say to yourself, *There's nothing left now*—because there might be. You might be about to discover firsthand that parallel universes really do exist.

Multiverses Unified

All animals are equal, but some animals are more equal than others.
—George Orwell, *Animal Farm*, 1945

I just couldn't get this nagging thought out of my mind: were the Level I and Level III multiverses somehow really one and the same? Could they somehow be unified, just as Maxwell had unified electricity and magnetism into electromagnetism, and Einstein had unified space and time into spacetime? On one hand, their natures seemed quite different: the Level I parallel universes from Chapter 6 are far away in our good old three-dimensional space, while the Level III parallel universes from this chapter can be right here as far as these three dimensions are concerned, but separated from us in Hilbert space, the abstract space with infinitely many dimensions where the wavefunction lives. On the other hand, the Level I and Level III multiverses have a lot in common. Jaume Garriga and Alex Vilenkin had written a paper showing that the Level I parallel universes that may have been created by cosmological inflation contain all the same sequences of events that Everett's quantum parallel universes do, and so had I. Figure 8.10 illustrates that if a quantum event causes two events to happen in quantum superposition, effectively splitting your

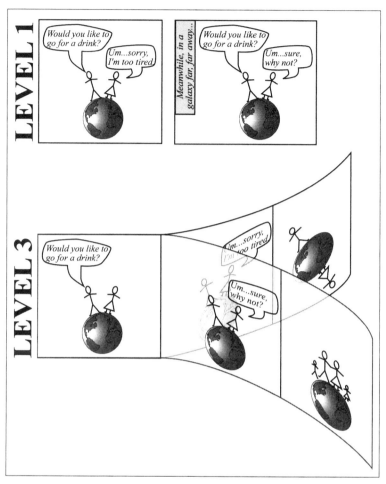

Figure 8.10: Comparison of Level I and Level III parallel universes. Whereas Level I parallel universes are far away in space, those of Level III are even right here, with quantum events causing classical reality to split and diverge into parallel storylines. Yet Level III adds no new storylines beyond levels I or II.

future into two parallel quantum branches, then the parallel quantum outcome that you're now unaware of is also occurring right here in your particular quantum branch—just really far away in space.

There was also another source of nagging: Anthony Aguirre. Anthony is one of my best friends, and our lives are parallel in many ways: we both try to balance our careers with two young sons, we're both obsessed with big questions, and together we've founded the Foundational Questions Institute, fqxi.org, a philanthropically funded organization that funds high-risk, high-reward physics research that conventional funding agencies shy away from. What was he nagging me about? "Are some parallel universes really more equal than others?" he'd ask.

What he was getting at was that the explanation I gave for quantum probabilities earlier in this chapter works great when you have outcomes with equal probability (like the quantum card whose chances of falling face-up and face-down were both 50%), but not when the probabilities were unequal. For example, suppose you start with the card tilted by an extremely small amount, so that the probability (square of the wavefunction) is 2/3 for it falling face-up and 1/3 for it falling face-down. Then Figure 8.2 would still look the same: there are still $2 \times 2 \times 2 \times 2 = 16$ outcomes after four trials, and the most typical outcome is the card falling face-up 50% of the time, not 2/3 of the time. The way that Everett saved the day and nonetheless managed to predict a probability of 2/3 from this was by arguing that some of these outcomes had a larger measure of existence than others and that, specifically, this measure of existence could be calculated as the square of the wavefunction. This worked, and many authors have since given more elaborate arguments for why squaring the wavefunction is the right thing to do, but Anthony convinced me that this was an ugly blemish on Everett's otherwise elegant argument. People often asked me if I believed that Everett's parallel universes were real. Answering, "Yeah, but . . . eh . . . hrm . . . some are more real than others" sounded really lame.

In March 2008, Anthony told me about an idea (which I'll explain in a moment) for a possible solution that his old Harvard professor David Layzer had suggested, and we spent two exciting hours in a Belmont café scribbling math symbols on the backs of napkins*—but in vain. We couldn't manage to make the math work. But I couldn't let go of the idea either. Two years later, I started obsessing about this again, and found a 1968 paper by the quantum-gravity theorist Jim Hartle that I felt contained another piece of the puzzle. But as I sat there in my Winchester apartment late in the evening of March 6, 2010, I just couldn't get the pieces to fit together. Frustrated, I decided to take a thinking walk through town. To my amazement, after five minutes in the wintry air, it finally clicked! I suddenly realized how to solve both

---

* I find it odd that there's so much talk about "back-of-the-envelope calculations" when, in my personal experience, most impromptu calculations are in fact done on napkins, despite their susceptibility to tearing and their generally inferior quality as a graphological medium.

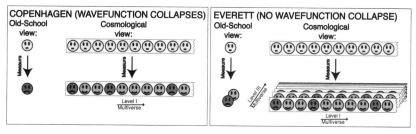

Figure 8.11: How the Level I and Level III multiverses are unified. Each circle represents a planet where you've bet money that your quantum card will land face-up. Before the measurement, you're in a neutral mood; afterward, you're either happy because you won or sad because you lost. The card starts just ever so slightly tilted, so that you expect to win with probability 2/3. These planets are typically very far apart, say, a googolplex meters in various directions, but I've drawn them side by side in a straight line to illustrate the key points.

problems in one fell swoop: unify the two multiverse levels and understand the unequal probabilities. It kept me up until about three a.m., and consumed me all of the next day in that wonderful trancelike way that you have to experience to fully understand. I felt that it was one of my most exciting clicks since rediscovering decoherence nineteen years earlier, and couldn't let go until I'd typed up a four-page skeleton paper for Anthony.

Figure 8.11 illustrates the key idea. Suppose that you're about to perform the Quantum Cards experiment with the card slightly tilted, so that you expect to see it fall face-up and win $100 with a probability of 2/3. In the old-school view (shown to the left in each rectangle in Figure 8.11), there's one copy of you to start with and either one or two copies after the experiment, depending on whether the wavefunction collapses or not: if the Copenhagen interpretation is correct, there will be one definite outcome generated at random, while if Everett is correct, there will be two parallel universes, each containing a single copy of you: one where you're happy about winning and one where you're sad.

Now let's instead assume that the Level I multiverse from Chapter 6 exists, as modern cosmology suggests. This means that an infinite number of indistinguishable copies of you are about to perform the exact same experiment on other planets far, far away in space, illustrated by the strip of neutral faces in the figure. In my calculation, I applied the Schrödinger equation to the wavefunction describing the entire collection of particles making up all these copies of you and your experiment.

What ends up happening? If the wavefunction collapses, then you get a single random outcome for the infinite space (the Level I multiverse), such that you're happy on 2/3 of the planets and sad on 1/3 of them—that's not surprising. If Everett is right in that there's no collapse, then you end up with the entire infinite space in a quantum superposition of different states, each of which has you happy on some planets and sad on others. Now here's the kicker: all of those states of space turn out to be indistinguishable from one another, with you being happy on exactly 2/3 of the infinitely many planets! Any finite sequence of planets with happy and sad outcomes in one of those states can be found somewhere else in space in each of the other states. You might think that there should also be states of space that are different, say, one where you're happy on every single planet. However, using the Schrödinger equation and the mathematics of Hilbert space, I was able to prove that the wavefunction you actually get is equal to simply a superposition of infinitely many indistinguishable states. Anthony and I found this striking for several reasons.

First of all, the great debate about whether the wavefunction collapses ends in a grand anticlimax: it simply doesn't matter! Figure 8.11 illustrates that regardless of whether Everett is right or not, you're happy on 2/3 of the planets. Indeed, both sides of the collapse debate emerge a bit bruised. The Copenhagen interpretation introduced this controversial collapse business to get rid of pesky parallel universes and obtain a unique outcome, yet you can see in the figure that this no longer helps: even with collapse, you still end up with parallel universes with both outcomes. The Everett interpretation had Level III (quantum) parallel universes as its hallmark, but you can see in the figure that you can safely ignore them, because they're all indistinguishable. In this sense, the Level I and Level III multiverses are unified: as long as you have an infinite space with a Level I multiverse, you can ignore all its Level III parallel universes, since they're in practice all just identical copies. Perhaps Level III can be unified with Level II as well, but we haven't yet been able to prove that.

Second, Figure 8.11 illustrates the origin of unequal probabilities by bringing Everett's many worlds to our good old three-dimensional space: the different outcomes aren't merely happening somewhere else in this hard-to-grasp mathematical Hilbert space, but also far away in our own space that we study with telescopes. Here, the key point is that after your card has fallen down but before you've opened your eyes and looked at it, you have no way of knowing which of your many copies you are, since they all feel subjectively indistinguishable up to this point. You therefore have to consider yourself to be a random member of this group of copies. Since you know that 2/3 of them will see the card face-up when they open their eyes, you'll consider what you see as random, with a 2/3 probability of seeing face-up. This is analogous to the way that French noblemen originally introduced the notion of probability to optimize their gambling strategies: if, in a game, all you know is that you'll be in one of many equally likely situations (corresponding to different ways that your cards could have been dealt, say), then you say that your probability of winning is simply the fraction of all the situations where you win.

Third, this allowed us to propose what we called the *cosmological interpretation of quantum mechanics*. Here we interpret the wavefunction for an object as describing not some funky imaginary ensemble of possibilities for what the object might be doing, but rather the actual spatial collection of identical copies of the object that exist in our infinite space. Moreover, quantum uncertainty that you experience simply reflects your inability to self-locate in the Level I multiverse, i.e., to know which of your infinitely many copies throughout space is the one having your subjective perceptions.

In some fields, coauthors on a paper traditionally list their names in alphabetical order. In cosmology, however, we usually let the author ordering reflect who has contributed most to the paper. In most cases, it's quite obvious who's done most of the work, but this time it was unusually hard to say. By the time we were ready to submit this paper for publication, both Anthony and I had worked quite hard on it and made arguably equally important contributions. We had an amusing phone conversation about this where we both lauded the other one's contributions while stubbornly holding back from offering each other to go first. I finally suggested a solution that we both liked: deciding the author order with a quantum random-number generator. In this particular universe, he's the first author (http://arxiv.org/pdf/1008.1066.pdf), but if our paper is correct, then I'm first author not only in half of the Level III parallel universes where we used this procedure, but in half of the Level I parallel universes as well.

In 2010, Alex Vilenkin invited me to give a talk about this paper at Tufts, and just as in the opening of Chapter 5, Alan Guth was in the audience. I kept getting flashbacks from thirteen years earlier of Alan's head slumping toward his chest, and tried to mentally brace for the inevitable, since I couldn't recall a single talk ever where he hadn't fallen asleep. Then a miracle happened, which felt like the best endorsement that our paper could ever have received, and like the pinnacle of my scientific career: Alan stayed awake through my entire talk!

Shifting Views: Many Worlds or Many Words?

So what should you make of all this quantum business? Should you believe in wavefunction collapse or in quantum parallel universes? Although quantum mechanics is arguably the most successful physical theory ever invented, the century-old debate about how it fits into a coherent picture of physical reality shows no sign of abating. A veritable zoo of interpretations of what's going on has cropped up over the years, including the ensemble, Copenhagen, instrumental, hydrodynamic, consciousness, Bohm, quantum logic, Many-Worlds, stochastic mechanics, many-minds, consistent histories, objective collapse, transactional, modal, existential, relational, Montevideo and cosmological interpretations.* Moreover, different proponents of a particular interpretation often disagree about its detailed definition. Indeed, there isn't even consensus on which ones should be called interpretations. . . .

You might figure that since the experts are still arguing about this about a century after quantum mechanics was discovered, with no consensus in sight, they'll probably be arguing for another century as well. However, the whole context of the debate has changed in three important ways, involving theory, cosmology and technology, causing sociological changes that I find quite interesting.

First of all, we've seen how the theoretical discoveries by Everett, Zeh and others have shown that even if you drop the controversial wavefunction-collapse postulate and keep only the simple bare-bones quantum mechanics where the Schrödinger equation always holds, then you'll still subjectively *feel* like the wavefunction collapses when you make observations, obeying all the right probability rules, and you'll remain blissfully unaware of any quantum parallel universes.

Second, the cosmology discoveries that we covered in Chapters 5 and 6 have suggested that we're stuck with parallel universes even if Everett is wrong. Moreover, we saw earlier how these Level I parallel universes elegantly merge with the quantum ones.

* You'll find references for all these interpretations in http://arxiv.org/abs/1008.1066.

Third, support for the idea that quantum gravity somehow collapses the wavefunction has itself collapsed, because of a string-theory breakthrough known as the AdS/CFT correspondence. The details of this acronym don't matter for our discussion: the key point is that a mathematical transformation has been found showing that certain quantum-field theories with gravity can be reinterpreted as other quantum-field theories without gravity. Gravity clearly isn't causing wavefunction collapse if the very presence of gravity is merely a matter of interpretation.

Fourth, ever more accurate experiments have ruled out many attempts to explain away the quantum weirdness. For example, could the apparent quantum randomness be replaced by some kind of unknown quantity stored inside particles, so-called hidden variables? The Irish physicist John Bell showed that, in this case, quantities that could be measured in certain difficult experiments would inevitably disagree with the standard quantum predictions. After many years, technology finally improved to the point that these experiments could be done, and the hidden-variable explanation was ruled out.

Could it be that there's a small correction to the Schrödinger equation that we haven't discovered yet, but which causes quantum superpositions to break down for sufficiently large objects? Back when quantum mechanics was born, there were indeed many physicists who believed that quantum mechanics would prove to work only on the atomic scale. Well, no longer! The simple double-slit interference experiment (Figure 7.7), hailed by Feynman as the mother of all quantum effects, has been successfully repeated for objects larger than individual elementary particles: atoms, small molecules and even the soccer ball–shaped carbon-60 "Bucky Ball" molecule. Back in grad school, I asked my classmate Keith Schwab if he thought one could experimentally demonstrate that a macroscopic object was in two places at once. Amazingly, two decades later, he runs his own lab at Caltech working on doing exactly this, with a metal rod containing many billions of atoms. Indeed, his Santa Barbara colleague Andrew Cleland has already done this with a metal paddle large enough to see with the naked eye. Anton Zeilinger's group in Vienna has even started discussing doing it with a virus. If we imagine, as a thought experiment, that this virus has some primitive kind of consciousness, then the Many-Worlds interpretation seems unavoidable: extrapolation to superpositions involving other sentient beings such as humans would then be merely a quantitative rather than a qualitative one. Zeilinger's group has also demonstrated that counterintuitive quantum properties of photons persisted while they traveled 89 kilometers through space—hardly a microscopic distance. So I feel that the experimental verdict is in: the world *is* weird, and we just have to learn to live with it.

Indeed, many people have warmed up to quantum weirdness, for reasons that aren't philosophical but financial: this very weirdness may offer useful new technologies. According to a recent estimate, more than a quarter of the U.S. gross national product is now based on inventions made possible by quantum mechanics, from lasers to computer chips. Indeed, fledgling technologies such as quantum cryptography and quantum computing explicitly exploit the Level III multiverse and work only if the wavefunction doesn't collapse.

These breakthroughs in theory, cosmology and technology have caused a notable shift in views. When I give talks, I like to know what the people in my audience think. When I asked them which interpretation of quantum mechanics they identified most closely with, here's what they said, first at a 1997 quantum-mechanics conference at UMBC in Maryland, then at a 2010 talk I gave at the Harvard Physics Department:

Interpretation	Maryland 1997	Harvard 2010
Copenhagen	13	0
Everett	8	16
Bohm	4	0
Consistent histories	4	2
Modified dynamics	1	1
None of the above/undecided	18	16
Total votes	48	35

Although these polls were highly informal and unscientific, and clearly don't survey a representative sample of all physicists, they nonetheless indicate a rather striking shift in opinion: after reigning supreme for decades, the Copenhagen interpretation saw its approval rating drop below 30% in 1997 and to 0% (!) in 2010. In contrast, after being proposed in 1957 and going virtually unnoticed for about a decade, Everett's Many-Worlds interpretation survived twenty-five years of fierce criticism and occasional ridicule to top the 2010 poll. It's also worth noting that there's a large fraction of undecided voters, suggesting that the quantum-mechanics debate is still in full swing.

The Austrian animal behaviorist Konrad Lorenz mused that important scientific discoveries go through three phases: first they're completely ignored, then they're violently attacked, and finally they're brushed aside as well known. The poll suggests that after spending the 1960s in phase 1, Everett's parallel universes have now shifted to somewhere between phase 2 and phase 3.

To me, this shift means that it's time to update the quantum textbooks to mention decoherence (many still don't) and to make clear that the Copenhagen interpretation is better thought of as the *Copenhagen approximation:* even though the wavefunction probably doesn't collapse, it's a very useful approximation to do the calculations as if it does collapse when you make an observation.

All physics theories have two parts: mathematical equations and words that tell us what they mean. Although above I rattled off the names of over a dozen interpretations of quantum mechanics, many of them differ only in the "words" part. To me, the most interesting question is what the math part is, and specifically whether the simplest math of all (just the Schrödinger equation with no exceptions) is enough. So far, there isn't a shred of experimental evidence to the contrary, yet many of the quantum interpretations add a lengthy "words" part to talk away the parallel universes. So when you pick your own favorite interpretation, it really comes down to what bothers you most: a profusion of worlds or a profusion of words. When the time came to write a paper for the proceedings of that 1997 Maryland conference, I called it "The Interpretation of Quantum Mechanics: Many Worlds or Many Words?" in an attempt to tease some of my colleagues. I was expecting to get flamed with plenty of hate mail from them as a result, but I have to hand it to them: even though I think they're wrong about quantum mechanics, they do have a good sense of humor. . . .

In Chapter 7, we talked about how everything is made of particles, and how particles are in a sense purely mathematical objects. In this chapter, we've seen that in quantum mechanics, there's something that is arguably even more fundamental: the wavefunction and the infinite-dimensional place called Hilbert space where it lives. The particles can be created and destroyed, and can be in several places at once. In contrast, there is, was and always will be only one wavefunction, and it's the object that moves through Hilbert space as dictated by Schrödinger's equation. But if the ultimate physical reality corresponds to the wavefunction, then what sort of beast is a wavefunction? What's it made of? What's Hilbert space made of? As far as we know, nothing: they seem to be purely mathematical objects! So once again, as we attempt to dig deeper in search of the underlying physical reality, we've found a hint that the bedrock itself is purely mathematical. We'll take this idea much further in Chapter 10.

THE BOTTOM LINE

- In the mathematically simplest quantum theory, there's something more fundamental than our three-dimensional space and the particles within it: the wavefunction and the infinite-dimensional place called Hilbert space where it lives.
- In this theory, particles can be created and destroyed, and can be in several places at once, but there is, was, and always will be only one wavefunction, moving through Hilbert space as determined by the Schrödinger equation.
- This mathematically simplest quantum theory, where the Schrödinger equation always rules, predicts the existence of parallel universes where you live out countless variations of your life.
- It also implies that quantum randomness is an illusion, caused by quantum cloning of you.
- There's nothing quantum about apparent randomness, which happens even if you're classically cloned.
- This mathematically simplest quantum theory also predicts a censorship effect called decoherence, which hides most such weirdness from us, mimicking wavefunction collapse.
- Decoherence happens constantly in your brain, debunking popular suggestions about "quantum consciousness."
- This quantum multiverse is unified with the spatial multiverse from Chapter 6, so that a wavefunction for a system describes its infinite copies throughout space, and quantum uncertainty reflects your ignorance about which particular copy you're observing.
- If we live in an infinite uniform space as in the cosmological standard model, then it doesn't matter whether the wavefunction ultimately collapses: all of Everett's many worlds are indistinguishable, and collapse doesn't prevent all quantum outcomes from actually happening.
- The quantum multiverse arguably makes you subjectively immortal, in which case you'll eventually find yourself the oldest person on the planet, and this may not even require quantum mechanics, merely the Level I multiverse in an infinite space. But I don't think so, as I'll explain in Chapter 11.
- The wavefunction and Hilbert space, which constitute arguably the most fundamental physical reality, are purely mathematical objects.

Part Three

STEPPING BACK

9

Internal Reality, External Reality and Consensus Reality

Sweet exists by convention, bitter by convention, color by convention; atoms and void [alone] exist in reality.

—Democritus, ca. 400 B.C.

"Nooooo! My suitcase!"

They were already boarding my flight from Boston to Philadelphia, where I was supposed to help with a BBC documentary about Hugh Everett, when I realized that my hand wasn't holding a suitcase. I ran back to the security checkpoint.

"Did someone just forget a black roll-on bag here?"

"No," said the guard.

"But there it is—that's my suitcase, right there!"

"That's not a black suitcase," said the guard. "That's a teal suitcase."

Until then, I'd never realized how color-blind I was, and it was quite humbling to realize that many assumptions I'd previously made about reality—and my wardrobe—were dead wrong. How could I ever trust what my senses told me about the outside world? And if I couldn't, then how could I hope to ever know anything with certainty about the external reality? After all, everything I know about the outside world and my untrustworthy senses, I've learned from my senses. This puts me on the same shaky epistemological footing as a prisoner who's spent his whole life in solitary confinement, whose only information about the outside world and his untrustworthy prison guard is what his prison guard has told him. More generally, how can I trust what my conscious perceptions tell me about the world if I don't understand how my mind works?

This basic dilemma has been eloquently explored by philosophers throughout the ages, including titans such as Plato, René Descartes, David Hume and Immanuel Kant. Socrates said: "The only true wisdom is in knowing you know nothing." So how can we make further progress in our quest to understand reality?

So far in this book, we've taken a physics approach to exploring our external physical reality, zooming out to the transgalactic macrocosm and zooming in to the subatomic microcosm, attempting to understand things in terms of their basic building blocks such as elementary particles. However, all we have direct knowledge of are instead *qualia*, the basic building blocks of our conscious perception,* exemplified by the redness of a rose, the sound of a cymbal, the smell of a steak, the taste of a tangerine or the pain of a pinprick. So don't we also need to understand consciousness before we can fully understand physics? I used to answer "yes," thinking that we could never figure out the elusive "theory of everything" for our external physical reality without first understanding the distorting mental lens through which we perceive it. But I've changed my mind, and in this brief interlude chapter, I want to tell you why.

External Reality and Internal Reality

Perhaps you're thinking, *Okay, Max, but I'm not color-blind. And I'm looking at the external reality right now with my own eyes, and I'd have to be paranoid to think it's not the way it looks.* But please try these simple experiments:

> **Experiment 1:** Turn your head from left to right a few times.
> **Experiment 2:** Move your eyes from left to right a few times, without moving your head.

Did you notice how the first time, the external reality appeared to rotate, and the second time, it appeared to stay still, even though your eyeballs rotated both times? This proves that what your mind's eye is looking at isn't the external reality, but a reality model stored in your brain! If you looked at the image recorded by a rotating video camera, you'd clearly see it move as it did in Experiment 1. But your eyes are a form of biological video camera, so Experiment 2 shows that your consciousness isn't directly perceiving the images formed on their retinas. Rather, as neuroscientists have now studied in great detail, the information recorded by your retinas gets processed in highly complex ways and is used to continually update an elaborate model of the outside

---

* For introductions to the vast literature on consciousness by psychologists, neuroscientists, philosophers and others, I recommend the books about the mind in the "Suggestions for Further Reading" section.

world that's stored in your brain. Take another look in front of you, and you'll see that, thanks to this advanced information processing, your reality model is three-dimensional even though the raw images from your retinas are two-dimensional.

I don't have a light switch near my bed, so I'll often take a good look at my bedroom and all the obstacles littering the floor, then turn off the light and walk to my bed. Try it yourself: put down this book, stand up, look around, and then walk a few steps with your eyes closed. Can you "see"/"feel" the objects in the room moving relative to you? That's your reality model being updated, this time using information from your leg movements rather than from your eyes. Your brain continuously updates its reality model using any useful information it can get hold of, including sound, touch, smell and taste.

Let's call this reality model your *internal reality*, because it's the way you subjectively perceive the external reality from the internal vantage point of your mind. This reality is internal also in the sense that it exists only internally to you: your mind feels as if it's looking at the outside world, while it's actually looking only at a reality model inside your head—which in turn is continually tracking what's outside your brain via elaborate but automatic processes that you're not consciously aware of.

It's absolutely crucial that we don't conflate this internal reality with the external reality that it's tracking, because the two are very different. My brain's internal reality is like the dashboard of my car: a convenient summary of the most useful information. Just as my car's dashboard tells me my speed, fuel level, motor temperature, and other things useful for a driver to be aware of, my brain's dashboard/reality model tells me my speed and position, my hunger level, the air temperature, highlights of my surroundings and other things useful for the operator of a human body to be aware of.

The Truth, the Whole Truth and Nothing but the Truth

Once my car's dashboard malfunctioned and sent me to the garage with its "CHECK ENGINE" indicator illuminated even though nothing was wrong. Similarly, there are many ways in which a person's reality model can malfunction and differ from the true external reality, giving rise to illusions (incorrect perceptions of things that do exist in the external reality), omissions (nonperception of things that do exist in the external reality) and hallucinations (perceptions of things that don't exist in the external reality). If we swear under oath to tell the truth, the whole truth and nothing but the truth, we should be aware that our perceptions might violate all three with illusions, omissions and hallucinations, respectively.

So metaphorically speaking, the "CHECK ENGINE" incident was my car hallucinating—or experiencing phantom pain. I recently discovered that my car also suffers from an illusion: based on its speedometer reading, it thinks it's always driving two miles per hour faster than it really is. That's not bad compared to the vast list of human illusions that cognitive scientists have discovered, which afflict all our senses and distort our internal reality. If your version of this figure is in color rather than black and white, you'll probably see the lower dot in the left panel as orange and the upper dot as somewhat brown. Figure 9.1 shows two examples of optical illusions, where our visual system creates an internal reality different from the external reality. In the external reality, the light from both of them has identical properties, with a wavelength around 600 nanometers. If a spotlight beamed out such light, it would be orange light. What about brown? Have you ever seen a spotlight or a laser pointer produce a brown beam? Well, you never will, because there's no such thing as brown light! The color brown doesn't exist in the external reality, but only in your internal reality: it's simply what you perceive when seeing dim orange light against a darker background.

For fun, I sometimes compare how the same news story is reported online by MSNBC, FOX News, the BBC, Al Jazeera, *Pravda* and elsewhere. I find that when it comes to telling the truth, the whole truth,

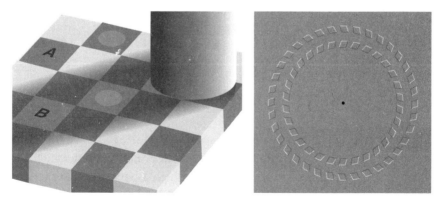

Figure 9.1: Optical illusions. In the left panel, squares A and B have the same shade of gray, and the two dots have identical color. In the right panel, look at the black dot while moving your head forward and backward, and see the circles move.

and nothing but the truth, it's the second part that accounts for most of the differences in how they portray reality: what they omit. I think the same holds for our senses: although they can produce hallucinations and illusions, it's their omissions that account for most of the discrepancy between the internal and external realities. My visual system omitted the information that distinguishes between black and teal suitcases, but even if you're not color-blind, you're missing out on the vast majority of the information that light carries. When I was taught in elementary school that all colors of light can be made up by mixing three primary colors red, green, and blue, I thought that this number three told us something fundamental about the external reality. But I was wrong: it teaches us only about the omissions of our visual system. Specifically, it tells us that our retina has three kinds of cone cells, which take the thousands of numbers that can be measured in a spectrum of light (see Figure 2.5 in Chapter 2) and keeps only three numbers, corresponding to the average light intensity across three broad ranges of wavelengths.

Moreover, wavelengths of light outside of the narrow range 400–700 nanometers go completely undetected by our visual system, and it came as quite a shock when human-built detectors revealed that our external reality was vastly richer than we'd realized, teeming with radio waves, microwaves, x-rays, and gamma rays. And vision isn't the only one of our senses that's guilty of omissions: we can't hear the ultrasound chirping of mice, bats and dolphins; we're oblivious to most faint scents that dominate the olfactory inner reality of dogs, and so on. Although some animal species capture more visual, auditory, olfactory, gustatory or other sensory information than we humans do, they're all unaware of the subatomic realm, the galaxy-spangled cosmos, and the dark energy and dark matter that, as we saw in Chapter 4, makes up 96% of our external reality.

Consensus Reality

In the first two parts of this book, we've seen how our physical world can be remarkably well described by mathematical equations, fueling the hope that one day equations can be found for a "theory of everything," perfectly describing our external reality on all scales. The ultimate triumph of physics would be to start with the external reality from the "bird perspective" of a mathematician studying these equations (which are ideally simple enough to fit on her T-shirt) and to derive from them her internal reality, the way she subjectively perceives it from her "frog perspective" inside the external reality. To accomplish this would clearly require a detailed understanding of how consciousness works, including illusions, omissions, hallucinations and other complications.

However, between the external reality and the internal reality, there's also a third and intermediate *consensus reality*, as illustrated in Figure 9.2. This is the version of reality that we life-forms here on Earth all agree on: the 3-D positions and motions of macroscopic objects, and other everyday attributes of the world for which we have a *shared description* in terms of familiar concepts from classical physics. Table 9.1 summarizes these reality descriptions and perspectives and how they're interrelated.

Each of us has our own personal inner reality, perceived from the subjective perspective of our own position, orientation and state of mind, and distorted by our personal cognitive biases: in your inner reality, dreams are real and the world turns upside down when you stand on your head. In contrast, the consensus reality is shared. When you give your friend driving directions to your place, you do your best to transform your description from one involving subjective concepts from your inner reality (such as "here" and "in the direction I'm facing") to shared concepts from the consensus reality (such as "on 70 Vassar

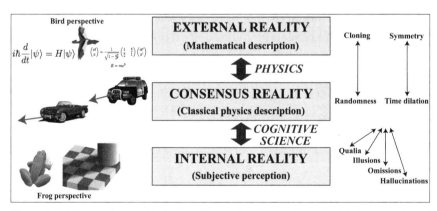

Figure 9.2: We can view reality in three interrelated ways: from the bird perspective of a mathematician studying the equations describing it, from the subjective frog perspective of a self-aware observer in it, and from the intermediate consensus perspective in which we usually describe it to one another (as classical objects moving in 3-D, say). The ultimate quest for understanding splits conveniently into two parts, which can be tackled separately: for physics to reveal how the external reality relates to the consensus reality (including complications such as observer cloning appearing as randomness and rapid motion appearing as time slowdown) and for cognitive science to reveal how the consensus reality relates to the internal reality (including qualia and complications such as illusions, omissions and hallucinations).

Street" and "north"). Since we scientists need to be precise and quantitative when we refer to our shared consensus reality, we try extra-hard to be objective: we say that light has a "600-nanometer wavelength" instead of "orange color" and that something has "$CH_3COOC_5H_{11}$ molecules" instead of "banana flavor." The consensus reality isn't free from some shared illusions relative to the external reality, as we'll elaborate on below: for example, cats, bats and robots also experience the same quantum randomness and relativistic time dilation. However, it's by definition free from illusions that are unique to biological minds, and therefore decouples from the issue of how our human consciousness works. The internal reality may feel teal deficient to me, black and white to a seal, iridescent to a bird seeing four primary colors, and still more different to a bee seeing polarized light, a bat using sonar, a blind person with keener touch and hearing, or the latest robotic vacuum cleaner, but we all agree on whether the door is open.

This is why I've changed my mind: although understanding the detailed nature of human consciousness is a fascinating challenge in its own right, it's *not* necessary for a fundamental theory of physics, which need "only" derive the consensus reality from its equations. In other words, what Douglas Adams called "the ultimate question of life, the universe and everything" splits cleanly into two parts that can be tackled separately: the challenge for physics is deriving the consensus reality from the external reality, and the challenge for cognitive science is to derive the internal reality from the consensus reality. These are two great challenges for the third millennium. They're each daunting in their own right, and I'm relieved that we need not solve them simultaneously.

Reality Cheat Sheet	
External reality	The physical world, which I believe would exist even if we humans didn't
Consensus reality	The shared description of the physical world that self-aware observers agree on
Internal reality	The way you subjectively perceive the external reality
Reality model	Your brain's model of the external reality; this is the internal reality that you perceive
Bird perspective	Your perspective on the external reality when studying the abstract mathematical equations that describe it
Frog perspective	Your subjective perspective of the physical world (your internal reality)

Table 9.1: Key terms introduced in this chapter that we'll use later on

Physics: Linking External Reality to Consensus Reality

We've seen above that the consensus reality is quite different from the internal reality, and that the challenge of linking the two is as hard as understanding consciousness. However, as we've seen in the earlier parts of this book, the consensus reality is also quite different from the external reality, making it crucial not to conflate the two. Indeed, in my opinion, the history of modern physics shows that in several of the greatest breakthroughs, the hardest part wasn't doing the math, but understanding how these two realities are related.

When Einstein discovered special relativity in 1905, many of the key equations had already been written down by Hendrik Lorentz and others. However, what required Einstein's genius was figuring out the relation between the mathematics and the measurements. He realized that the lengths and durations appearing in the mathematical description of the external reality differ from those measured in the consensus reality, and that the difference depends on motion: if an airplane flies over a group of people, then in their consensus reality, it will be shorter than before it took off, and its onboard clocks will run slower.*

When Einstein discovered general relativity a decade later, Bernhard Riemann and others had already developed key parts of the mathematical formalism. However, the crowning achievement was again so difficult that it required Einstein's insight: understanding that curved space in the mathematical description of the external reality corresponded to gravitation in the consensus reality. To appreciate how hard this was, imagine that on the eve of his death, Isaac Newton was approached by a genie who granted him one last wish. After some contemplation, Newton made up his mind:

"Please tell me what the state-of-the-art equations of gravity will be in three hundred years."

The genie scribbled down the complete equations of general relativity on a sheet of paper, and being a kind genie, it also explained how

---

* Einstein realized that whereas observers sharing the same location and motion will share a common consensus reality, two groups moving relative to one another will have different consensus realities. In other words, there can be many different consensus realities, but their differences are explained by physical effects that have nothing to do with consciousness or the internal structure of the observers.

to translate them into the old-fashioned mathematical notation of the time. Would it be obvious to Newton how to interpret this as a generalization of his own theory?

The difficulty of linking external reality to consensus reality reached a new record high with the discovery of quantum mechanics, manifested in the fact that we physicists still argue about how to interpret the theory today, about a century after its inception. As we saw in Chapter 8, the external reality is described by a Hilbert space where a wavefunction changes deterministically over time, whereas the consensus reality is one where things happen seemingly at random, with probability distributions that can be computed to great accuracy from the wavefunction. It took over thirty years from the birth of quantum mechanics before Everett showed how these two realities could be reconciled, and the world had to wait another decade for the discovery of decoherence, which was crucial for reconciling the presence of macrosuperpositions in the external reality with their absence in the consensus reality.

Today, the grand challenge of theoretical physics is unifying quantum mechanics with gravitation. Based on this historical progression of examples, I predict that the correct mathematical theory of quantum gravity will break all previous records in being difficult to interpret. Suppose that on the eve of the next quantum-gravity conference, our friend the genie broke into the lecture hall and scribbled the equations of the ultimate theory on a blackboard. Would any of the participants realize what was being erased the next morning? I doubt it!

In summary, our quest to understand reality splits into two parts that can be tackled separately: the grand challenge for cognitive science is to link our consensus reality with our internal reality, and the grand challenge for physics is to link our consensus reality with our external reality. We've seen that, although the former challenge is daunting, so is the latter. Our consensus reality appears to have impenetrably solid and stationary objects, but all except a quadrillionth of the volume of a rock is empty space between particles in restless schizophrenic vibration. Our consensus reality feels like a three-dimensional stage where events unfold over time, but as we'll explore in Chapter 11, Einstein's work suggests that change is an illusion, time being merely the fourth dimension of an unchanging spacetime that just is, never created and never destroyed, containing our cosmic history as a DVD contains a movie. The quantum world feels random, but as we saw in the last chapter, Everett's work suggests that randomness, too, is an illusion, being simply the way our minds feel when cloned into diverging parallel universes. The quantum-gravity world feels—well, here we physicists still have a loooooooong way to go.

In the remainder of this book, we're going to focus on the physics quest, and push it to its logical extreme: given what we know about our consensus reality, what's the external reality like? What's its ultimate nature?

THE BOTTOM LINE

- I've argued that, although there's only one true reality, there are several complementary perspectives on it.
- In the internal reality of your mind, the only information you have about the external reality is the small sample transmitted through your senses.
- This information is distorted in many ways, and arguably tells you as much about how your senses and your brain work as it tells you about the external reality.
- The mathematical description of the external reality that theoretical physics has uncovered appears very different from the way we perceive this external reality.
- Midway between the internal and external realities lies the "consensus reality," the shared description of the physical world that all self-aware observers agree on.
- This cleanly splits what Douglas Adams jocularly called "the ultimate question of life, the universe and everything" into two parts that can be tackled separately: the challenge for the physical sciences is deriving the consensus reality from the external reality, and the challenge for the cognitive sciences is to derive the internal reality from the consensus reality.
- The rest of this book is focused on the first of these two challenges.

10

Physical Reality and Mathematical Reality

Philosophy is written in this grand book, the universe, which stands continually open to our gaze. But the book cannot be understood unless one first learns to comprehend the language and read the characters in which it is written. It is written in the language of mathematics, and its characters are triangles, circles, and other geometric figures without which it is humanly impossible to understand a single word of it; without these one is wandering in a dark labyrinth.
—Galileo Galilei, *The Assayer,* 1623

The enormous usefulness of mathematics in the natural sciences is something bordering on the mysterious and . . . there is no rational explanation for it.
—Eugene Wigner, 1960

Whoa! It's Friday morning in Princeton, and I've just finished reading emails about a book project, a broken oven and a quantum-suicide debate, when I find this gem in my inbox from a senior professor I know:

Date: *December 4, 1998, 7:17:42 EST*
Subject: *Not an easy e-mail to write . . .*

Dear Max,

. . . . your crackpot papers are not helping you. First, by submitting them to good journals and being unlucky so that they get published, you remove the "funny" side of them. . . . I am the Editor of the leading journal . . . and your paper would have never passed. This might not be that important except that colleagues perceive this side of your personality as a bad omen on future development. . . . You must realize that, if you do not fully separate these activities from your serious research, perhaps eliminating them altogether, and relegate them to the pub or similar places, you may find your future in jeopardy.

I'd had cold water poured on me before, but this was one of those great moments when I realized I'd set a new personal record, the new high score to try to top. When I forwarded this email to my dad, who's greatly inspired my scientific pursuits, he responded with a Dante quote: *"Segui il tuo corso et lascia dir le genti!"* Italian for "Follow your own path and let people talk!"

I find it amusing how strong the conformist herd mentality is among many physicists, given that we all pay lip service to thinking outside the box and challenging authority. I'd become acutely aware of this sociological situation already back in grad school: for example, Einstein's revolutionary relativity theory never won the Nobel Prize,* Einstein himself dismissed Friedmann's expanding-universe discovery, and Hugh Everett never even got a job in physics. In other words, much more important discoveries than I could realistically hope to make were being dismissed. So I faced a dilemma back in grad school: I'd fallen in love with physics precisely because I was fascinated with the biggest questions, yet it seemed clear that if I just followed my heart, then my next job would be at McDonald's.

I didn't want to choose between my passion and my career, so I developed a secret strategy that ended up working surprisingly well, letting me have my cake and eat it, too. I called it my "Dr. Jekyll/Mr. Hyde Strategy," and it exploited a sociological loophole. Whereas Giordano Bruno was burned at the stake in 1600 for his views (which included heresies such as space being infinite) and Galileo was condemned to lifelong house arrest for arguing that the Earth orbits the Sun, today's sanctions are milder. If you're interested in big philosophical-sounding questions, most physicists will treat you in much the same way as if

---

* At nobelprize.org, you can read that the Nobel Prize in Physics was awarded to Albert Einstein in 1922 "for his services to Theoretical Physics, and especially for his discovery of the law of the photoelectric effect." However, a Swedish colleague of mine on the Nobel Committee once showed me the less-publicized full version of the award text. In my translation of it below, I've boldfaced a hilarious caveat that some curmudgeons presumably inserted to reflect their misgivings about relativity theory, today universally hailed as one of the greatest triumphs of the human mind:

THE ROYAL SWEDISH ACADEMY OF SCIENCE has at its meeting on November 9, 1922, in accordance with the regulations in the November 27, 1895, will of ALFRED NOBEL, decided to, **independently of the value that, after possible confirmation, may be attributed to the relativity and gravitation theory**, award the prize that for 1921 is given to the person who within the domain of physics has made the most important discovery or invention, to ALBERT EINSTEIN for his contributions to Theoretical Physics, especially his discovery of the photoelectric effect.

you're captivated by computer games: what you do after work is your own business and won't be held against you as long as it doesn't distract you from your day job, and as long as you don't talk too much about it at work. So whenever authority figures asked what I worked on, I transformed into the respectable Dr. Jekyll and told them that I worked on mainstream topics in cosmology, such as those of Chapter 4, involving lots of measurements and numbers and blah blah blah. But secretly, when nobody was watching, I'd transform into the evil Mr. Hyde and do what I *really* wanted to do: pursue the ultimate nature of reality as in Chapters 6, 8 and most of the rest of this book. To allay fears, I put a blurb on my website about having some "side interests," joking that every time I'd written ten mainstream papers, I'd allow myself to indulge in writing one wacky one. This was very convenient, since I was the only one who kept count. . . . When I graduated from Berkeley, I'd published eight papers, but half of them were written by Mr. Hyde, so I omitted these from my Ph.D. thesis. I really liked my Berkeley thesis advisor, Joe Silk, but to be on the safe side, I made sure he was far from the laser printer before printing the Hyde papers, and I showed them to him only after he'd officially signed my thesis. . . .* And I stuck with this strategy: whenever I applied for jobs and research grants, I only mentioned Dr. Jekyll's work, while on the side, I kept doing research on these big questions that set me on fire—in a good, non-Bruno kind of way.

This devious strategy worked beyond my wildest expectations, and I'm extremely grateful that I get to work at a university with brilliant colleagues and students without having to stop thinking about my greatest interests. But now I feel that I have a debt to pay to the science community, and that the time has come to pay my dues! If we imagine all research topics arranged in front of us, in a metaphorical space, then there's a border delimiting what's mainstream physics from what's not. The amazing thing about this border is that, as illustrated in Figure 10.1, it's continually shifting! In some places, it has contracted, with theories from alchemy to astrology leaving the mainstream. In other places, it has expanded to reclassify ideas such as relativity theory and

---

* I also timed Mr. Hyde's papers strategically. Just as politicians like to discreetly reveal unpopular news on a Friday afternoon to give people time to forget before next week's news cycle, I wrote that alleged crackpot paper during the summer of 1996, right after I'd been offered my Princeton postdoc job, because I knew that this would give people the maximum time to forget before I had to apply for jobs again.

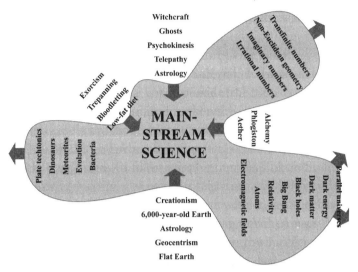

Figure 10.1: The boundary of what's considered mainstream keeps shifting.

the germ theory of disease from speculative minority views to main-stream science. I've long believed that there are additional topics that physicists can usefully contribute to even though they at first sound rather philosophical, and I've now had tenure long enough that I can't make excuses: I feel that I now have a moral obligation to more junior scientists to bring Mr. Hyde out of the academic closet and do my part to push the boundary a little. That's why Anthony Aguirre and I started the Foundational Questions Institute I mentioned in Chapter 8, http://fqxi.org. And that's why I'm writing this book.

So what paper of mine was it that triggered that "stop or you'll ruin your career"? What topic was it about that was so far outside the current mainstream boundary of Figure 10.1 that this professor felt the need to bring me back into the fold? It was about the core idea of this book: that our physical world is a giant mathematical object. And this is the chapter where we're going to start exploring it.

Math, Math Everywhere!

What's the answer to the ultimate question of life, the universe, and everything? In Douglas Adams's science-fiction spoof *The Hitchhiker's Guide to the Galaxy*, the answer was found to be 42; the hardest part turned out to be finding the real question. Indeed, although our inquisitive ancestors undoubtedly asked such big questions, their search for a "theory of everything" evolved as their knowledge grew. As the ancient Greeks replaced myth-based explanations with mechanistic models of the Solar System, their emphasis shifted from asking *why* to asking *how*.

Since then, the scope of our questioning has dwindled in some areas and mushroomed in others, as illustrated in Figure 10.1. Some questions were abandoned as naive or misguided, such as explaining the sizes of planetary orbits from first principles, which was popular during the Renaissance. The same may happen to currently trendy pursuits such as predicting the amount of dark energy in the cosmos, if it turns out that the amount in our neighborhood is a historical accident as we discussed in Chapter 6. Yet our ability to answer other questions has surpassed earlier generations' wildest expectations: Newton would have been amazed to know that we'd one day measure the age of our Universe to an accuracy of 1%, and comprehend the microworld well enough to make an iPhone.

I find it very appropriate that Douglas Adams joked about 42, because mathematics has played a striking role in these successes.* The idea that our Universe is in some sense mathematical goes back at least to the Pythagoreans of ancient Greece, and has spawned centuries of discussion among physicists and philosophers. In the seventeenth century, Galileo famously stated that our Universe is a "grand book" written in the language of mathematics. More recently, the physics Nobel laureate Eugene Wigner argued in the 1960s that "the unreasonable effectiveness of mathematics in the natural sciences" demanded an explanation.

Shapes, Patterns and Equations

Below we're going to explore a really extreme explanation. However, first we need to clear up what exactly it is that we're trying to explain. Please stop your reading for a few moments and look around

---

* I've switched from collecting stamps to collecting cool questions whose answer is 42. Here are my favorites so far:

1. At what latitude was this book written?
2. What's the radius of the rainbow, in degrees?
3. At most how many percent of the gas around it can a black hole gobble up?

Feeding a black hole turns out to be a lot like feeding a baby: most of the material comes flying back at great speeds. . . . Black holes can eat at most $1-1/\sqrt{3} \approx 42\%$ of the gas around them.

you. Where's all this math that we're going on about? Isn't math all about numbers? You can probably spot a few numbers here and there, for example the page numbers of this book, but these are just symbols invented and printed by people, so they can hardly be said to reflect our Universe being mathematical in any deep way.

Because of our education system, many people equate mathematics with arithmetic. Yet like physics, mathematics has evolved to ask broader questions. For example, when I quoted Galileo above, he spoke of geometric figures such as circles and triangles being mathematical, so when you look around you, do you see any geometric patterns or shapes? Here again, human-made designs like the rectangular shape of this book don't count. But try throwing a pebble and watch the beautiful shape that nature makes for its trajectory! Galileo made a remarkable discovery illustrated in Figure 10.2: the trajectories of anything you throw have the *same* shape, called an upside-down *parabola*. Moreover, the shape of a parabola can be described by a very simple mathematical equation: $y = x^2$, where x is the horizontal position and y is the vertical position (the height). Depending on the initial speed and direction, the shape can be stretched both vertically and horizontally, but it always remains a parabola.

When we observe how things move around in orbits in space, we discover another recurring shape, as illustrated in Figure 10.3: the *ellipse*. The equation $x^2 + y^2 = 1$ describes the points on a circle, and an ellipse is simply a stretched circle. Depending on the initial speed and direction of the orbiting object and the mass of the thing it's orbiting around,

Figure 10.2: When you throw something up into the air, its trajectory always has the same shape, called an upside-down *parabola*, if it doesn't collide with anything and air resistance is unimportant.

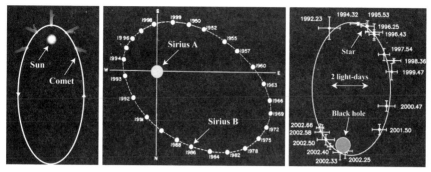

Figure 10.3: When something is orbiting something else due to gravity, its orbit always has the same shape, called an *ellipse*, which is simply a circle that's stretched in one direction (that's if there's no source of friction, and you ignore Einstein's corrections to Newton's gravity theory, which are usually tiny unless you're near a black hole). The orbit is an ellipse even for dramatically different objects, say, for a comet orbiting the Sun (left), a white dwarf stellar corpse orbiting Sirius A, the brightest star in our night sky (center), and a star orbiting the monster black hole at the center of our Galaxy (right), which is four million times more massive than our Sun. *(Right panel courtesy of Reinhard Genzel and Rainer Schödel)*

the shape of the orbit can be both stretched and tilted, but it always remains an ellipse. Moreover, these two shapes are related: the tip of a very elongated ellipse is shaped almost exactly like a parabola, so in fact, all of these trajectories are simply parts of ellipses.*

We humans have gradually discovered many additional recurring shapes and patterns in nature, involving not only motion and gravity, but also areas as disparate as electricity, magnetism, light, heat, chemistry, radioactivity and subatomic particles. These patterns are summarized by what we call our *laws of physics*. Just as the shape of an ellipse, all these laws can be described using mathematical equations, as illustrated in Figure 10.4. Why is that?

Numbers

Equations aren't the only hints of mathematics that are built into nature: there are also *numbers*. As opposed to human creations such as the page numbers in this book, I'm now talking about numbers that are basic properties of our physical reality. For example, how many pencils can you arrange so that they're all perpendicular (at 90 degrees) to each other?

---

* Indeed, if you prevented the basketball in Figure 10.2 from hitting the ground by compressing our entire planet into a black hole at the center, then the parabolic part of the ball's trajectory would remain the same, and would extend into a full ellipse around the black hole.

Figure 10.4: Just as art and poetry can capture a lot in just a few symbols, so can the equations of physics. From left to right, top to bottom, these masterpieces describe electromagnetism, near–light speed motion, gravity, quantum mechanics and our expanding Universe. We still haven't found equations for a unified theory of everything.

The answer is 3—for instance, by placing them along the 3 edges emanating from a corner of your room. Where did that number 3 come sailing in from? We call this number the dimensionality of our space, but why are there 3 dimensions rather than 4 or 2 or 42? And why are there, as far as we can tell, exactly 6 kinds of quarks in our Universe? As we saw in Chapter 7, there are many additional whole numbers (so-called integers) built into nature that describe what types of elementary particles exist.

As if that weren't enough mathematical goodies, there are also quantities encoded in nature that aren't whole numbers, but require decimals to write out. Nature encodes 32 such fundamental numbers according to my latest count. Does the number shown when you stand on your bathroom scale count as such a number? No, that number doesn't count, because it's measuring something (your mass) that changes from day to day and therefore isn't a basic property of our Universe. What about the mass of a proton, 1.672622×10^{-27}kg, or the mass of an electron, 9.109382×10^{-31}kg, which seem to stay perfectly constant over time? They don't count either, because they're measuring the number of kilograms, and that's just a rather arbitrary unit of mass that we humans have made up. But if you divide one of these last two numbers by the other, then you get something truly fundamental: the proton is about 1836.15267 more massive than the electron.* 1836.15267 is a *pure number*, just as π or $\sqrt{2}$, in the sense that it's a quantity that doesn't involve any human units of measurement such as grams, meters, seconds or volts. Why is it close to 1836? Why not 2013? Why not 42? The short answer is that we don't know, but that we think we can in principle calculate this number and every other fundamental constant of nature ever measured from just the 32 numbers listed in Table 10.1.

Don't worry about the intimidating-sounding technical names of the numbers in this table, which are irrelevant for what we're getting at here. The point is that there's something very mathematical about our Universe, and that the more carefully we look, the more math we seem to find. Apropos constants of nature, there are hundreds of thousands of pure numbers that have been measured across all areas of physics, ranging from ratios of masses of elementary particles to ratios of characteristic wavelengths of light emitted by different molecules, and using sufficiently powerful computers to solve the equations describing the laws of nature, it seems that every single one of these numbers can be computed from the 32 in Table 10.1. Some of the computations and some of the measurements are really difficult and haven't been done yet, and perhaps when that happens, some of the decimals won't match between theory and experiment. That sort of discrepancy has happened repeatedly in the past, and has typically been resolved in one of three ways:

---

* If you wonder why the ratio can be measured more accurately than the two masses separately, the reason is that the two measurement errors are very strongly related (correlated).

1. Someone discovered a mistake in the experiment.
2. Someone discovered a mistake in the calculation.
3. Someone discovered a mistake in our laws of physics.

In the third case, a more fundamental law of physics was usually found, as when Newton's equations for gravity were replaced by Ein-

Parameter	Meaning	Measured value
g	Weak-coupling constant at m_Z	0.6520 ± 0.0001
θ_W	Weinberg angle	0.48290 ± 0.00005
g_s	Strong-coupling constant at m_Z	1.220 ± 0.004
μ^2	Quadratic Higgs coefficient	$\approx -2 \times 10^{-34}$
λ	Quartic Higgs coefficient	≈ 0.5
G_e	Electron Yukawa coupling	0.000002931 ± 10^{-9}
G_μ	Muon Yukawa coupling	0.0006060 ± 0.0000002
G_τ	Tauon Yukawa coupling	0.01022
G_u	Up quark Yukawa coupling	0.000014 ± 0.000003
G_d	Down quark Yukawa coupling	0.000029 ± 0.000003
G_c	Charm quark Yukawa coupling	0.0073 ± 0.0001
G_s	Strange quark Yukawa coupling	0.00054 ± 0.00003
G_t	Top quark Yukawa coupling	0.995 ± 0.008
G_b	Bottom quark Yukawa coupling	0.0230 ± 0.0002
$\sin \theta_{12}$	Quark CKM–matrix angle	0.2243 ± 0.0016
$\sin \theta_{23}$	Quark CKM–matrix angle	0.0413 ± 0.0015
$\sin \theta_{13}$	Quark CKM–matrix angle	0.0037 ± 0.0005
δ_{13}	Quark CKM–matrix phase	1.05 ± 0.24
θ_{qcd}	CP-violating QCD vacuum phase	$< 10^{-9}$
G_{ve}	Electron neutrino Yukawa coupling	$< 1.3 \times 10^{-11}$
$G_{v\mu}$	Muon neutrino Yukawa coupling	$< 9.8 \times 10^{-7}$
$G_{v\tau}$	Tau neutrino Yukawa coupling	< 0.00009
$\sin^2 2\theta'_{12}$	Neutrino MNS–matrix angle	0.857 ± 0.024
$\sin^2 2\theta'_{23}$	Neutrino MNS–matrix angle	≥ 0.95
$\sin^2 2\theta'_{13}$	Neutrino MNS–matrix angle	$\leq 0.098 \pm 0.013$
δ'_{13}	Neutrino MNS–matrix phase	?
$\rho\Lambda$	Dark-energy density	$(1.16 \pm 0.07) \times 10^{-123}$
ξ_b	Baryon mass per photon ρ_b/n_γ	$(4.66 \pm 0.06) \times 10^{-29}$
ξ_c	Cold dark matter mass per photon ρ_c/n_γ	$(24.9 \pm 0.7) \times 10^{-29}$
ξ_v	Neutrino mass per photon $\rho_v/n_\gamma = \frac{3}{11}\Sigma\, m_{vi}$	$< 0.5 \times 10^{-29}$
Q	Scalar fluctuation amplitude δ_H on horizon	$(2.0 \pm 0.2) \times 10^{-5}$
n	Scalar spectral index	0.960 ± 0.007

Table 10.1: Every fundamental property of nature ever measured can be computed from the 32 numbers in this table—at least in principle. Some of these numbers have been measured very accurately, while others haven't yet been experimentally determined. The detailed meaning of these numbers doesn't matter for our discussion, but if you're interested, you'll find it explained in my paper at http://arxiv.org/abs/astro-ph/0511774. But what determines these numbers?

stein's, explaining why the motion of Mercury around the Sun isn't quite a perfect ellipse. In all cases, the feeling that there's something mathematical about nature was further strengthened.

If you discover an even more accurate law of physics in the future, it might decrease the number of parameters from the 32 in Table 10.1 by allowing you to compute some of the numbers from others in the table, or it might increase them by adding new ones, say, involving the masses of new kinds of particles that might be discovered by the Large Hadron Collider outside Geneva.

More Clues

So what do we make of all these hints of mathematics in our physical world? Most of my physics colleagues take them to mean that nature is for some reason described by mathematics, at least approximately, and leave it at that. In his book *Is God a Mathematician?*, the astrophysicist Mario Livio concludes that *"scientists have selected what problems to work on based on those problems being amenable to a mathematical treatment."* But I'm convinced that there's more to it than that.

First of all, *why* does math describe nature so well? I agree with Wigner that it demands an explanation. Second, throughout this book, we've come across clues suggesting that nature isn't just *described* by mathematics, but that some aspects of it *are* mathematical:

1. In Chapters 2–4, we saw that the very fabric of our physical world, space itself, is a purely mathematical object in the sense that its only intrinsic properties are mathematical properties—numbers such as dimensionality, curvature and topology.

2. In Chapter 7, we saw that all the "stuff" in our physical world is made of elementary particles, which in turn are purely mathematical objects in the sense that their only intrinsic properties are mathematical properties—numbers listed in Table 7.1 such as charge, spin and lepton number.

3. In Chapter 8, we saw that there's something that's arguably even more fundamental than our three-dimensional space and the particles within it: the wavefunction and the infinite-dimensional place called Hilbert space where it lives. Whereas particles can be created and destroyed, and can be in several places at once, there is, was and always will be only one wavefunction, moving through Hilbert

space as determined by the Schrödinger equation—and the wavefunction and Hilbert space are purely mathematical objects.

What does this all mean? Now let me share with you what I think it means, and let's see if it makes more sense to you than to that professor who said it would ruin my career.

The Mathematical Universe Hypothesis

I was quite fascinated by all these mathematical clues back in grad school. One Berkeley evening in 1990, while my friend Bill Poirier and I were sitting around speculating about the ultimate nature of reality, I suddenly had an idea for what it all meant: that our reality isn't just *described* by mathematics—it *is* mathematics, in a very specific sense that I'll describe below. Not just aspects of it, but all of it, including you.*
This idea sounds rather crazy and far-fetched, so after telling Bill about it, I mulled it over for many years before writing that first paper about it.

Before delving into the details, here's my logical framework for thinking about this business. First there are two hypotheses, one seemingly innocuous and one seemingly radical:

> **External Reality Hypothesis (ERH):** *There exists an external physical reality completely independent of us humans.*

> **Mathematical Universe Hypothesis (MUH):** *Our external physical reality is a mathematical structure.*

Second, I have an argument that, with a sufficiently broad definition of mathematical structure, the former implies the latter.
My starting assumption, the External Reality Hypothesis, isn't too controversial: I'd guess that the majority of physicists favor this long-standing idea, though it's still debated. Metaphysical solipsists reject it flat out, and supporters of the Copenhagen interpretation of quantum mechanics may reject it on the grounds that there's no reality without observation. Assuming that an external reality exists, physics

---

* Roger Penrose expresses similar sentiments in his book *The Road to Reality*.

theories aim to describe how it works. Our most successful theories, such as general relativity and quantum mechanics, describe only parts of this reality: gravity, for instance, or the behavior of subatomic particles. In contrast, the Holy Grail of theoretical physics is a theory of everything—a complete description of reality.

Reducing the Baggage Allowance

My personal quest for this theory begins with an extreme argument about what it's allowed to look like: *If we assume that reality exists independently of humans, then for a description to be complete, it must also be well defined according to nonhuman entities—aliens or supercomputers, say—that lack any understanding of human concepts. Put differently, such a description must be expressible in a form that's devoid of any human baggage like "particle," observation or other English words.*

In contrast, all physics theories that I've been taught have two components: mathematical equations and "baggage" —words that explain how the equations are connected to what we observe and intuitively understand. When we derive the consequences of a theory, we introduce new concepts and words for them, such as *protons, atoms, molecules, cells* and *stars*, because they're convenient. It's important to remember, however, that it's we humans who create these concepts; in principle, everything could be calculated without this baggage. A hypothetical ideal supercomputer could calculate how the state of our Universe changes over time without interpreting what's happening in human terms, simply figuring out how all the particles would move or how the wavefunction would change.

For example, suppose the basketball trajectory in Figure 10.2 is that of a beautiful buzzer beater that wins you the game, and that you later want to describe what it looked like to a friend. Since the ball is made of elementary particles (quarks and electrons), you could in principle describe its motion without making any reference to basketballs:

- *Particle 1 moves in a parabola.*
- *Particle 2 moves in a parabola.*
- *. . .*
- *Particle 138,314,159,265,358,979,323,846,264 moves in a parabola.*

That would be slightly inconvenient, however, because it would take you longer than the age of our Universe to say it. It would also be redundant, since all the particles are stuck together and move as a single unit. That's why we humans have invented a word *ball* to refer to the entire unit, enabling us to save time by simply describing the motion of the whole unit once and for all.

The ball was designed by humans, but it's quite analogous for composite objects that aren't man-made, such as molecules, rocks and stars: inventing words for them is convenient both for saving time, and for providing concepts or so-called shorthand abstractions in terms to understand the world more intuitively. Although useful, such words are all optional baggage: for example, I've used the word *star* repeatedly in this book, but you could in principle replace every occurrence of it by a definition in terms of its building blocks, say, *gravitationally bound clump of about 10^{57} atoms, some of which are undergoing nuclear fusion*. In other words, nature contains many types of entities that are almost begging to be named. Sure enough, virtually every human population on Earth has a word for *star* in its own language, often invented independently to reflect its own cultural and linguistic tradition. I suspect that most alien civilizations in distant solar systems have also invented a name or symbol for *star* even if they don't communicate using sounds.

Another striking fact is that you can often predict the existence of such name-worthy entities mathematically, from the equations governing their parts. In this way, the whole Lego-like hierarchy of structures that we discussed in Chapter 7 can be predicted, from elementary particles to atoms to molecules, and what we humans add are merely catchy names for the objects at each level. For example, if you solve the Schrödinger equation for five or fewer quarks, it turns out that there are only two fairly stable ways for them to be arranged: either as a clump of two up quarks and a down quark or as a clump of two down quarks and an up quark, and we humans have added the baggage of calling these two types of clumps "protons" and "neutrons" for convenience. Similarly, if you apply the Schrödinger equation to such clumps, it turns out that there are only 257 stable ways for them to be assembled together. We humans have added the baggage of calling these proton/neutron assemblies "atomic nuclei," and have also invented specific names for each kind: *hydrogen, helium*, etc. The Schrödinger equation also lets you calculate all the ways of putting atoms together into larger objects, but this time, there turn out to be so many different stable objects that it's inconvenient to name them all—instead, we've just named important classes of objects (such as "molecules" and "crystals"), and the most common or interesting objects in each class (e.g., "water," "graphite," "diamond").

I think of these composite objects as *emergent*, in the sense that they emerge as solutions of equations involving only more fundamental objects. This emergence is subtle and easy to miss because historically, the scientific process has mostly gone in the opposite direction: for example, we humans knew of stars before realizing that they were made of atoms, we knew of atoms before realizing that they were made of electrons, protons and neutrons, and we knew of neutrons before we discovered quarks. For every emergent object that's important to us humans, we create baggage in the form of new concepts.

The same pattern of emergence and human baggage creation can be seen in Figure 10.5. Here I've crudely organized scientific theories into a family tree where each might, at least in principle, be derivable from more fundamental ones above it. As mentioned, all these theories have two components: mathematical equations and words that explain how they're connected to what we observe. For example, we saw in Chapter 8 how quantum mechanics, as usually presented in textbooks, has both components: math such as the Schrödinger equation as well as fundamental postulates written out in plain English, such as the wavefunction-collapse postulate. At each level in the hierarchy of theories, new concepts (e.g., protons, atoms, cells, organisms, cultures) are introduced because they're convenient, capturing the essence of what is going on without recourse to the more fundamental theory above it. It's we humans who introduce these concepts and the words for them: in principle, everything could have been derived from the fundamental theory at the top of the tree, although such an extreme reductionist approach is often useless in practice. Crudely speaking, as we move down the tree, the number of words goes up while the number of equations goes down, dropping to near zero for highly applied fields such as medicine and sociology. In contrast, theories near the top are highly mathematical, and physicists are still struggling to understand the concepts, if any, in terms of which we can understand them.

The Holy Grail of physics is to find what's jocularly referred to as a "Theory of Everything," or ToE, from which all else can be derived—this would replace the big question mark at the top of the theory tree. As we discussed in Chapter 7, we know that something is missing here because we lack a consistent theory unifying gravity with quantum mechanics. This ToE would be a complete description of the external physical real-

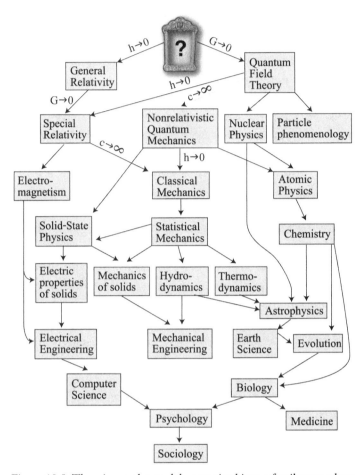

Figure 10.5: Theories can be crudely organized into a family tree where each might, at least in principle, be derivable from more fundamental ones above it. For example, special relativity can be obtained from general relativity in the approximation that Newton's gravitational constant G equals zero, classical mechanics can be derived from special relativity in the approximation that the speed of light c is infinite, and hydrodynamics with its concepts such as density and pressure can be derived from the classical physics of how particles bounce around. However, these cases where the arrows are well understood form a minority. Deriving biology from chemistry or psychology from biology appears unfeasible in practice. Only limited and approximate aspects of such subjects are mathematical, and it's likely that all mathematical models found in physics so far are similarly approximations of limited aspects of reality.

ity that the External Reality Hypothesis assumes. At the beginning of this section, I argued that such a complete description must be devoid of any human baggage. This means that it must contain no concepts at all! In other words, it must be a purely mathematical theory, with no explanations or "postulates" as in quantum textbooks (mathematicians are perfectly capable of—and often pride themselves on—studying abstract mathematical structures that lack any intrinsic meaning or connection with physical concepts). Rather, an infinitely intelligent mathematician should be able to derive the entire theory tree of Figure 10.5 from these equations alone, by deriving the properties of the physical reality that they describe, the properties of its inhabitants, their perceptions of the world, and even the words they invent. This purely mathematical theory of everything could potentially turn out to be simple enough to describe with equations that fit on a T-shirt.

All of this begs the question: is it actually possible to find such a description of the external reality that involves no baggage? If so, such a description of objects in this external reality and the relations between them would have to be completely abstract, forcing any words or symbols to be mere labels with no preconceived meanings whatsoever. Instead, the only properties of these entities would be those embodied by the relations between them.

Mathematical Structures

To answer this question, we need to take a closer look at mathematics. To a modern logician, a mathematical structure is precisely this: a set of abstract entities with relations between them. Take the integers, for instance, or geometric objects such as the dodecahedron, a favorite of the Pythagoreans. This is in stark contrast to the way most of us first perceive mathematics—either as a sadistic form of punishment, or as a bag of tricks for manipulating numbers. Like physics, mathematics has evolved to ask broader questions.

Modern mathematics is the formal study of structures that can be defined in a purely abstract way, without any human baggage. Think of mathematical symbols as mere labels without intrinsic meaning. It doesn't matter whether you write, "Two plus two equals four," "2 + 2 = 4," or "*Dos más dos es igual a cuatro.*" The notation used to denote the entities and the relations is irrelevant; the only properties of integers are those embodied by the relations between them. That is, we don't invent mathematical structures—we discover them, and invent only the notation for describing them. It is crucial not to conflate the *language* of mathematics (which we invent) with the *structures* of mathematics (which we discover). If an alien civilization gets interested in 3-D shapes with only flat identical faces, they might discover the five from Figure 7.2 that we Earthlings call Platonic solids. They are free to invent their own exotic names for them, but they can't invent a sixth one—it simply doesn't exist. It's in the same sense that the mathematical structures that are popular in modern physics are discovered rather than invented, from 3+1-dimensional pseudo-Riemannian manifolds to Hilbert spaces.

In summary, there are two key points to take away from our discussion above:

1. The External Reality Hypothesis implies that a "theory of everything" (a complete description of our external physical reality) has no baggage.
2. Something that has a complete baggage-free description is precisely a mathematical structure.

Taken together, this implies the Mathematical Universe Hypothesis, i.e., that the external physical reality described by the ToE is a mathematical structure.* So the bottom line is that if you believe in an external reality independent of humans, then you must also believe that our physical reality is a mathematical structure. Nothing else has a baggage-free description. In other words, we all live in a gigantic mathematical object—one that's more elaborate than a dodecahedron, and probably also more complex than objects with intimidating names such as Calabi-Yau manifolds, tensor bundles and Hilbert spaces, which appear in today's most advanced physics theories. Everything in our world is purely mathematical—including you.

What Is a Mathematical Structure?

"Waaaaaaaaaaaaaaaaaait a minute!" That's what my friend Justin Bendich used to cry out whenever a physics claim raised urgent unanswered questions, and the Mathematical Universe Hypothesis raises three:

---

* In the philosophy literature, John Worrall has coined the term *structural realism* as a compromise position between scientific realism and anti-realism; crudely speaking, stating that the fundamental nature of reality is correctly described only by the mathematical or structural content of scientific theories. This term has been interpreted and refined in different ways by different science philosophers, and Gordon McCabe has argued that the term *universal structural realism* should be used for my hypothesis that our physical Universe is isomorphic to a mathematical structure.

- What exactly is a mathematical structure?
- How exactly can our physical world be a mathematical structure?
- Does this make any testable predictions?

We'll tackle the second question in Chapter 11 and the third one in Chapter 12. Let's start by exploring the first question—we'll return to it in more detail in Chapter 12.

Baggage and Equivalent Descriptions

Earlier, we described how we humans add baggage to our descriptions. Now let's look at the opposite: how mathematical abstraction can remove baggage and strip things down to their bare essence. Consider the particular sequence of chess moves that have become known as the "Immortal Game," where white spectacularly sacrifices both rooks, a bishop and the queen to checkmate with the three remaining minor pieces as shown in Figure 10.6. Here on Earth, this game was first played in 1851 by Adolf Anderssen and Lionel Kieseritzky. However, the same game is replayed annually in the town of Marostica, Italy, with live players dressed as chess pieces, and it's regularly repeated by countless chess enthusiasts around the world. Some players (including my brother Per, his son Simon and my son Alexander in Figure 10.6) use pieces made of wood, while others use pieces of marble or plastic with different shapes and sizes. Some boards are brown and beige, some are black and white, and some are virtual, being mere 3-D or 2-D computer graphics as in

Less abstract / More baggage ←————————————→ More abstract / Less baggage

```
1.e4 e5 2.f4 exf4 3.Bc4 Qh4+
4.Kf1 b5 5.Bxb5 Nf6 6.Nf3 Qh6
7.d3 Nh5 8.Nh4 Qg5 9.Nf5 c6
10.g4 Nf6 11.Rg1 cxb5 12.h4 Qg6
13.h5 Qg5 14.Qf3 Ng8 15.Bxf4 Qf6
16.Nc3 Bc5 17.Nd5 Qxb2
18.Bd6 Bxg1 19. e5 Qxa1+
20. Ke2 Na6 21.Nxg7+ Kd8
22.Qf6+ Nxf6 23.Be7
```

Figure 10.6: An abstract game of chess is independent of the colors and shapes of the pieces, and of whether its moves are described on a physically existing board, by stylized computer-rendered images, or by so-called algebraic chess notation—it's still the same chess game. Analogously, a mathematical structure is independent of the symbols used to describe it.

Figure 10.6. Yet there's a sense in which none of these details matter: when chess aficionados call the Immortal Game beautiful, they're not referring to the attractiveness of the players, the board, or the pieces, but to a more abstract entity, which we might call the abstract game, or the sequence of moves.

Let's look in detail at how we humans go about describing such abstract entities. First of all, a description needs to be specific, so we invent objects, words or other symbols to correspond to the abstract ideas: for example, in the United States, we name the chess piece that can move diagonally a "bishop." Second, it's obvious that this name is arbitrary, and that other names would have worked just as well—indeed, this piece is call a *fou* (fool) in French, *strelec* (shooter) in Slovak, *löpare* (runner) in Swedish and *fil* (elephant) in Persian. However, we can reconcile the uniqueness of the Immortal Game with the multiplicity of possible descriptions of it by introducing the powerful idea of *equivalence*:

1. We define what we mean by two descriptions being equivalent.
2. We say that if two descriptions are equivalent, then they're describing one and the same thing.

For example, we agree that any two descriptions of a chess position are equivalent if the only difference between them lies in the sizes of the pieces, or in the names that the players give to the pieces in their native language.

Any word, concept or symbol that appears in some but not all of the equivalent descriptions is clearly optional and therefore baggage. So if we want to get down to the bare essence of the Immortal Game, then how much baggage can we strip away? Clearly a lot, since computers are able to play chess without having any notion of human language or human concepts such as the colors, textures, sizes and names of chess pieces. To fully understand how far we can go, we need to make a more rigorous definition of equivalence:

> **Equivalence:** *Two descriptions are equivalent if there's a correspondence between them that preserves all relations.*

Chess involves abstract entities (different chess pieces and different squares on the board) and relations between them. For example, one relation that a piece may have to a square is that the former is standing on the latter. Another relation that a piece may have to a square is that it's allowed to move there. For example, the two center panels in Figure 10.6 are equivalent by our definition: there's a correspondence between the three-dimensional and two-dimensional pieces and boards such that whenever a 3-D piece stands on a particular square, the corresponding 2-D piece stands on the corresponding square. Similarly, a description of a chess position given purely verbally in English is equivalent to a description given purely verbally in Spanish if you can provide a dictionary specifying the correspondence between the English and Spanish words, and if using it to translate the Spanish description produces the English description.

When newspapers and websites print chess games, they customarily use yet another equivalent description: so-called algebraic chess notation (Figure 10.6, right). Here pieces are represented not by objects or words, but by single letters; *bishop* is equivalent to *B*, for example, and squares are represented by a letter specifying the column and a number specifying the row. Since the abstract game description in Figure 10.6 (right) is equivalent to a description in the form of a movie of the game being played on a physical board, everything in the latter description that isn't in the former description is mere baggage—from the physical existence of a board to the shapes, colors, and names of the pieces. Even the specifics of algebraic chess notation are baggage: when computers play chess, they typically use other abstract chess-position descriptions, involving certain patterns of zeros and ones in their memory. So what is it that's left when you strip away all this baggage? What is it that's described by all these equivalent descriptions? The Immortal Game itself, 100% pure, with no additives.

Baggage and Mathematical Structures

Our case study involving abstract chess pieces, board squares and relations between them was an example of a much more general concept: a mathematical structure. This is a standard concept in modern mathematical logic. I'll give a more technical description in Chapter 12, but this nontechnical definition is all we need for now:

> **Mathematical structure:** *Set of abstract entities with relations between them*

To understand what this means, let's consider a couple of examples. Figure 10.7 (left) is a description of a mathematical structure with four entities, some of which are related by the relation *likes to*. In the figure, the entity *Philip* is represented by an image with many intrinsic properties, such as being brown-haired. In contrast, the entities of a mathematical structure are purely abstract, which means that they have no intrinsic properties whatsoever. This means that whatever symbols we use to represent them are mere labels whose properties are irrelevant: to avoid the mistake of attributing properties of the symbols to the abstract entities that they symbolize, let's consider the more spartan description in the middle panel. This description is equivalent to the first one, because if you apply the correspondence given by the dictionary *Philip* = 1, *Alexander* = 2, *ski* = 3, *skate* = 4 and *likes to* = R, all relations are preserved. For example, "Alexander likes to skate" translates to "2 R 4," which is indeed a relation that holds in the middle panel.

Just as chess games can be described using symbols alone, without any graphics, so can mathematical structures. For example, the right panel of Figure 10.7 gives a third equivalent description of our mathematical structure in terms of a four-by-four table of numbers. In this table, an entry of 1 means that the relation (*likes to*) holds between the element corresponding to that row and the element corresponding to that column, so the fact that there's a 1 in the third column of the first row means that "Philip likes to ski." There are clearly many more equivalent ways of describing this mathematical structure, but

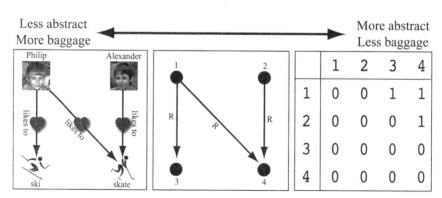

Figure 10.7: Three equivalent descriptions of the same mathematical structure, which mathematicians would call an "ordered graph with four elements." Each description contains some arbitrary baggage, but the structure that they all describe is 100% baggage-free: its four entities have no properties whatsoever except the relations that hold between them, and the relation has no properties whatsoever except the information about which elements it relates.

there's only one unique mathematical structure that's described by all these equivalent descriptions. In summary, any particular description of a mathematical structure contains baggage, but the structure itself doesn't. It is important not to confuse the description with that which is described: even the most abstract-looking description of a mathematical structure is still not the structure itself. Rather, the structure corresponds to the class of all equivalent descriptions of it. Table 10.2 summarizes the relations between these and other key concepts linked to the mathematical-universe idea.

Symmetry and Other Mathematical Properties

Some mathematicians enjoy debating what mathematics really is, and there's certainly no consensus. However, a quite popular definition of mathematics is "the formal study of mathematical structures." In this vein, mathematicians have identified large numbers of interesting mathematical structures, ranging from familiar ones such as the cube, the icosahedron (Figure 7.2), and the integers (the whole numbers) to ones with exotic names like Banach spaces, orbifolds and pseudo-Riemannian manifolds.

One of the most important things that mathematicians do when studying mathematical structures is prove theorems about their proper-

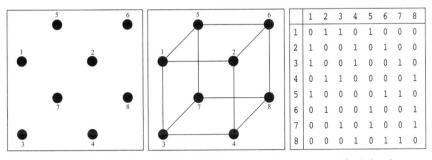

Figure 10.8: The center panel describes a mathematical structure with eight elements (symbolized by dots) and relations between them (symbolized by lines). You can interpret these elements as the corners of a cube and the relation as specifying which corners are connected by an edge, but this interpretation is completely optional baggage—the right panel gives an equivalent description of this same mathematical structure without any graphics or geometry—for example, the fact that there's a 1 in the fifth column of the sixth row means that a relation holds between elements 5 and 6. This mathematical structure has many interesting properties, including both mirror symmetry and certain rotation symmetries. In contrast, the mathematical structure described in the left panel has no relations and no interesting properties at all except its cardinality of 8, the number of elements it contains.

ties. But what properties can a mathematical structure have if its entities and relations aren't allowed to have any intrinsic properties at all?

Consider the mathematical structure described by the left panel of Figure 10.8. It has no relations at all between its entities, so there's nothing to distinguish any one entity from any other. This means that this mathematical structure has no properties whatsoever except its *cardinality*, the number of entities that it has. Mathematicians call this mathematical structure "the set of 8 elements," and its only property is having eight elements—a pretty boring structure!

The middle panel of Figure 10.8 describes a different and more interesting mathematical structure with eight elements, which includes a relation. One description of this structure is that the elements are the corners of a cube and the relation specifies which corners are connected by an edge. However, remember not to confuse the description with that which is described: the mathematical structure itself has no intrinsic properties such as size, color, texture or composition—it has only these eight related entities which you can optionally interpret as cube corners. Indeed, the right panel of Figure 10.8 gives an equivalent definition of this mathematical structure without making any reference to geometric notions such as cubes, corners or edges.

So if the entities of this structure have no intrinsic properties, does the structure itself have any interesting properties (besides having eight elements)? In fact, it does: *symmetries!* In physics, we say that something has a symmetry if it remains unchanged when you transform it in some way. For example, we say that your face has *mirror symmetry* if it looks the same after being reflected left to right. In the same way, the mathematical structure in Figure 10.8 (middle) has mirror symmetry: if you swap elements 1 and 2, 3 and 4, 5 and 6, and 7 and 8, the drawing of the relations will still look the same. It also has some *rotational symmetry*, corresponding either to rotating the cube in the drawing by 90 degrees around one of its faces, by 120 degrees around one of its corners, or by 180 degrees around an edge center. Although we intuitively think of symmetry as having to do with geometry, you can in fact discover these same symmetries just by messing around with the table in the right panel of Figure 10.8: if you renumber the eight elements in certain ways and then re-sort the table by increasing row and column numbers, you end up with the exact same table that you started with.

A famous thorny issue in philosophy is the so-called *infinite regress problem*. For example, if we say that the properties of a diamond can be explained by the properties and arrangements of its carbon atoms, that the properties of a carbon atom can be explained by the properties and arrangements of its protons, neutrons and electrons, that the properties of a proton can be explained by the properties and arrangements of its quarks, and so on, then it seems that we're doomed to go on forever trying to explain the properties of the constituent parts. The Mathematical Universe Hypothesis offers a radical solution to this problem: at the bottom level, reality is a mathematical structure, so its parts have no intrinsic properties at all! In other words, the Mathematical Universe Hypothesis implies that we live in a *relational reality*, in the sense that the properties of the world around us stem not from properties of its ultimate building blocks, but from the relations between these building blocks.* The external physical reality is therefore more than the sum of its parts, in the sense that it can have many interesting properties while its parts have no intrinsic properties at all.

Mathematical Universe Cheat Sheet	
Baggage	Concepts and words that are invented by us humans for convenience, which aren't necessary for describing the external physical reality
Mathematical structure	Set of abstract entities with relations between them; can be described in a baggage-independent way
Equivalence	Two descriptions of mathematical structures are equivalent if there's a correspondence between them that preserves all relations; if two mathematical structures have equivalent descriptions, they are one and the same
Symmetry	The property of remaining unchanged when transformed; for example, a perfect sphere is unchanged when rotated
External Reality Hypothesis	The hypothesis that there exists an external physical reality completely independent of us humans
Mathematical Universe Hypothesis	The hypothesis that our external physical reality is a mathematical structure; I argue that this follows from the External Reality Hypothesis
Computable Universe Hypothesis	Our external physical reality is a mathematical structure defined by computable functions (Chapter 12)
Finite-Universe Hypothesis	Our external physical reality is a finite mathematical structure (Chapter 12)

Table 10.2: Summary of key concepts linked to the mathematical-universe idea

* Our brain may provide another example of where properties stem mainly from relations. According to the so-called concept cell hypothesis in neuroscience, particular firing patterns in different groups of neurons correspond to different concepts. The main difference between the concept cells for "red," "fly" and "Angelina Jolie" clearly don't lie in the types of neurons involved, but in their relations (connections) to other neurons.

The particular mathematical structures illustrated in Figures 10.7 and 10.8 belong to the family of mathematical structures known as *graphs:* abstract elements, some of which are connected pairwise. You can use other graphs to describe the mathematical structures corresponding to the dodecahedron and the other Platonic solids from Figure 7.2. Another example of a graph is the network of friends on Facebook: here the elements correspond to all the Facebook users, and two users are connected if they're in a friend relation. Although mathematicians have studied graphs extensively, they constitute merely one of many different families of mathematical structures. We'll discuss mathematical structures in greater detail in Chapter 12, but let's first briefly look at a few more examples here, to get a sense for how diverse mathematical structures can be.

There are many mathematical structures corresponding to different types of numbers. For example, the so-called *natural numbers* 1, 2, 3, . . . together form a mathematical structure. Here the elements are the numbers and there are many different kinds of relations. Some relations (say, *equals, is greater than* and *is divisible by*) can hold between two numbers ("15 *is divisible by* 5," say), some relations hold between three numbers ("17 *is the sum of* 12 *and* 5," say) and some relations involve other numbers of numbers. Mathematicians have gradually discovered larger classes of numbers that form their own mathematical structures, such as *integers* (including negative numbers), *rational numbers* (including fractions), *real numbers* (including the square root of 2), *complex numbers* (including the square root of –1), and *transfinite numbers* (including infinite numbers). When I close my eyes and think of the number 5, it looks yellow to me. Yet in all these mathematical structures, the numbers themselves have no such intrinsic properties at all, and their only properties are given by their relations to other numbers—5 has the property that it's the sum of 4 and 1, say, but it's not yellow, and it's not made of anything.

Another large class of mathematical structures corresponds to different types of spaces. For example, the three-dimensional Euclidean space that we learned about in school is a mathematical structure. Here the elements are points in the 3-D space and real numbers that are interpreted as distances and angles. There are many different kinds of relations. For example, three points can satisfy the relation that they lie on a line. There's a different mathematical structure corresponding to Euclidean space with four dimensions and with any other number of dimensions. Mathematicians have also discovered many other types of more general spaces that form their own mathematical structures, like so-called Minkowski space, Riemann spaces, Hilbert spaces, Banach spaces and Hausdorff spaces. Many people used to think that our three-dimensional physical space was a Euclidean space. However, we saw in Chapter 2 that Einstein put an end to that. First his special relativity theory said that we live in a Minkowski space (including time as a fourth dimension), then his general relativity said that we instead live in a Riemann space, which meant that it could be curved. Then, as we saw in Chapter 7, quantum mechanics came along and said that we're really living in a Hilbert space. Again, the points in these spaces aren't made of anything, and have no color, texture or other intrinsic properties whatsoever.

Although the collection of known mathematical structures is large and exotic, and even more remain to be discovered, every single mathematical structure can be analyzed to determine its symmetry properties, and many have interesting symmetry. Intriguingly, one of the most important discoveries in physics has been that our physical reality also has symmetries built into it: for example, the laws of physics have rotational symmetry, which means that there's no special direction in our Universe that you can call "up." They also appear to have translation (sideways shifting) symmetry, meaning that there's no special place that we can call the center of space. Many of the spaces just mentioned have beautiful symmetries, some of which match the observed symmetries of our physical world. For example, Euclidean space has both rotational symmetry (meaning that you can't tell the difference if the space gets rotated) and translational symmetry (meaning that you can't tell the difference if the space gets shifted sideways). The four-dimensional Minkowski space has even more symmetry: you can't even tell the difference if you do a type of generalized rotation between the space and time dimensions—and Einstein showed that this explains why time appears to slow down if you travel near the speed of light, as mentioned in the last chapter. Many more subtle symmetries of nature have been discovered in the last century, and these symmetries form the foundations of Einstein's relativity theories, quantum mechanics, and the standard model of particle physics.

Note that these symmetry properties that are so important in physics come precisely from the lack of intrinsic properties of the building blocks of reality, that is, from the very essence of what it means for it to be a mathematical structure. If you take a colorless sphere and paint part of it yellow, then its rotational symmetry gets destroyed. Similarly, if the points in a three-dimensional space had any properties that made some points intrinsically different from others, then the space would lose its rotational and translational symmetry. "Less is more," in the sense that the less properties the points have, the more symmetry the space has.

If the Mathematical Universe Hypothesis is correct, then our Universe is a mathematical structure, and from its description, an infinitely intelligent mathematician should be able to derive all these physics theories. How exactly would she do this? We don't know, but I'm quite sure about what her first step would be: to calculate the symmetries of the mathematical structure.

At the beginning of this chapter, you saw the grim prognostication that my publications on the relation between mathematics and physics were too crazy and would ruin my career. I've now told you about the first part of these ideas, arguing that our external physical reality is a mathematical structure, which sounds quite crazy indeed. However, that was just the warm-up—it's going to get much crazier later, when we examine the implications and testable predictions of the Mathematical Universe Hypothesis! Among other things, we'll be led inexorably to a new multiverse so large that it makes even the Level III Multiverse of quantum mechanics pale in comparison. But before that, we need to answer a burning question. Our physical world is changing over time, whereas mathematical structures don't change—they just exist. So how can our world possibly be a mathematical structure? We'll tackle that in the next chapter.

THE BOTTOM LINE

- Since antiquity, people have puzzled over why our physical world can be so accurately described by mathematics.
- Ever since, physicists have kept discovering more shapes, patterns and regularities in nature that are describable by mathematical equations.
- The fabric of our physical reality contains dozens of pure numbers, from which all measured constants can in principle be calculated.
- Some key physical entities such as empty space, elementary particles and the wavefunction appear to be purely mathematical in the sense that their only intrinsic properties are mathematical properties.
- The External Reality Hypothesis (ERH)—that there exists an external physical reality completely independent of us humans—is accepted by most but not all physicists.
- With a sufficiently broad definition of mathematics, the ERH implies the Mathematical Universe Hypothesis (MUH) that our physical world is a mathematical structure.
- This means that our physical world not only is described by mathematics, but that it is mathematical (a mathematical structure), making us self-aware parts of a giant mathematical object.
- A mathematical structure is an abstract set of entities with relations between them. The entities have no "baggage": they have no properties whatsoever except these relations.
- It is crucial not to conflate the *language* of mathematics (which we invent) with the *structure* of mathematics (which we discover).
- A mathematical structure can have many interesting properties—for example, symmetries—even though neither its entities nor its relations have any intrinsic properties whatsoever.
- The MUH solves the infamous infinite regress problem where the properties of nature can only be explained from the properties of its parts, which require further explanation, ad infinitum: the properties of nature stem not from properties of its ultimate building blocks (which have no properties at all), but from the relations between these building blocks.

11

Is Time an Illusion?

The distinction between past, present, and future is only a stubbornly persistent illusion.

—Albert Einstein, 1955

Time is an illusion, lunchtime doubly so.
—Douglas Adams, *The Hitchhiker's Guide to the Galaxy*

If you're like me, then you're disturbed by unanswered questions. The last chapter raised many, so it's valid for you to question what I've said. For example, I argued that our external physical reality is a mathematical structure, but what does this really mean? This physical reality is constantly changing—leaves move in the wind and planets orbit the Sun—while mathematical structures are static: an abstract dodecahedron always has had and always will have exactly twelve pentagonal faces. How can something changing possibly be something unchanging? Another urgent question concerns how you personally fit into this supposed mathematical structure—how can your self-awareness, thoughts and feelings be part of a mathematical structure?

How Can Physical Reality Be Mathematical?

Timeless Reality

Einstein can help us answer these questions. He taught us that there are two equivalent ways of thinking about our physical reality: either as a three-dimensional place called *space*, where things change over time, or as a four-dimensional place called *spacetime* that simply exists, unchanging, never created and never destroyed.* These two perspectives cor-

---

* This idea of time as the fourth dimension as an unchanging reality has been promoted and explored by many, including H. G. Wells in his 1895 novel *The Time Machine*. Julian Barbour gives an interesting account of the idea and its history in his book *The End of Time*.

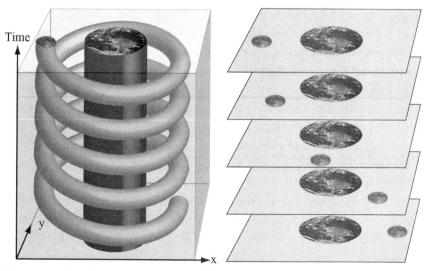

Figure 11.1: The Moon's orbit around the Earth. We can equivalently think of this either as a position in space that changes over time (right), or as an unchanging spiral shape in spacetime (left), corresponding to a mathematical structure. The snapshots of space (right) are simply horizontal slices of spacetime (left).

respond to the frog and bird perspectives on reality that we discussed in Chapter 9: the latter is the outside overview of a physicist studying its mathematical structure, like a bird surveying a landscape from high above; the former is the inside view of an observer living in this structure, like a frog living in the landscape surveyed by the bird.

Mathematically, spacetime is a space with four dimensions, the first three being our familiar dimensions of space, and the fourth dimension being time. Figure 11.1 illustrates this idea. Here, I've drawn it so that the time dimension is in the vertical direction and the space dimensions are in the horizontal directions. To avoid confusion, I've plotted only two of the three space dimensions, labeled x and y, because smoke starts pouring out of my ears if I try to visualize four-dimensional objects. . . . The figure shows the Moon moving around Earth in a circular orbit—to keep things legible, I've drawn the orbit much smaller than to scale and made several simplifications.* The right panel shows the frog perspective: five snapshots of space with the Moon in different positions, while

---

* To keep things simple, Figure 11.1 ignores the fact that both the Earth and the Moon are rotating, that the Moon's orbit is slightly oblong (it's an ellipse rather than a perfect circle), and that the Moon's gravitational pull causes Earth to undergo circular motion, too, with a radius that's about 74% of Earth's radius.

Earth remains in the same place. The left panel shows the bird perspective: here the motion of the frog perspective is replaced by unchanging *shapes* in spacetime. Since Earth isn't moving, it's at the sample place in space for all time, and therefore makes a vertical cylinder in spacetime. The Moon is more interesting, manifesting itself as a spiral in spacetime that encodes where it is at different times. Please look at the left and right panels until you've figured out how they're related, since this is crucial for the rest of our discussion. To get snapshots of space (right) from spacetime (left), you simply make horizontal slices through spacetime at the times you're interested in.

Note that spacetime doesn't exist within space and time—rather, space and time exist within it. I'm arguing that our external physical reality is a mathematical structure, which is by definition an abstract, immutable entity existing outside of space and time. This mathematical structure corresponds to the bird perspective of our reality, not the frog perspective, so it should contain spacetime, not just space. The mathematical structure contains additional elements as well, as we'll get to below, corresponding to the stuff contained in our spacetime. However, this doesn't alter its timeless nature: if the history of our Universe were a chess game, the mathematical structure would correspond not to a single position, but to the entire game (Figure 10.6). If the history of our Universe were a movie, the mathematical structure would correspond not to a single frame but to the entire DVD. So from the bird's perspective, trajectories of objects moving in four-dimensional spacetime resemble a tangle of spaghetti. Where the frog sees something moving with constant velocity, the bird sees a straight strand of uncooked spaghetti. Where the frog sees the Moon orbit Earth, the bird sees the rotini-like spiral of Figure 11.1. Where the frog sees hundreds of billions of stars moving around in our Galaxy, the bird sees hundreds of billions of intertwined spaghetti strands. To the frog, reality is described by Newton's laws of motion and gravitation. To the bird, reality is the geometry of the pasta.

Past, Present and Future

"Excuse me, but what's the time?" I'm guessing that you, like me, have asked this question, as if there were such a thing as *the* time at a fundamental level. Yet you've probably never approached a stranger and asked, "Excuse me, but what's the place?" If you were really hopelessly lost, you'd probably instead have said something like "Excuse me, but where am I?"—thereby acknowledging that you're not asking about a property of space, but rather about a property of yourself: your location in space while you're asking the question. Similarly, when you ask for the time, you're not really asking about a property of time, but rather about your location in time. Spacetime contains all places and all times, so there's no *the* time any more than there's *the* place. It would therefore be more appropriate (scientifically if not socially) to ask, "When am I?" Spacetime is like a map of cosmic history without a "You are here" marker. If you need such a marker to orient yourself, I recommend a phone with both a clock and a GPS.

When Einstein wrote that "The distinction between past, present, and future is only a stubbornly persistent illusion," he was referring to the fact that these concepts have no objective meaning in spacetime. Figure 11.2 illustrates that when we think about the "present," we mean the time slice through spacetime corresponding to the time when we're having that thought. We refer to the "future" and "past" as the parts of spacetime above and below this slice. This is analogous to your use of the terms *here*, *in front of me* and *behind me* to refer to different parts of spacetime relative to your present position. The part that's in front of you is clearly no less real than the part behind you—indeed, if you're walking forward, some of what's presently in front of you will be behind

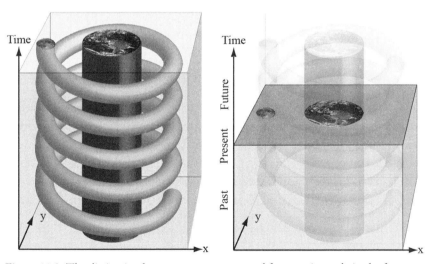

Figure 11.2: The distinction between past, present and future exists only in the frog perspective (right), not in the bird perspective of the mathematical structure (left)—in the latter, you can't ask, "What time is it?"; merely, "When am I?"

you in the future, and is presently behind various other people. Analogously, in spacetime, the future is just as real as the past—parts of spacetime that are presently in your future will in your future be in your past. Since spacetime is static and unchanging, no parts of it can change their reality status, and all parts must be equally real.*

In summary, time is not an illusion, but the flow of time is. So is change. In spacetime, the future exists and the past doesn't disappear. When we combine Einstein's classical spacetime with quantum mechanics, we get quantum parallel universes as we saw in Chapter 8. This means that there are many pasts and futures that are all real—but this in no way diminishes the unchanging mathematical nature of the full physical reality.

This is how I see it. However, although this idea of an unchanging reality is venerable and dates back to Einstein, it remains controversial and subject to vibrant scientific debate, with scientists I greatly respect expressing a spectrum of views. For example, in his book *The Hidden Reality*, Brian Greene expresses unease toward letting go of the notions that change and creation are fundamental, writing, "I'm partial to there being a process, however tentative . . . that we can imagine generating the multiverse." Lee Smolin goes further in his book *Time Reborn*, arguing that not only is change real, but that indeed time may be the only thing that's real. At the other end of the spectrum, Julian Barbour argues in his book *The End of Time* not only that change is illusory, but that one can even describe physical reality without introducing the time concept at all.

How Spacetime and "Stuff" Can Be Mathematical

Earlier, we've seen how spacetime can be viewed as a mathematical structure. But what about all the stuff *in* spacetime, say, the book you're reading right now? How can that be part of a mathematical structure?

In recent years, we've seen that many things that seemed totally unrelated to mathematics, such as texts, sounds, images and movies, can be represented mathematically by computers and transmitted over the

---

* In his book *The Hidden Reality*, Brian Greene further hammers home this conclusion by pointing out that, according to Einstein's relativity theory, the horizontal slice delimiting past from future in Figure 11.2 gets tilted if you start moving; there clearly can't be any fundamental distinction between past and future if you can reclassify a distant supernova explosion from already having happened to not yet having happened simply by walking faster.

Internet as a bunch of numbers. Let's take a closer look at how computers do this—because as we'll soon see, nature is doing something rather analogous to represent all the stuff around us.

I just typed the word *word*, and my laptop represented it in its memory as the four-number sequence "119 111 114 100." It represents each lowercase letter by a number that's 96 plus its order in the alphabet, so $a = 97$, $w = 119$. At the same time, my laptop is playing *De Profundis* by the Estonian composer Arvo Pärt, which it's also representing as a sequence of numbers—these numbers are interpreted not as letters, but as the positions in which it puts the loudspeaker membranes at 44,100 different instances each second, which in turn causes the air vibrations that my ears and my brain interpret as sound. As soon as I hit the *w* key on my keyboard, my laptop displayed a visual image of a *w* on my screen, and this image is also represented by numbers. Although all my screen images look smooth and continuous, my screen is in fact made up of $1,920 \times 1,200$ pixels in a rectangular grid, as illustrated in Figure 11.3, and the color of each pixel is represented by three numbers, each between 0 and 255, specifying the intensities of red, green and blue light coming from the pixel; suitable combinations of these three colors can then produce all intensities of all the colors of the rainbow. Last night, when my kids and I watched a YouTube video, my laptop similarly divided not only the two spatial dimensions of my screen into pixels, but also the time dimension, slicing time into 30 frames per second.

At work, we physicists often simulate some event in 3-D, such as a hurricane, a supernova explosion or the formation of a solar system. To do this, we divide three-dimensional space into 3-D pixels (*voxels*). We also divide the 4-D spacetime into 4-D voxels. Each such 4-D voxel represents what's happening at that place at that time by a group of

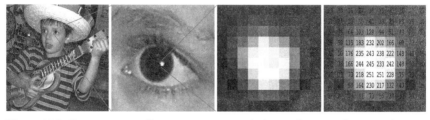

Figure 11.3: Computers usually represent gray-scale images by a number at each point (pixel) of the image (rightmost panel). The larger the number, the more intense the light from the pixel, with 0 representing black (no light at all) and 255 representing white. Similarly, so-called *fields* in classical physics are represented by numbers at each point in spacetime, which, loosely speaking, specify the amount of "stuff" existing at each point.

numbers encoding everything that's relevant, such as the temperature, the pressure, and the densities and velocities of different substances in the voxel. For example, in a simulation of our Solar System, a voxel corresponding to the center of the Sun will have an extremely large temperature number, and a voxel outside the Sun containing almost empty space will have a pressure number close to zero. The numbers in neighboring voxels satisfy certain relations that are captured by mathematical equations, and when a computer is performing a simulation, it's using these relations to fill in missing numbers like a Sudoku player. If a computer is making a weather forecast, then the spacetime voxels corresponding to right now are filled in with measured numbers for air pressure, air temperature, etc. The computer then uses the relevant equations to calculate the numbers that go in the spacetime voxels corresponding to tomorrow and the rest of the week.

Although such simulations represent aspects of our external physical reality mathematically, they do so only approximately. Spacetime certainly isn't made of the crude voxels we use to simulate tomorrow's weather, which is one of the reasons why weather forecasts are often inaccurate. Yet this idea that there's a bunch of numbers at each point in spacetime is quite deep, and I think it's telling us something not merely about our *description* of reality, but about reality itself. One of the most fundamental concepts in modern physics is that of a *field*, which is just this: something represented by numbers at each point in spacetime. For example, there's a *temperature field* corresponding to the air around you: there's a well-defined temperature at each point, totally independent of any human-invented voxels, and you can measure the temperature number by holding a thermometer there—or your finger, if you don't need great accuracy. There's also a *pressure field:* at each point, there's a pressure number which you can measure with a barometer—or with your ear, which will hurt if the number is too extreme and which can detect sound if the pressure is fluctuating over time.

We now know that neither of these two fields are truly fundamental: they're merely different measures of how fast the air molecules are moving on average, so these numbers stop being well defined if you try to measure them on subatomic scales. However, there are other fields that seem to be quite fundamental, forming part of the very fabric of our external physical reality. As a first example, let's look at the *magnetic field*. It's represented by not one (like temperature) but three numbers at each point in spacetime, encoding both a strength and a direction.

You've probably measured the magnetic field using a compass, watching its magnetic pointer align itself with Earth's magnetic field, which points north. The pointer aligns itself faster if the magnetic field is stronger, such as near an MRI machine. A second example is the *electric field*, which is also represented by a triplet of numbers encoding strength and direction. An easy way to measure it is by the force it exerts on a charged object—like when your hair gets electrically attracted to a plastic comb. These electric and magnetic fields can be elegantly unified into what's known as the *electromagnetic field*, represented by six numbers at each point in spacetime. As we discussed in Chapter 7, light is simply a wave rippling through the electromagnetic field, so if our physical world is a mathematical structure, then all the light in our Universe (which feels quite physical) corresponds to six numbers at each point in spacetime (which feels quite mathematical). These numbers obey the mathematical relations that we know as Maxwell's equations, shown in Figure 10.4.

There's a caveat here: what I've just described was our understanding of electricity, magnetism and light in classical physics. Quantum mechanics complicates this picture, but without making it any less mathematical, replacing classical electromagnetism with *quantum field theory*, the bedrock of modern particle physics. In quantum field theory, the wavefunction specifies the degree to which each possible configuration of the electric and magnetic fields is real. This wavefunction is itself a mathematical object, an abstract point in Hilbert space.

As we saw in Chapter 7, quantum field theory says that light is made of particles called photons, and, crudely speaking, the numbers constituting the electric and magnetic fields can be thought of as specifying how many photons there are at each time and place. Just as the strength of the electromagnetic field corresponds to the number of photons at each time and place, there are other fields corresponding to all the other elementary particles known. For example, the strengths of the *electron field* and the *quark field* relate to the numbers of electrons and quarks at each time and place. In this way, all motions of all particles in all of spacetime correspond, in classical physics, to a bunch of numbers at each point in a four-dimensional mathematical space—a mathematical structure. In quantum field theory, the wavefunction specifies the degree to which each possible configuration of each of these fields is real.

As we discussed in Chapter 7, we physicists still haven't found a mathematical structure that can describe *all* aspects of reality, including gravity, but so far, there's no indication that string theory or any of the other most actively pursued candidates for such a description are any less mathematical than quantum field theory.

Description Versus Equivalence

Before moving on, there's an important semantic issue that we need to sort out. Whereas most of my physics colleagues would say that our external physical reality is (at least approximately) *described by* mathematics, I'm arguing that it *is* mathematics (more specifically, a mathematical structure). In other words, I'm making a much stronger claim. Why?

Everything I've said so far in this chapter suggests that our external physical reality can be *described* by a mathematical structure. If a future physics textbook contains the coveted Theory of Everything (ToE), then its equations are a complete description of the mathematical structure that is the external physical reality. I'm writing *is* rather than *corresponds to* here, because if two structures are equivalent, then there's no meaningful sense in which they're not one and the same, as emphasized by the Israeli philosopher Marius Cohen.* Recall the powerful mathematical notion of equivalence that we described in Chapter 10, which embodies the very essence of mathematical structures: if two complete descriptions are equivalent, then they're describing one and the same thing.† This means that if some mathematical equations completely describe both our external physical reality and a mathematical structure, then our external physical reality and the mathematical structure are one and the same, and then the Mathematical Universe Hypothesis is true: our external physical reality is a mathematical structure.

Remember that two mathematical structures are equivalent if you can pair up their entities in a way that preserves all relations. If you can thus pair up every entity in our external physical reality with a corresponding one in a mathematical structure ("This electric-field strength here in physical space corresponds to this number in the mathematical

---

* Marius Cohen, "On the Possibility of Reducing Actuality to a Pure Mathematical Structure" (master's thesis, Ben Gurion University of the Negev, Israel, 2003).
† If you have a mathematics background and are familiar with the notion of isomorphism, you can restate this argument as follows. From the definition of a mathematical structure, it follows that if there's an isomorphism between a mathematical structure and another structure (a one-to-one correspondence between the two that respects the relations), then they're one and the same. If our external physical reality is isomorphic to a mathematical structure, it therefore fits the definition of being a mathematical structure.

structure," for example), then our external physical reality meets the definition of being a mathematical structure—indeed, that same mathematical structure.

We saw in Chapter 10 that if someone wishes to avoid accepting the Mathematical Universe Hypothesis, they can do so by rejecting the External Reality Hypothesis that there's an external physical reality completely independent of us humans. They could then argue that our Universe is somehow made of stuff perfectly described by a mathematical structure, but which also has other properties that aren't described by it, and can't be described in an abstract, human-independent, baggage-free way. However, I think this viewpoint would make the famous science philosopher Karl Popper from Chapter 6 turn in his grave, since he emphasized that scientific theories must have observable effects. In contrast, since the mathematical description is supposedly perfect, accounting for everything that can be observed, those additional bells and whistles that would make our Universe nonmathematical would by definition have no observable effects whatsoever, rendering them 100% unscientific.

What Are You?

We've now seen how both spacetime and the stuff in it can be viewed as being part of a mathematical structure. But what about *us*? Our thoughts, our emotions, our self-awareness, and that deep existential feeling *I am*—none of this feels the least bit mathematical to me. Yet we too are made of the same kinds of elementary particles that make up everything else in our physical world, which we've argued is purely mathematical. How can we reconcile this?

In my opinion, we don't yet fully understand what we are. Moreover, as we discussed in Chapter 9, we don't really need to fully understand the mysteries of consciousness to understand our external physical reality. Nonetheless, I feel that modern physics has provided some tantalizing hints about fruitful ways of viewing ourselves, so let's explore this topic further.

The Braid of Life

George Gamow, the cosmology pioneer whom we encountered in Chapter 3, titled his autobiography *My World Line,* a phrase also used by Einstein to refer to paths through spacetime. However, your own world line strictly speaking isn't a line: it has a non-zero thickness and it's not straight. Let's first consider the roughly 10^{29} elementary particles (quarks and electrons) that your body is made of. Together, they form a tubelike shape through spacetime, analogous to the spiral shape of the Moon's orbit (Figure 11.1) but more complicated, reflecting the fact that your motion from birth to death is more complicated than the Moon's. For example, if you're swimming laps in a pool, that part of your spacetime tube has a zigzag shape, and if you're using a playground swing, that part of your spacetime tube has a serpentine shape.

However, the most interesting property of your spacetime tube isn't its bulk shape, but its internal structure, which is remarkably complex. Whereas the particles that constitute the Moon are stuck together in a rather static arrangement, many of your particles are in constant motion relative to one another.

Consider, for example, the particles that make up your red blood cells. As your blood circulates through your body to deliver the oxygen you need, each red blood cell traces out its own unique tube shape through spacetime, corresponding to a complex itinerary through your arteries, capillaries and veins with regular returns to your heart and lungs. These spacetime tubes of different red blood cells are intertwined to form a braid pattern (Figure 11.4, middle panel) which is more elaborate than anything you'll ever see in a hair salon: whereas a classic braid consists of three strands with perhaps thirty thousand hairs each, intertwined in a simple repeating pattern, this spacetime braid consists of trillions of strands (one for each red blood cell), each composed of trillions of hairlike elementary-particle trajectories, intertwined in a complex pattern that never repeats. In other words, if you imagine spending a year giving a friend a truly crazy hairdo, braiding his hair by separately intertwining not strands but all the individual hairs, the pattern you'd get would still be very simple in comparison.

Yet the complexity of all this pales in comparison to the patterns of information processing in your brain. As we discussed in Chapter 8 and illustrated in Figure 8.7, your roughly hundred billion neurons are constantly generating electrical signals ("firing"), which involves shuffling around billions of trillions of atoms, notably sodium, potassium and calcium ions. The trajectories of these atoms form an extremely elaborate braid through spacetime, whose complex intertwining corresponds to storing and processing information in a way that somehow gives rise

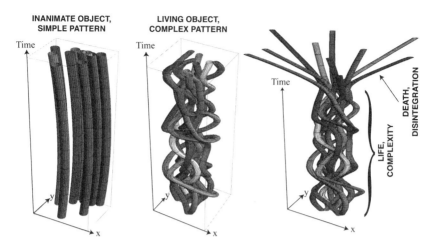

Figure 11.4: Complexity is a hallmark of life. The motion of an object corresponds to a pattern in spacetime. An inanimate clump of ten particles accelerating toward the left constitutes a simple pattern (left), while the particles that make up a living organism constitute a complex pattern (middle), corresponding to the complex motions that accomplish information processing and other vital processes. When a living organism dies, it eventually disintegrates and its particles separate from each other (right). These crude illustrations show merely ten particles; your own spacetime pattern involves about 10^{29} particles and is mind-blowingly complex.

to our familiar sensation of self-awareness. There's broad consensus in the scientific community that we still don't understand how this works, so it's fair to say that we humans don't yet fully understand what we are. However, in broad brushstrokes, we might say this: *You're a pattern in spacetime*. A mathematical pattern. Specifically, you're a braid in spacetime—indeed one of the most elaborate braids known.

Some people find it emotionally displeasing to think of themselves as a collection of particles. Indeed, I got a good laugh back in my twenties when my friend Emil addressed my friend Mats as an *atombhög*, Swedish for "atom heap," in an attempt to insult him. However, if someone says, "I can't believe I'm just a heap of atoms!" I object to the use of the word *just*: the elaborate spacetime braid that corresponds to their mind is, hands down, the most beautifully complex type of pattern we've ever encountered in our Universe. The world's fastest computer, the Grand Canyon or even the Sun—their spacetime patterns are all simple in comparison.

Whereas many of the particles inside you are in the constant complex motion that corresponds to your being alive, others move only in less-elaborate ways, such as many of the ones that make up your skin and help keep the other particles from flying apart. This means that your spacetime tube is a bit like those electrical cables where the inner strands are braided together and the shared insulation on the outside resembles a hollow tube. Moreover, most of your particles get regularly replaced. For example, about three-quarters of your body weight is water molecules, which get replaced every month or so, and your skin cells and red blood cells are replaced every few months. In spacetime, the trajectories of these particles joining and then leaving your body make a pattern reminiscent of the familiar silk strands attached to a corncob. At both ends of your spacetime braid, corresponding to your birth and death, all the threads gradually separate, corresponding to all your particles joining, interacting and finally going their own separate ways (Figure 11.4, right). This makes the spacetime structure of your entire life resemble a tree: at the bottom, corresponding to early times, is an elaborate system of roots corresponding to the spacetime trajectories of many particles, which gradually merge into thicker strands and culminate in a single tubelike trunk corresponding to your current body (with a remarkable braidlike pattern inside as we described above). At the top, corresponding to late times, the trunk splits into ever-finer branches, corresponding to your particles going their own separate ways once your life is over. In other words, the pattern of life has only a finite extent along the time dimension, with the braid coming apart into frizz at both ends.

All of the patterns we've discussed of course exist in four dimensions rather than three, and the metaphors about braids, cables and trees, shouldn't be taken too literally. The key point is simply that you can be an unchanging pattern in spacetime—the specific details of this pattern are less important for the points we're making. This pattern is part of the mathematical structure that is our Universe, and the relations between different parts of the pattern are encoded in mathematical equations. As we saw in Chapter 8, Everett's quantum mechanics endows you with an even more interesting—but no less mathematical—structure, since a single you (the tree trunk) can split into many branches, each feeling that they're the one and only you—we'll return to this later.

Living in the Moment

Now we've discussed how space itself, the stuff in space and even you yourself can be a part of a mathematical structure. But this came at a price: we had to abandon the familiar feeling that time flows as a mere illusion, and instead think of time as a fourth dimension in an unchanging mathematical structure. So how can we reconcile this with our subjective experience that things change from one moment to the next?

All your subjective perceptions exist in spacetime, just as every scene of a movie exists on its DVD. Specifically, spacetime contains a large number of braidlike patterns corresponding to subjective perceptions both at different places, corresponding to different people, and at different times. Let's refer to each such perception as an "observer moment." I coined a different name for this in my 1996 mathematical-universe paper, but I like *observer moment* better, and Nick Bostrom and other philosophers have established it as the standard term in recent years. You know from experience that some of these observer moments feel connected and fused together into a seemingly seamless sequence, corresponding to what you call your life. However, this feeling raises tough questions. *How does the connecting work?* Specifically, is there some sort of rule for which observer moments feel connected, and why does this connected sequence of observer moments subjectively feel like time flowing?

An obvious guess could be that the connecting has to do with continuity: that two observer moments feel connected if they're adjacent in spacetime and part of the same pattern. However, Figure 11.5 illus-

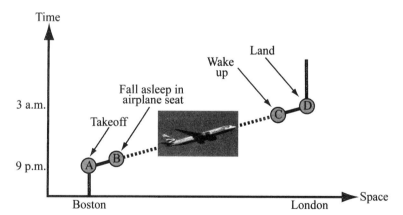

Figure 11.5: My world line when flying to London. I take off (A), fall asleep shortly thereafter (B) and wake up (C) shortly before landing (D). Even though my conscious perception at (C) is at a different point than (B) in both space and time, it appears to connect seamlessly with my last conscious perception at (B), but not with the many other conscious perceptions (of fellow passengers) that are much closer to (C) than to (B) in both space and time.

trates that the question is trickier than it first seems, and that the answer can't be this simple. First of all, the observer moment (labeled C) corresponding to my waking up feels connected to the one (labeled B) corresponding to my falling asleep. Specifically, it feels to me as if C is the continuation of B, even though these two observer moments are nowhere near each other in spacetime. Second, there are many other observer moments (corresponding to perceptions of other people on my flight) that are much closer to C in both space and time, so why doesn't C instead feel connected to one of those observer moments? Third, imagine a perfect clone of me being assembled while I'm sleeping, with all the particles in the same configurations, except located in another identical-looking airplane. Then the subjective perception of my clone upon awakening will be subjectively identical to the one I have at C, so by definition, it too feels connected to B even though its spacetime pattern isn't.*

This suggests that the continuity business is a red herring, and that there simply is no new physical process to be discovered that somehow makes certain observer moments feel connected, thereby explaining our familiar feeling that time flows. Fortunately, there's a simpler explanation that doesn't require any new physics, which we'll now explore. The Mathematical Universe Hypothesis combined with our subjective experience tells us that there are very complex braidlike structures in spacetime that are self-aware and subjectively feel like observer moments. We know that these structures can be quite localized both in space and in time: your brain occupies just over a liter of volume, and the time it takes for your brain to have individual thoughts or sensations is typically about a tenth of a second, give or take a factor of ten. This means that how an observer moment subjectively feels depends only on what's right there in that localized region of spacetime—not on what's elsewhere in space (such as the external reality you see around you), and not on what's elsewhere in time (such as what you experienced a few seconds ago). Yet crucial components of your conscious perceptions involve both of those: right now, you feel aware both of the book in front of you and of the sentence that you read five seconds ago, even

---

* If the assembly instructions for my clone were transmitted wirelessly from the body scanner that analyzed the original me, then the spacetime braids of me and my clone would still be connected by a very elaborate pattern in the electromagnetic field. But an identical copy of me waking up in the Level I multiverse of Chapter 6 would feel connected to C as well, without there being any information transfer between the two copies.

Figure 11.6: The subjective perceptions in spacetime (observer moments) of a diver and a skier at four separate times. Each film strip corresponds to a single observer moment, including both a clear image of what's currently happening, and progressively hazier memories of what happened in the past. If I'd rearranged the eight strips in a random order, you could easily reconstruct the sequences because of relations between them: the current visual impressions (right frame) in some observer moments match memories in others.

though neither belong to the small spacetime region constituting your present observer moment. In other words, it appears that the way your observer moment subjectively feels involves what's elsewhere in both space and time—even though it wasn't supposed to involve either. How can this be?

We discussed the spatial part of this paradox in Chapter 9, and concluded that your consciousness is actually observing not the outside world, but rather an elaborate reality model contained in your brain which is continually updated via input from your sensory organs to track what's actually taking place in the outside world.* So the space-

---

* For a detailed discussion of time experience and the rich philosophical literature on the subject during the past two millennia, see http://plato.stanford.edu/entries/time-experience. In particular, the idea that key aspects of time perception such as duration can only be explained as perceptions of our memory were explored about 1,600 years ago by Saint Augustine; the Mathematical Universe Hypothesis gives such questions a new urgency.

time pattern corresponding to your current observer moment includes the state of your reality model right now. As illustrated in Figure 11.6, it's quite analogous for the temporal part: your world model includes not merely information about the present state of your surroundings, but also memories of how your surroundings were in the past. Each of the eight film strips represents a single observer moment. For each one, there's a clear image of what's currently happening, and progressively hazier memories of what happened in the past. You're therefore aware of an entire time sequence of events right now at this very moment. Just as your spatial-reality model gives you the subjective feeling of looking at a three-dimensional space even though your mind is actually looking at the reality model in your brain, this temporal-reality model with its sequence of memories gives you the subjective feeling of time flowing through a sequence of events even while your mind is actually looking at the reality model in your brain in a single observer moment.

In other words, your subjective feeling that time is flowing comes from the relations between these memories that you have right now. Imagine a thought experiment where a perfect clone of me is built asleep, complete with all my memories, and is only woken up long enough to perceive a single observer moment. He'd still feel that time flowed from a complex and interesting past, even though he got to experience only that one moment. This means that the subjective perceptions of duration and change are qualia, basic instantaneous perceptions just as redness, blueness or sweetness.

This implication of the Mathematical Universe Hypothesis is pretty radical, so please pause your reading for a moment to take it in and think about it. What you're aware of right at this moment feels not like a photo but like a movie clip. This movie isn't reality—it exists only in your head, as part of your brain's reality model. It contains lots of information about the actual external physical reality—as long as you aren't dreaming or hallucinating—but still constitutes only a very heavily edited version of reality, akin to the evening news on TV, mainly featuring certain highlights of patterns nearby in space and time that your brain thinks are useful for you to be aware of.

Just as when you watch news on TV, you're not watching distant parts of space directly: you're watching merely an edited movie *about* these parts of space. Similarly, you're not watching the past, but an edited movie *about* the past. As opposed to watching the news during several minutes, you watch your internal newsreel *all at once*, thus being simultaneously aware of present and past events. A second later, you watch your internal newsreel once again, all at once, and it's mostly unchanged like a TV rerun, but has been slightly re-edited to add another second of material at the end and shorten the remainder. In other words, even though an observer moment objectively occupies less than a liter of volume and a second of time, it subjectively feels as if it occupies all the space you're aware of and all the time you remember. You feel as if you're observing this space and time from here and now, but all that space and time are just part of the reality model that you're experiencing. This is why you subjectively feel that time flows even though it doesn't.

Self-Awareness

Moreover, you too are in the movie, since your reality model includes a model of yourself—that's why you're not merely aware but also self-aware. This means that when you feel that you're looking at this book, what's really going on is that your brain's reality model has its model of you looking at its model of the book, as illustrated in Figure 11.7.

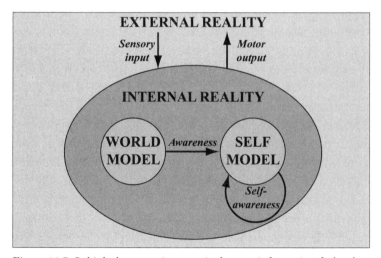

Figure 11.7: I think that consciousness is the way information feels when being processed in certain complex ways, and that the particular kind of consciousness that we humans subjectively perceive arises when your brain's model of you is interacting with your brain's model of the world. The arrows above indicate information flow. For example, information input from your senses continually helps your world model track key aspects of what's actually going on in the external reality, and information output via your motor cortex controls your muscles to affect the external reality, say, by turning a page in this book.

Which leads to the ultimate consciousness question: who's looking at your brain's reality model, to give rise to subjective consciousness? Here's my guess: *nobody!* If there were another part of your brain that really looked at the whole reality model and became aware of all the information in it, then this brain region would need to physically transfer all that information into its own local copy. This would be a huge waste of resources from an evolutionary perspective, and there's no evidence from neuroscience research of such wasteful duplication. Moreover, it wouldn't answer the question: if a spectator is really needed, then this duplicate reality model would in turn need a spectator to be subjectively perceived, leading to another infinite regress problem.

Rather, my guess is that the answer is beautifully simple: no spectator is needed, because your consciousness basically *is* your reality model. *I think that consciousness is the way information feels when being processed in certain complex ways.* Since the different parts of your brain interact with each other, different parts of your reality model can interact with each other, so the model of you can interact with your model of the outside world, giving rise to the subjective sensation of the former perceiving the latter. When you're looking at a strawberry, your brain's model of the color red feels subjectively very real—and so does your brain's model of your mind's eye as an observing vantage point. We already know that our brain is astonishingly creative in interpreting the same basic types of electrical signals in a bundle of neurons as qualia that subjectively feel completely different: we perceive them as colors, sounds, smells, tastes or touches, depending on whether the neuron bundle comes from our eyes, ears, nose, mouth or skin. The key difference lies not in the neurons that carry this information, but in the patterns whereby they're connected. Although your perception of yourself and your perception of the strawberry are extremely different, it's therefore plausible that they're both fundamentally the same kind of thing: complex patterns in spacetime. In other words, I'm arguing that your perceptions of having a self, that subjective vantage point that you call "I," are qualia just as your subjective perceptions of "red" or "green" are. In short, redness and self-awareness are both qualia.

Predicting Your Future

One of the key purposes of science, and indeed one of the key purposes of having a brain, is predicting our future. But if time doesn't flow, then what do we even mean by predicting our future?

Figure 11.6 illustrates how we can reformulate this as a sensible question even without the notions of change or the flow of time. The eight observer moments illustrated belong to two different people, one diving and one skiing, each corresponding to a long braidlike pattern in spacetime. Comparing the eight observer moments reveals some interesting relations between them, where the current visual impressions (right frame on each film strip) of some observer moments match fresh memories (middle frame) in others, and fresh memories in some match older memories (left frame) in others. This uniquely defines two separate time sequences of observer moments, corresponding to the left and right columns of strips, with later times corresponding to higher up in the figure.

Consider all observer moments in all of spacetime. The ones that are natural to call *your future perceptions* are the ones that can be similarly matched with your current observer moment, fitting together like pieces of a puzzle. Specifically, they should share your current memories in the correct order (with some reasonable allowance for forgetting and misremembering), and with additional memories added to the sequence. Suppose, for example, that you're the diver who's just seen the giant turtle swim up toward the right (Figure 11.6, left column, second observer moment from the top) and want to predict your future. As a thought experiment, suppose also that you're infinitely intelligent and have figured out which mathematical structure our Universe is, and have calculated what all of its observer moments are and how they subjectively feel. You realize that the only one that matches with your current observer moment and has an extra second's worth of perceptions at the end is the top-left observer moment in the figure. You therefore make the prediction that this is what you'll perceive in one second: in one second, you're going to see the giant turtle turn and start swimming toward you. In this way, you recover the traditional scientific notion of causality: that you can predict the future from the present.

Where Are You? (And What Do You Perceive?)

We've now seen how our physical reality can be a mathematical structure, including space, time, stuff and even you yourself. We've also seen how you, at least in principle, might be able to make predictions about your future by analyzing observer moments, matching them up like pieces in a puzzle. To predict things in practice, this observer-moment approach often reduces to "business as usual" physics. Suppose, for example, that you do the experiment illustrated in Figure 10.2, throwing a basketball up into the air and studying its motion. If you assume (1) that Einstein's equations of gravity describe this motion, and (2) that there's no other person who subjectively feels exactly like you, with the exact same life memories, then you know that the only future observer moments that smoothly match with your present one are ones where you see the ball fly in a parabolic trajectory as in the figure, so that's what you predict that you'll perceive. How did you know that it was going to be a parabola and not another shape, say, a spiral? By solving Einstein's equations and getting a parabola as the solution.

Predicting Your Future, Revisited

We've seen, however, that the second assumption is likely to be false: if either the Level I or the Level III multiverse exists, then there are other people who subjectively feel exactly like you, and the problem of predicting your future gets much more interesting! I was sneaky when I chose the heading "Where Are You (And What Do You Perceive)?" because I want to ask the question also when the word *you* is interpreted in the plural sense. As we'll see, it gets particularly tricky when the number of yous increases or decreases.

Let's continue our thought experiment where you know every detail about the mathematical structure that we inhabit. Then predicting your future boils down to three steps:

1. Find all self-aware entities in it.
2. Figure out what they subjectively perceive so that you know which ones might be you, and what they perceive in the future.
3. Predict what you'll subjectively perceive in the future (probabilities for different options).

Amusingly, as we'll see below, all three of these steps involve daunting unsolved problems!

Finding Self-Awareness

Let's start with the first step. Given some mathematical structure that is our external physical reality, perhaps including a multiverse, how do

we find all self-aware entities in it? We discussed how we humans correspond to certain complex braidlike patterns in spacetime. However, we
don't want to limit our exploration of self-awareness to our own human
form of life, so let's use the more general term *self-aware substructure* (or
"SAS" for short) to refer to any part of a mathematical structure that has
subjective perceptions. We'll also use *observer* as a synonym from time
to time, but will stick with SAS whenever we need to remind ourselves
to avoid anthropocentrism.

So how do we find SASs in a mathematical structure? The short
answer is clearly that we don't yet know—science simply hasn't advanced
to that point. We can't even answer the question in the particular case
we're most familiar with: our own spacetime. First of all, we don't know
what mathematical structure we inhabit, since a self-consistent model
of quantum gravity remains conspicuous with its absence. Second, even
if we knew our mathematical structure, we wouldn't know what to do
with it to find its SASs.

Imagine that a friendly visiting alien gives you an "SAS-buster," a convenient handheld device that looks a bit like a metal detector but which
makes a loud beeping noise whenever it detects an SAS. You play around
with it and find that it beeps quietly when you point it at a goldfish, more
loudly when you point it at a cat and with ear-piercing volume when you
point it at yourself, but that it remains dead silent when you point it at a
cucumber, a car or a corpse. How might this SAS-buster work?

Although the minimalistic user manual that came with the SAS-
buster merely refers to "a proprietary algorithm," my guess is that part
of what it does is measure both the complexity and the information content of the object you point it toward. The *complexity* of something is
usually defined as the smallest number of bits required to fully describe
it (a bit is a zero or a one). For example, a diamond describable as 10^{24}
carbon atoms arranged in a perfectly regular lattice pattern has very low
complexity compared to a hard drive with a terabyte of random numbers, since the latter can't be described with less than a terabyte (about
8×10^{12} bits) of information. Yet that hard drive is much less complex
than your brain, where more than a hundred quadrillion (10^{17}) bits of
information are needed just to describe the state of its synapses alone.

However, a hard drive wouldn't be self-aware no matter how big it
was, so complexity alone clearly isn't enough to make an SAS. I suspect
that another quantity that the SAS-buster measures is the *information
content* of the object you point it at. There are rigorous mathematical

definitions of information content in mathematics and physics, tracing back to the work of Claude Shannon and John von Neumann over half a century ago. Whereas the complexity of an object measures how complicated it is to describe, its information content* measures the extent to which it describes the rest of the world. In other words, information is a measure of how much *meaning* complexity has. If you fill your hard drive with random numbers, then it contains no information about the outside world, but if you fill it with history books or with movie clips of your family, then it does. Your brain contains a vast amount of information about the outside world, both in the form of memories of distant times and places and in the form of its continually updated model of what's happening around you right now. When a person dies, the information content of the electrical firing patterns of their neurons vanishes as this entire electrical system shuts down, and before long, the information content stored chemically and biologically in their synapses begins to disappear as well.

Yet complexity and information content still aren't sufficient to guarantee self-awareness—for example, a video camera has both without being self-aware in any meaningful sense. This means that the SAS-buster needs to look for additional ingredients of self-awareness that are harder to understand. For example, Figure 11.7 suggests that an SAS needs to be able not only to store information, but also to process it in some form of computation, and that a high degree of interconnectedness may be required in the information processing. The neuroscientist Giulio Tononi has made an intriguing proposal for how to quantify the required interconnectedness, described in the publications by Koch and Tononi in the "Suggestions for Further Reading" section. The core idea is that for an information processing system to be conscious, it needs to be integrated into a unified whole that can't be decomposed into nearly independent parts.[†] This means that all parts need to compute jointly

---

* What I'm casually calling the information content of an object is in technical terms called the mutual information between the object and the rest of the world.
† This is closely linked to so-called redundancy and error-correcting codes used in bar codes, hard drives, mobile telephony and other modern information technology: you use more bits than the minimum needed, which encode your information in a clever collective way such that none of your information is lost even if you lose any modest fraction of your bits. Our brain appears to use a similarly redundant architecture, since it doesn't seem to depend critically on any single neuron, and keeps functioning well even if a modest number of neurons die. Perhaps part of the reason that consciousness evolved is that such redundancy is evolutionarily useful.

with lots of information about each other—otherwise there would be more than one independent consciousness, such as in a room full of people or, perhaps, in the two brain halves of a patient whose connecting corpus callosum has been cut. If there are fairly independent parts that are too simple, then these won't be conscious at all, like the independent pixels of a video camera.

Generations of physicists and chemists have studied what happens when you group together vast numbers of atoms, finding that their collective behavior depends on the pattern in which they're arranged: the key difference between a solid, a liquid and a gas lies not in the types of atoms, but in their arrangement. My guess is that we'll one day understand consciousness as yet another phase of matter. I'd expect there to be many types of consciousness just as there are many types of liquids, but in both cases, they share certain characteristic traits that we can aim to understand.

As a baby step toward consciousness, let's first consider *memory*—what traits does it have? For a substance to be useful for storing information, it clearly needs to have a large repertoire of possible long-lived states. Solids do, whereas liquids and gases don't: if you engrave someone's name on a gold ring, the information will still be there years later, but if you engrave it on the surface of a pond, it will be lost within a second as the water surface changes its shape. Another desirable trait of a memory substance is that it's not only easy to read from (as a gold ring), but also easy to write to: altering the state of your hard drive or your synapses requires less energy than engraving gold.

What traits should we ascribe to *"computronium,"* the most general substance that can process information as a computer? Rather than just remain immobile as a gold ring, it must exhibit complex dynamics so that it's future state depends in some complicated (and hopefully controllable/programmable) way on the present state. Its atom arrangement must be less ordered than a rigid solid where nothing interesting changes, but more ordered than a liquid or a gas. At the microscopic level, computronium doesn't need to be very complicated, because computer scientists have shown that as long as a device can perform certain basic logic operations, it's *universal*: it can be programmed to perform the same computation as any other computer with enough time and memory.

What about *"perceptronium,"* the most general substance that feels subjectively self-aware? If Tononi is right, then it should not merely have the traits of computronium, but also the property that its information is indivisible, forming a unified whole. So when our SAS-buster analyzes a room full of atoms, it will first discover which ones are strongly connected to others and classify the connected atom groups as objects, say, a bench with two people on it. It will then identify parts of these objects that meet the criteria for computronium: say two brains and two cell phone CPUs. Finally, it will determine that there's only perceptronium in the two brains, and that these are two separate pieces that are rather disconnected from one another, one corresponding to the consciousness of each person.

Computing the Internal Reality: What Has History Taught Us?

Once you've found a self-aware entity with your SAS-buster, the next step is to calculate what it subjectively perceives. In the language of Chapter 9, we wish to compute its internal reality from the external reality. This is a tough challenge with which we have limited experience, since physics has historically tended to focus on the opposite problem: given our subjective perceptions, we've looked for mathematical equations that could describe them. For example, Newton observed the motion of the Moon and came up with a law of gravity that explained it. Nonetheless, I feel that the history of physics has taught us many valuable lessons about how the internal and external realities are related: below are seven examples.

Don't panic

Although the problem is unsolved and very hard, we saw in Chapter 9 that we can conveniently split it into two parts: we physicists can limit ourselves to starting with the external reality and predicting the consensus reality that all reasonable observers agree on, leaving the quest for the internal reality to neuroscientists and psychologists. For most of the tricky predict-the-future questions that we'll encounter below, we'll see that the distinction between the consensus reality and the internal reality doesn't matter. Moreover, the history of physics has provided useful case studies such as classical mechanics, general relativity and quantum mechanics, where we know both the key equations and how it feels to be governed by them.

We perceive that which is stable

We humans replace the bulk of both our "hardware" (e.g., our cells) and our "software" (e.g., our memories) many times in our life span. Nonetheless, we perceive ourselves as stable and permanent. Likewise, we perceive objects other than ourselves as permanent. Or rather, what we perceive as objects are those aspects of the world that display a certain permanence. For instance, when observing the ocean, we perceive the moving waves as objects because they display a certain permanence, even though the water itself is only bobbing up and down. Similarly, as we saw in Chapter 8, we perceive only those aspects of the world that are fairly stable against quantum decoherence.

We perceive ourselves as local

Both relativity and quantum mechanics illustrate that you perceive yourself as being "local" even if you aren't. Although in the external reality of general relativity, you're an extended braidlike pattern in a static four-dimensional spacetime, you nonetheless perceive yourself as localized at a particular place and time in a three-dimensional world where things happen. As we discussed previously, your basic perceptions are observer moments, each of which corresponds to a particular localized part of your braid pattern rather than to the whole thing, your whole life.

Quantum mechanics teaches us the same lesson: if you enter a quantum superposition of being in two separate places at once in the external reality (the mathematical Hilbert space where the Schrödinger equation rules), then as we saw in Chapter 8, both of these copies of you will perceive an inner reality where they're in a well-defined location.

We perceive ourselves as unique

In Chapter 8, we also saw that we perceive ourselves as unique and isolated systems even if we aren't. We saw that even if quantum mechanics effectively clones us so that we end up in several macroscopically different places at once, intricately entangled with other systems, we perceive ourselves as remaining unique and isolated, retaining an independent and distinct identity. What appears as "observer branching" in the external reality is perceived as merely a slight randomness in the internal reality.

The same thing happens with classical cloning as in Figure 8.3: cloning with determinism is perceived as uniqueness with randomness. In other words, our well-defined local and unique identity exists only in our internal reality; at a fundamental level, it's an illusion.

We perceive ourselves as immortal(?)

In Chapter 8, we also discussed the possibility that the Level I and/or Level III multiverses make us feel immortal. In summary, the relation between the external and internal realities is quite subtle when the number of copies of you increases or decreases:

- When the number of yous increases, you perceive subjective randomness.
- When the number of yous decreases, you perceive subjective immortality.

The latter is particularly controversial, and whether it's a correct inference or not may hinge on the resolution of the so-called measure problem that we'll describe further on.

We perceive that which is useful

Why do we perceive the world as stable and ourselves as local and unique? Here's my guess: because it's useful. It appears that we humans have evolved self-awareness in the first place because certain aspects of our world are somewhat predictable, so that being good at modeling the world, predicting the future and making smart decisions increases our reproductive success. Self-awareness would then be a side effect of this advanced information processing. More generally, any SAS that's either evolved or engineered with a purpose might have self-awareness as a by-product of having an internal model of the world and itself.

It's then quite natural that the SAS will perceive only those aspects of the external reality that are useful for attaining its goals. For example, migratory birds perceive Earth's magnetic field because it's useful for navigation, whereas star-nosed moles are blind because visual perception isn't useful for their underground lifestyle. Although what's useful and thus perceived varies among species, certain basic considerations appear to be shared among all life-forms. For instance, it's only useful to perceive aspects of the world that exhibit enough stability and regularity that information about them can help predict the future. If you're looking out over a stormy ocean, perceiving the exact motions of trillions of water molecules would be rather useless because they tend to collide with each other and change directions within less than a trillionth of a second. On the other hand, perceiving that a humongous wave is headed your way is quite useful, because you can predict its future motion several seconds in advance and use this prediction to avoid getting flushed out of the gene pool.

In the same way, it's useful for an SAS to perceive itself as being localized and unique, because information can be processed only locally. Even if there exists an identical copy of you a googolplex meters away, or in a decohered part of the quantum Hilbert space, no information can be transferred between the two of you, so both of you might as well keep things simple and act as if the other copy doesn't exist.

We perceive that for which awareness is needed

Because the parts of our brain that model the world and our place in it (and give rise to consciousness) are very useful and in high demand, their use is mostly reserved for computations/decisions that really require them. Just as you wouldn't use a supercomputer to run a word processor, your brain doesn't use its consciousness module for mundane tasks such as regulating your heartbeat—they're instead outsourced to other brain regions whose workings you're not consciously aware of. This suggests that if a future robot becomes self-aware, it might remain unaware of self-contained rote tasks that don't require access to its reality model (multiplying numbers together, say). The consciousness framework envisioned by Giulio Tononi explains how such unconscious cognitive outsourcing can work.

For us humans, I find it interesting that our bodily defense against microscopic enemies (our highly complex immune system) doesn't appear to be self-aware even though our defense against macroscopic enemies (our brain controlling various muscles) does. This is presumably because the aspects of our world that are relevant in the former case are so different (e.g., smaller length scales, longer time scales) from that of the latter that sophisticated, logical thinking and the accompanying self-awareness aren't needed.

When Are You?

Previously, we discussed how a mathematical structure can contain self-aware observer moments, such as the one you're having right now, and we explored the challenges of finding these observer moments and figuring out how they subjectively feel. You exist in a mathematical structure containing some sort of spacetime, so to make physical predictions, you should try to learn what kind of mathematical structure you're in and your current observer moment's location in it: where in space and when in time are you? As we'll see, the "when" part is even more subtle than the "where" part, particularly when the number of yous changes over time.

Beyond Popper's Two-Timing

To me, science is all about understanding reality and our place in it. From a pragmatic perspective, it's about building a model of reality that lets us predict our future as successfully as possible, so that we can choose to do what we predict will have the best outcome—my guess is that it's to help accomplish this task that we've been fortunate enough to evolve consciousness. Thinkers throughout the ages have tried to formalize this scientific process, and I think most contemporary scientists agree that it boils down to this:

1. Make predictions from assumptions.
2. Compare observations with predictions, update assumptions.
3. Repeat.

We scientists often call a collection of assumptions a *theory*. In the MUH context, the key assumptions that go into the model of reality are what mathematical structure we inhabit and which particular observer moment therein is the one you're experiencing right now. Karl Popper emphasized the second item on the list, arguing that assumptions that can't make testable predictions aren't scientific. Although he placed particular emphasis on falsifiability, i.e., that it should in principle be able to test whether scientific assumptions are false, there's a beautiful mathematical toolkit known as Bayesian decision theory which generalizes the true/false dichotomy to allow shades of gray: each possible assumption gets assigned a number between zero and one, the probability with which you think it's correct, and there's a simple formula for how to update these probabilities whenever you make new observations.

As elegant and well accepted as it is, there's a problem with this approach to science: it requires *two* connected observer moments. In the first, you make your prediction, and in the second you contemplate what you've observed. This works well in the conventional situation where there is, was and always will be at most one copy of you (Figure 11.8, left), but breaks down for any parallel-universe scenario where you have alter egos. As we saw in Chapters 6 and 8, this breakdown can lead to novel effects such as subjective immortality and subjective randomness (Figure 11.8).

In the MUH context, we've argued that the perceptions of time flowing and of assumptions and observations having been made, exist in every *single* observer moment that we experience. This means that we must transcend Popper's two-time approach to science with a one-time approach that can be applied to a single observer moment. I like to imagine that I have this awesome pocket-sized remote control for reality itself. When attending a dull meeting, I can press the Fast-Forward button. When I experience something amazing, I can rewind and re-play it as many times as I want. And to transcend Popper, I simply press Pause. Now I can, in the spirit of Horace, truly seize the moment,

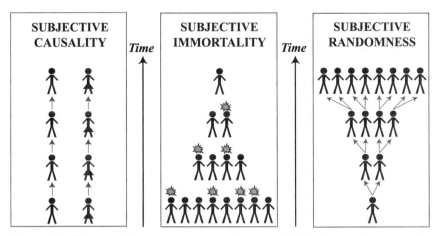

Figure 11.8: When each observer moment can be uniquely linked to a predecessor and a successor, we perceive subjective causality (left). When some but not all successors disappear, we may perceive subjective immortality. When several subjectively distinguishable successors share the same predecessor, we perceive subjective randomness.

taking it in, absorbing it and reflecting on it without feeling rushed on toward the future. In particular, I can reflect on what I assume and what I observe. If my brain is working well, then I'll find that my internal-reality model agrees well with the latest news that my senses are reporting from the outside world. And if my scientific-reasoning algorithm is good, then I'll find that the predictions I remember making for this moment are in decent agreement with what's actually happening. While my senses are hard at work recording new information to be consciously perceived in future observer moments, the conscious part of my mind is busy using my scientific-reasoning algorithm to update my assumptions about more subtle and abstract aspects of reality.

Why Aren't You an Ant?

So how should you reason in your observer moment, once you've pressed the Pause button? You need a good framework for this not only to get by in a multiverse, but also, as we'll see, for making sense of the so-called doomsday argument and other famous philosophical conundrums. If you believe in the Mathematical Universe Hypothesis, then you should try to figure out which mathematical structure you inhabit. If that structure contains many observer moments that subjectively feel like yours, then you could be any one of them. Unless there's something in the mathematics that somehow breaks the symmetry and favors some over others, you're equally likely to be any one of them. Therefore, as I argued in my 1996 mathematical universe paper, you reach the following conclusion:

> *You should reason as if your observer moment is a random one among those it could be.*

The past two decades have seen a vigorous and fascinating discussion of various alternative modes of reasoning in the philosophy literature, triggered in part by the doomsday argument (which we'll explore shortly) and related puzzles. The basic idea that we should expect to find our consciousness not in a random *place* (as per the Copernican principle) but in a random *observer* has a long history; we saw in Chapter 6 that Brandon Carter formulated it as his weak anthropic principle, and Alex Vilenkin from Chapter 5 has formulated it as the *principle of medi-*

ocrity. Contemporary philosophers such as Nick Bostrom, Paul Almond and Milan Ćirković have explored it extensively, and in 2002, Bostrom coined the now standard terminology of the *Strong Self-Sampling Assumption.*

> **Strong Self-Sampling Assumption (SSSA):** *Each observer moment should reason as if it were randomly selected from the class of all observer moments in its reference class.*

The subtlety here is how *reference class* should be interpreted, and philosophers who accept the SSSA often argue about this. If you use the maximally restrictive option and limit the reference class to other yous' observer moments that feel subjectively indistinguishable from your own, you recover my old approach. However, we'll see that you can often reach additional interesting conclusions by being more liberal: you'll still reach correct conclusions even if distinguishable observer moments are allowed, as long as the way they subjectively feel different doesn't bias the answer you're seeking. To get a feeling for how this works, let's consider an example of the SSSA in action—Nick Bostrom's Sleeping Beauty puzzle:

> *Sleeping Beauty volunteers to undergo the following experiment and is told all of the following details. On Sunday she's put to sleep. A fair coin is then tossed. If the coin comes up heads, Beauty is awoken and interviewed on Monday only. If the coin comes up tails, she's awoken and interviewed on Monday and Tuesday, but when she's put to sleep again on Monday, she's given a dose of an amnesia-inducing drug that ensures she can't remember her previous awakening. Any time Beauty is awakened and interviewed, she's asked, "What odds would you give that the coin landed heads?"*

After a large number of publications on the subject, the philosophy community is now split between the "halfers" and the "thirders," who feel that she should assign probabilities of 1/2 and 1/3, respectively. In the MUH framework, there's no such thing as true randomness, so let's replace the coin by a quantum measurement that realizes both outcomes equally in two Level III parallel universes. There are now three subjectively indistinguishable observer moments in the mathematical structure that correspond to her being interviewed, and they're all equally real:

1. The coin landed heads and it's Monday.
2. The coin landed tails and it's Monday.
3. The coin landed tails and it's Tuesday.

Since only one of the three corresponds to the heads option, she should assign a probability of 1/3 to this and will experience the corresponding subjective randomness once she finds out.

Now suppose the experimenters secretly decided to repaint her fingernails in a color depending on the quantum-measurement outcome. This means that the observer moments aren't all indistinguishable, but as long as she doesn't know the color code, the odds she gives shouldn't change. In other words, we're free to broaden the reference class as long as it won't bias the results.

This conclusion has radical implications: it suggests that no matter how vast and crazy a multiverse may exist out there, we humans are likely to be *rather typical* among all observers asking this sort of question! For example, it's extremely unlikely that typical solar systems contain quadrillions of hominids similar to us, because if that were the case, we'd be about a million times more likely to find ourselves in such a populous solar system rather than in our own with its measly 7 billion denizens. In other words, the SSSA allows us to make statements about what's going on even in places that we can't observe.

However, like any powerful tool, the SSSA must be used with caution. For example, why aren't you an ant? If we take carbon-based life-forms on Earth as our reference class, our over ten quadrillion (10^{16}) six-legged friends outnumber us bipeds by more than a million to one, so doesn't that imply that your current observer moment is a million times more likely to be that of an ant than that of a human? If so, that would rule out your basic reality framework with 99.9999% confidence. Okay, we've neglected that humans live about a hundred times longer than ants, but that doesn't change the troubling conclusion.

Instead, the resolution lies in the choice of reference class. As Figure 11.9 illustrates, you have many different choices of reference class, the most inclusive being all observer moments of all self-aware substructures and the most exclusive being only the ones that subjectively feel exactly like you do right now. If you ask the question "What sort of entity should I expect to be?," then your reference class clearly needs to be restricted to entities that ask such questions—and ants don't!

The business of using the right reference class corresponds to correctly using what statisticians call conditional probabilities, and botching this can cause epic failures. In 2010, a major poll failed to predict that U.S. Senate Majority Leader Harry Reid, would get reelected in Nevada because the robo-calling software hung up when the target didn't speak English, thus losing the responses from pro-Reid Hispanic voters. In Chapter 6, we saw that a typical region of space might expect itself to be in a universe with too much dark energy to form any galaxies, and that a typical hydrogen atom in our particular Universe should expect to find itself in an intergalactic gas cloud or a star. But that's not where you should expect to find yourself: "all points" or "all atoms" are irrelevant reference classes to you, because neither points nor atoms ask questions.

Why Aren't You a Boltzmann Brain?

If you think it sounds crazy to have an extraterrestrial alien classmate in your reference class, you'll be amused to know that some of my colleagues are busy arguing about even-more-exotic classmates: simulations and Boltzmann brains.

We're living proof that atoms can be put together in an elaborate pattern that subjectively feels self-aware. So far, our physics research has turned up no evidence whatsoever suggesting that ours is the only

Figure 11.9: *What's the probability that [INSERT YOUR FAVORITE QUESTION HERE] given that I'm a . . . ?* What you replace the ellipsis by is your *reference class*, illustrated above. Under the Mathematical Universe Hypothesis, it's always valid to reason as if you're a random member of the most restrictive reference class, corresponding to all observer moments who subjectively feel like you do, but in some cases you can draw additional valid and interesting conclusions from broadening the reference class, say, to humans or other self-aware entities who are capable of asking the same question.

possible path to consciousness. We therefore need to consider the possibility that there may be other kinds of atom arrangements that feel self-aware as well, and that some life-forms (perhaps even we or our descendants) will one day build such entities. They might be reminiscent of intelligent robots by having actual physical bodies that can interact with the world around them, or they might be simulations like characters in *Star Trek: The Next Generation* holodeck episodes or Agent Smith in *The Matrix** whose bodies are purely virtual, with lives playing out in the virtual reality of an extremely powerful computer. Some such simulations could have observer moments that subjectively feel exactly like you feel right now.

If this is the case, you clearly need to include the simulated yous in your reference class. Nick Bostrom and others have published extensively on this topic, concluding that there's a reasonable probability that we're in fact simulated. I'll give a counterargument in the next chapter, but if you want to play it safe in the meantime, in the spirit of Pascal's Wager, my advice is that you should live life to its fullest and do novel and interesting things. That way, in case you're a simulation, whoever created you will be less likely to get bored and switch you off. . . .

Whereas simulations are purposefully created, so-called Boltzmann brains are created by coincidences. After pioneering the field known as statistical mechanics about 150 years ago, the Austrian physicist Ludwig Boltzmann realized that if you leave a warm object alone for enough time, even most unlikely arrangements of atoms will occur by chance. The time it will take for the particles to spontaneously rearrange themselves into a self-aware brain is extremely long, but if you wait long enough, it will happen.

Now fast-forward to today's Universe, and let's consider its long-term fate. The accelerated expansion will eventually dilute away all the matter that currently fills our Universe, but if the cosmic dark-energy density remains constant (as current measurements suggest), then it will forever provide a very slight amount of heat energy. This heat comes from the same kind of quantum fluctuations that generated the cosmic microwave–background fluctuations in Chapter 5, and Stephen Hawking famously discovered that the faster our Universe expands, the higher this so-called Hawking temperature will be. The dark energy

---

* Note that many of the characters in *The Matrix* have simulated experiences in human brains; in contrast, the simulated people in the movie *The Thirteenth Floor* involve no human hardware whatsoever.

makes our Universe expand much more slowly than during inflation, so the temperature it provides is merely a millionth of a trillionth of a trillionth (10^{-30}) of a degree above absolute zero.

This is hardly balmy, even by Swedish standards, but it isn't absolute zero, which means that if you wait long enough, this heat energy will rearrange itself into anything you want. In the standard cosmological model, this random rearranging goes on *forever*, so it will randomly produce an exact replica of you who subjectively feels exactly like you do, complete with false memories of having lived your entire life. Much more often, it will replicate merely your disembodied brain, surviving just long enough to replicate your current observer moment. And then it will do it again, infinitely many times over, so that for every copy of you that has evolved and lived a real life, there are infinitely many delusional disembodied Boltzmann brains who think that they've lived that same real life.

This is deeply troubling. If our spacetime really contains these Boltzmann brains, then you're basically 100% certain to be one of them! After all, the observer moment of the evolved you is in the same reference class as those of these brains, since they subjectively feel the same, so you should reason as if you're a random one of these observer moments—and the disembodied ones outnumber the embodied one by infinity to one. . . .

Before you get too worried about the ontological status of your body, here's a simple test you can do to determine whether you're a Boltzmann brain. Pause. Introspect. Examine your memories. In the Boltzmann-brain scenario, it's indeed more likely that any particular memories that you have are false rather than real. However, for every set of false memories that could pass as having been real, very similar sets of memories with a few random crazy bits tossed in (say, you remembering Beethoven's Fifth Symphony sounding like pure static) are vastly more likely, because there are vastly more disembodied brains with such memories. This is because there are vastly more ways of getting things almost right than getting them exactly right. Which means that if you really are a Boltzmann brain who at first thinks you're not, then when you start jogging your memory, you should discover more and more utter absurdities. And after that, you'll feel your reality dissolving, as your constituent particles drift back into the cold and almost empty space from which they came.

In other words, if you're still reading this, you're *not* a Boltzmann brain. This means that something is fundamentally wrong with what we assumed about the future of our Universe, and that there's a lesson to be learned. We'll shortly explore that in the "measure problem" section.

The Doomsday Argument: Is the End Nigh?

We've seen that the idea that you should be a typical observer is a powerful one, with surprising consequences. Another much debated consequence is the *doomsday argument*, which was first given by Brandon Carter in 1983.

During World War II, the Allied forces successfully estimated the number of German tanks from their serial numbers. If the first captured tank had the serial number 50, then this ruled out the hypothesis that there were more than a thousand tanks with 95% confidence, since the probability of capturing one of the first fifty ones built was less than 5%. The key assumption is that the first tank captured can be thought of as a random one from the reference class of all tanks.

Carter pointed out that if we assign each human a serial number at birth, then we can make exactly the same argument to estimate the total number of humans who will ever live. When I arrived on the scene in 1967, I was roughly the fifty-billionth person born, so if I'm a random human out of all people who'll ever live, then I can rule out the hypothesis that more than a trillion humans will be born with 95% confidence. In other words, it's highly unlikely that there'll be more than a trillion humans born, because this would place me within the first 5% of humans to exist—something which we could explain only by invoking an unlikely fluke coincidence. Moreover, if the world population stabilizes at 10 billion with an eighty-year life expectancy, then humanity as we know it will with 95% certainty end before the year 10,000 AD.

If I believe that our doomsday will be caused by nuclear weapons (or computer technology, biotech or any other technology that has existed only since after 1945), then my forecast gets gloomier: my birth rank since the dangers began is 1.6 billion, and I can rule out with 95% confidence that there will be another 32 billion births after me, around the year 2100. And that's the 95% confidence limit—a more likely end date for humanity is right around now. To escape this pessimistic conclusion, I'd need to come up with some a priori reason for why I should be among the first 5% of all humans to be born under the shadow of these technologies. We'll return to the existential risk posed by technology in Chapter 13.

Some people take the doomsday argument very seriously. For example, when I had the pleasure of meeting Brandon Carter at a conference, he excitedly told me about the latest evidence that the population explosion was slowing, saying that he'd predicted that this would happen, and that this meant we should expect humanity to survive for longer. Others have criticized the argument on various grounds. For example, things get more subtle if there are other planets with people similar to us. Figure 11.10 illustrates such an example, where the total number of people ever born varies sharply from planet to planet. If you know this to be the case, then you should be more optimistic about the future than the standard doomsday argument suggests. Indeed, if I believed the more extreme theory that there are only two inhabited planets in spacetime, supporting a total of 10 billion and 10 quadrillion people from beginning to end, then the probability is 50% that I'm now on the planet that will eventually enjoy a quadrillion people.

Unfortunately, this counterargument gives only false hope. I have no such information, and I have very good reason to believe that this two-planet theory is false: the observation that my birth rank is about 50 billion rules out the theory at more than 99.9999% confidence, since the probability of a random person being within the first 50 billion born is only 0.00005%.

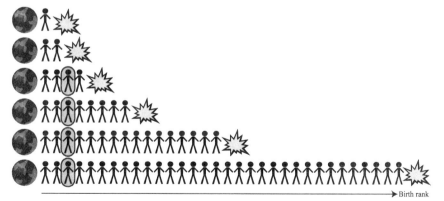

Figure 11.10: If you know that your birth rank is 3 billion, you might think there's only a 10% chance that more than 30 billion will ever live on your planet. But suppose you know that there are six planets similar to ours, where the total number of people born from the beginning to the end of their civilizations is 1, 2, 4, 8, 16 and 32 billion, respectively (each stick figure above represents a billion people). Then the probability that more than 30 billion will ever live on your planet is actually 25%; there are exactly four people who have your birth rank, you're equally likely to be any of them, and 25% of them live on the highly successful bottom planet in the image above.

Why Is Earth So Old?

In March 2005, I had the pleasure of meeting Nick Bostrom at a conference in California, and we soon discovered that we share not only Swedish childhoods but also a fascination with big questions. After some good wine, our conversation turned to doomsday scenarios. Could the Large Hadron Collider create a miniature black hole that would end up gobbling up Earth? Could it create a "strangelet" that could catalyze the conversion of Earth into strange quark matter? MIT colleagues of mine whose calculations I trust have concluded that there's negligible risk, but what if we've overlooked something? What used to reassure me the most was the fact that nature is vastly more violent than any of our human-made machines: for example, cosmic-ray particles created near monster black holes routinely slam into Earth with over a million times more energy than our accelerators can deliver, and 4.5 billion years after its formation, Earth is still alive and well. So Earth is clearly very robust, and I needn't worry. For the same reason, I shouldn't worry about other cosmic doomsday scenarios, such as space "freezing" into another lower-energy phase as per Chapter 5, with a cosmic death bubble containing this uninhabitable new kind of space expanding with the speed of light, destroying all people in its path at exactly the same instant they saw it come: if we're still here after all this time, such events must be nonexistent or very rare.

Then a terrible thought hit me: my reassuring argument was flawed! Suppose each planet has a 50% chance of getting destroyed each day. Then the vast majority will be gone within weeks, but in an infinite space with infinitely many planets, there'll always be an infinite number remaining, whose inhabitants can be blissfully unaware of the grim fate that awaits them. And if I'm simply a random observer in spacetime, then I expect myself to be one of these naive people who don't realize that they're like lambs about to be slaughtered. In other words, the fact that my region of space hasn't yet been destroyed tells me *nothing*, because *all* living observers are in regions of space that haven't been destroyed. I got really nervous. I felt as if I were in a zoo in front of a pack of hungry lions, and had just realized that the fence I thought protected me was an optical illusion—and one that the lions couldn't see.

Nick and I agonized about this for a while, until we managed to come up with a different anti-doomsday argument that wasn't flawed. Earth formed about 9 billion years after our Big Bang, and it's now fairly clear that our Galaxy (and other similar galaxies elsewhere) harbors a large number of Earth-like planets that formed several billion years earlier. This means that when we consider all observers similar to us in all of spacetime, a significant fraction of them exist long before us. Now, in a scenario where planets get randomly destroyed with some short half-life (say, a day, a year or a millennium), then almost all observer moments will happen very early on, and it's extremely unlikely for us to find ourselves on a planet that formed at such a leisurely pace relatively late in the game. We decided to write a paper about it, and worked on it late into the night in a hotel lounge. When I finally drifted off to sleep, I did so knowing that with 99.9% confidence, neither death bubbles, black holes, nor strangelets would get us for another billion years.

Unless, of course, we humans do something stupid of a kind that nature hasn't already tried. . . .

Why Aren't You Younger?

We just saw that if there were some terrible instability built into physics that made most planets short-lived, then we should expect to find ourselves on one of the first habitable planets to form, not on this slowpoke planet of ours. So that depressing theory is ruled out. Unfortunately for inflation, Alan Guth realized that under some reasonable-sounding assumptions, it predicts the same thing! Bothered by his brainchild predicting a much younger Earth, he called this the *youngness paradox*. Around when I became his colleague at MIT back in 2004, I spent a lot of time worrying about how to make predictions in a multiverse. I wrote a paper on this topic that painfully broke all my past length records, and was surprised to discover that the youngness paradox was even more extreme than we'd thought.

As we saw in Chapter 5, inflation typically goes on forever doubling the volume of space every 10^{-38} seconds or so, creating a messy spacetime with countless Big Bangs occurring at different times and countless planets forming at different times. We saw that an observer on any given planet will consider her Big Bang to be the moment when inflation ended in her part of space; for me personally, the delay between my Big Bang and my current observer moment is about 14 billion years. Now let's consider all simultaneous observer moments: for some, the time since their Big Bang is 13 billion years, for some it's 15, etc. Because of the frenetic volume doubling, there will be $2^{10^{38}}$ times as many Big Bangs happening one second later, because the volume doubled 10^{38} times during that extra second. Similarly, there are $2^{10^{38}}$ times more observers in the galaxies they form. This means that if I'm a random observer moment out of all currently occurring ones, then I'm $2^{10^{38}}$ times more probable to find myself in a one-second-younger universe, whose Big Bang happened one second more recently! That's about one with a hundred trillion trillion trillion zeros times more likely. My planet should be younger, my body should be younger, and everything should appear to have formed and evolved in haste.

A part of space that experienced its Big Bang more recently will be hotter, because it's had less time to cool off, so finding ourselves in a relatively cool universe is highly unlikely and we have a *coolness problem*: when I worked out the probability of measuring the cosmic microwave–background temperature to be less than three degrees above absolute zero, I got $10^{-10^{56}}$, so when the COBE-satellite measured this temperature to be 2.725 Kelvin, this measurement ruled out our whole inflation-based story with 99.999 . . . 999% confidence, where there are a hundred million trillion trillion trillion trillion nines after the decimal point. Not good . . . In the hall of shame for disagreements between theory and experiment, this crushes even the hydrogen-atom stability problem from Chapter 7 (28 nines) and the dark-energy problem from Chapter 4 (123 nines). Welcome to the *measure problem*!

The Measure Problem: Physics in Crisis

Something just went terribly wrong, but what exactly? Did this really rule out eternal inflation? Let's take a closer look. We asked a reasonable question about what a typical observer should expect to measure—we picked the particular example of the cosmic microwave temperature. Because we considered eternal inflation, we analyzed a spacetime containing many observer moments measuring many different temperatures, so we couldn't predict just one unique answer, merely probabilities for different temperature ranges. This, per se, isn't the end of the world: we saw in Chapter 7 how quantum mechanics predicts only probabilities, not definite answers, and is nonetheless a perfectly testable and successful scientific theory. Rather, the problem was that the probabilities we computed told us that what we in fact observe is ridiculously unlikely, so that the underlying theory is ruled out.

Could there be a mistake in our probability calculation? The math is straightforward in principle: the probabilities are simply the fractions of all observer moments in our reference class that measure various temperatures. If there are only five such observer moments and they observe 1, 2, 5, 10 and 12 degrees above absolute zero, then the fraction measuring less than three is two out of five, 2/5 = 40%—easy! But what if, as eternal inflation predicts, there are *infinitely* many such observer moments, and the fraction measuring less than three degrees is infinity divided by infinity? How do we make sense of that?

Mathematicians have developed an elegant scheme called *taking limits*, which in many cases can make sense of ∞/∞. For example, what fraction of all the counting numbers 1, 2, 3, . . . are even? There are infinitely many numbers and infinitely many of them are even, so the fraction is ∞/∞. But if we count only the first n numbers, we get a sensible answer that depends slightly on our counting cutoff n. If we keep increasing n, we find that the fraction jiggles around less and less as n grows. If we now take the limit where n approaches infinity, we get a well-defined answer that doesn't depend on n at all: exactly half of the numbers are even.

This seems like a sensible answer, but infinities are treacherous: the fraction of the numbers that's even depends on the order in which we count them! If we instead order the numbers 1, 2, 4, 3, 6, 8, 5, 10, 12, 7, 14, 16 and so on, then the same limit scheme gives the answer that 2/3 of the numbers are even! Because as we proceed down this list of numbers, we encounter two even numbers for every odd number. We didn't cheat, since all even and odd numbers eventually show up in our list; we merely reordered them. In the same way, by reordering the numbers appropriately, I can prove to you that the even fraction is one divided by your phone number. . . .

Analogously, the fraction of all the infinitely many observers in spacetime who make a particular observation depends on the order in which you count them! We cosmologists use the term *measure* to refer to an observer-moment ordering scheme, or, more generally, to a method for calculating probabilities from annoying infinities. The crazy probabilities I computed for the coolness problem corresponded to a particular measure, and most of my colleagues guess that the problem isn't with inflation but with the measure: somehow, it appears flawed to talk about the reference class of all observer moments at a fixed time.

The last few years have seen an avalanche of interesting papers proposing alternative measures. It's proven remarkably difficult to find one that works with eternal inflation: some measures flunk the coolness problem; others fail by predicting that you're a Boltzmann brain; yet others predict that we should see our sky warped by giant black holes. Alex Vilenkin recently told me that he was getting disheartened: a few years ago, he'd hoped that only one measure would avoid all these pitfalls, and that it would be so simple and elegant that it convinced us all. Instead, we now have a number of different measures that appear to give different but reasonable predictions, with no obvious way to choose between them. If the probabilities we predict depend on the measure we assume, and we can assume a measure giving almost any answer we want, then we really haven't predicted anything at all.

I share Alex's concern. In fact, I view the measure problem as the greatest crisis in physics today. The way I see it, inflation has logically self-destructed. As we saw in Chapter 5, we started taking inflation seriously because it made correct predictions: it predicted that typical observers should measure space around them to be flat rather than curved (the flatness problem); they should measure their cosmic microwave–background temperature to be similar in all directions (the horizon problem); they should measure a power spectrum similar to what the WMAP satellite saw, etc. But then it predicted infinitely many observers measuring different things with probabilities depending on a measure that we don't know. Which in turn means that inflation, strictly speaking, isn't predicting anything at all about what typical observers should see. All predictions are revoked, including those predictions that made us take inflation seriously in the first place! Self-destruction complete. Our inflationary baby Universe has grown into an unpredictable teenager.

In fairness to inflation, I don't feel that there's any competing cosmological theory on the market that does any better, so I don't view this as an argument against inflation per se. I simply feel strongly that we need to solve the measure problem, and my guess is that once we solve it, some form of inflation will still remain. Moreover, the measure problem isn't limited to inflation, but crops up in *any* theory with infinitely many observers. As an example, let's revisit collapse-free quantum mechanics. The quantum-immortality argument from Chapter 8 hinges crucially on there being infinitely many observers, so that some always survive, which means that we can't trust any of the conclusions until the measure problem has been solved.

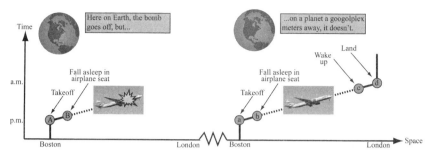

Figure 11.11: In Figure 11.5, we saw how observer moment (c) feels like the continuation of observer moment (b) because it shares all its memories. However, (c) also feels like the continuation of (B), an observer moment belonging to a doppelgänger whose flight is identical except for a terrorist bomb that kills all passengers before they wake up. If there are no other doppelgängers, then the correct prediction for both (B) and (b) is that they'll next perceive (c).

As Figure 11.11 illustrates, subjective immortality doesn't require quantum mechanics, merely parallel universes—it doesn't matter whether the two airplanes in the figure are in different parts of our 3-D space (Level I multiverse) or different parts of our Hilbert space (Level III multiverse). So let's quite generally consider any multiverse scenario where some mechanism kills half of all copies of you each second. After twenty seconds, only about one in a million (1 in 2^{20}) of your initial doppelgängers will still be alive. Up until that point, there have been $2^{20} + 2^{19} + \ldots + 4 + 2 + 1 \approx 2^{21}$ second-long observer moments, so only one in two million observer moments remembers surviving for twenty seconds. As Paul Almond has pointed out, this means that those surviving that long should rule out the entire premise (that they're undergoing the immortality experiment) at 99.99995% confidence. In other words, we have a philosophically bizarre situation: you start with a correct theory for what's going on, you make a prediction for what will happen (that you'll survive), you observationally confirm that your prediction was correct, and you then nonetheless turn around and declare that the theory is ruled out! Moreover, as we discussed in Chapter 8, you'd start experiencing increasingly bizarre fluke coincidences the longer you waited, which kept saving your life in ever more unlikely-seeming ways—getting saved by power failures, asteroid impacts, etc., would probably suffice to make most people start questioning their assumptions about reality. . . .

Infinite Problems

What's the measure problem telling us? Here's what I think: that there's a fundamentally flawed assumption at the very foundation of modern physics. The failures of classical mechanics required switching to quantum mechanics, and I think that today's best theories similarly need a major shakeup. Nobody knows for sure where the root of the problem lies, but I have my suspicions. Here's my prime suspect: ∞.

In fact, I have two suspects: "infinitely big" and "infinitely small." By *infinitely big*, I mean the idea that space can have infinite volume, that time can continue forever, and that there can be infinitely many physical objects. By *infinitely small*, I mean the continuum: the idea that even a liter of space contains an infinite number of points, that space can be stretched out indefinitely without anything bad happening, and that there are quantities in nature that can vary continuously. The two are closely related: we saw in Chapter 5 that inflation created an infinite volume by stretching continuous space indefinitely.

We have no direct observational evidence for either the infinitely big or the infinitely small. We speak of infinite volumes with infinitely many planets, but our observable Universe contains only about 10^{89} objects (mostly photons). If space is a true continuum, then to describe even something as simple as the distance between two points requires an infinite amount of information, specified by a number with infinitely many decimal places. In practice, we physicists have never managed to measure anything to more than about sixteen decimal places.

I remember distrusting infinity already as a teenager, and the more I've learned, the more suspicious I've become. Without infinity, there'd be no measure problem—we'd always calculate the same fractions regardless of what order we counted in. Without infinity, there'd be no quantum immortality.

Among physicists, my skepticism toward infinity places me in a very small minority. Among mathematicians, infinity and the continuum used to be viewed with considerable suspicion. Carl Friedrich Gauss, sometimes referred to as "the greatest mathematician since antiquity," had this to say two centuries ago: "I protest against the use of infinite magnitude as something completed, which is never permissible in mathematics. Infinity is merely a way of speaking, the true meaning being a limit which certain ratios approach indefinitely close, while others are permitted to increase without restriction." Criticizing the continuum and related ideas, his younger colleague Leopold Kronecker once said: "God made integers; all else is the work of man." In the past century, however, infinity has become mathematically mainstream, with only a few vocal critics remaining—for example, the Canadian-Australian mathematician Norman Wildberger has posted an essay arguing that "real numbers are a joke."

So why are today's physicists and mathematicians so enamored with infinity that it's almost never questioned? Basically, because infinity is an extremely convenient approximation, and we haven't discovered good alternatives. For example, consider the air in front of you. Keeping track of the positions and speeds of octillions of atoms would be hopelessly complicated. But if you ignore the fact that air is made of atoms and instead approximate it as a continuum, a smooth substance that has a density, pressure and velocity at each point, you find that this idealized air obeys a beautifully simple equation that explains almost everything we care about from how sound waves propagate through air to how winds work. Yet despite all that convenience, air isn't truly continuous. Could it be the same way for space, time and all the other building blocks of our physical word? We'll explore that in the next chapter.

THE BOTTOM LINE

- Mathematical structures are eternal and unchanging: they don't exist in space and time—rather, space and time exist in (some of) them. If cosmic history were a movie, then the mathematical structure would be the entire DVD.
- The Mathematical Universe Hypothesis (MUH) implies that the flow of time is an illusion, as is change.
- The MUH implies that creation and destruction are illusions, since they involve change.
- The MUH implies that it's not only spacetime that is a mathematical structure, but also all the stuff therein, including the particles that we're made of. Mathematically, this stuff seems to correspond to "fields": numbers at each point in spacetime that encode what's there.
- The MUH implies that you're a self-aware substructure that is part of the mathematical structure. In Einstein's theory of gravity, you're a remarkably complex braidlike structure in spacetime, whose intricate pattern corresponds to information processing and self-awareness. In quantum mechanics, your braid pattern branches like a tree.
- The movielike subjective reality that you're perceiving right now exists only in your head, as part of your brain's reality model, and it includes not merely edited highlights of here and now, but also a selection of prerecorded distant and past events, giving the illusion that time flows.
- You're self-aware rather than just aware because your brain's reality model includes a model of yourself and your relation to the outside world: your perceptions of a subjective vantage point you call "I" are qualia, just as your subjective perceptions of "red" and "sweet" are.
- The theory that our external physical reality is perfectly described by a mathematical structure while still not being one is 100% unscientific in the sense of making no observable predictions whatsoever.
- You should expect your current observer moment to be a typical one among all observer moments that feel like you. Such reasoning leads to controversial conclusions regarding the end of humanity, the stability of our Universe, the validity of cosmological inflation, and whether you're a disembodied brain or simulation.
- It also leads to the so-called measure problem, a serious scientific crisis that calls into question the ability of physics to predict anything at all.

12

The Level IV Multiverse

What is it that breathes fire into the equations and makes a universe for them to describe?

—Stephen Hawking

Why I Believe in the Level IV Multiverse

Why These Equations, Not Others?

Suppose that you're a physicist, and that you discover how to unify all physical laws into a "Theory of Everything." Using its mathematical equations, you're able to answer the tough questions that keep today's physicists awake at night, such as how quantum gravity works and how to solve the measure problem. A T-shirt with these equations becomes a best-seller, and you're awarded the Nobel Prize. You're elated, but the night before the award ceremony, you can't sleep because you struggle with an embarrassing question from my hero John Wheeler that still remains unanswered: *Why these particular equations, not others?*

In the last two chapters, I've argued for the Mathematical Universe Hypothesis (MUH), according to which our external physical reality is a mathematical structure, and this sharpens Wheeler's question. Mathematicians have discovered a large number of mathematical structures, and Figure 12.1 illustrates some of the simplest ones as boxes. None of the ones in the figure match our physical reality, even though some of them may describe certain limited aspects of our world. In 1916, the box labeled "GENERAL RELATIVITY" was a serious candidate for being an exact match, containing within it not only space and time but also various forms of matter, but the discovery of quantum mechanics soon made clear that our own physical reality had features that this particular mathematical structure lacked. Fortunately, you can now extend the figure by adding the mathematical structure that you've discovered and will get your prize for, knowing that this new box in the figure is *the* box, the one that corresponds to our physical reality.

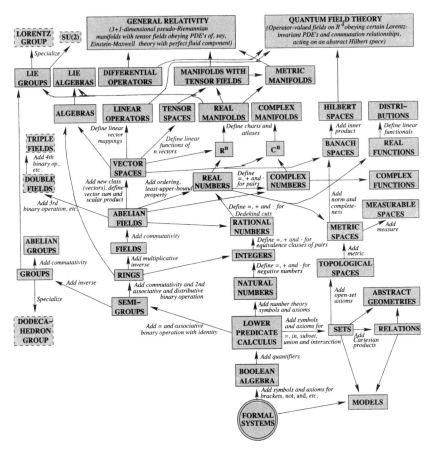

Figure 12.1: Relationships between various basic mathematical structures. The arrows generally indicate addition of new symbols and/or axioms. Arrows that meet indicate the combination of structures—for instance, an algebra is a vector space that's also a ring, and a Lie group is a group that's also a manifold. The full family tree is probably infinite in extent—the figure shows merely a small sample near the bottom.

At this point, I can hear John Wheeler's friendly voice interject: *But what about the other boxes?* If your box corresponds to a physically existing reality, then why don't they?

All boxes are on an equal mathematical footing, corresponding to different mathematical structures, so why should some be more equal than others when it comes to physical existence? Could there really be a fundamental, unexplained existential asymmetry built into the very heart of reality, splitting mathematical structures into two classes—those with and without physical existence?

Mathematical Democracy

This question really bothered me that Berkeley evening back in 1990, when I first had the mathematical universe idea and told my friend Bill Poirier about it in the fifth-floor hallway outside our dorm rooms in International House. Until a lightbulb went off in my head and I realized that there's a way out of this philosophical conundrum. I argued to Bill that complete mathematical democracy holds: that mathematical existence and physical existence are equivalent, so that *all* structures that exist mathematically exist physically as well. Then each other box in Figure 12.1 also describes a physically real universe—just a different one from the one we happen to inhabit. This can be viewed as a form of radical Platonism, asserting that all the mathematical structures in Plato's "realm of ideas" exist "out there" in a physical sense.

In other words, the idea is that there's a fourth level of parallel universes that's vastly larger than the three we've encountered so far, corresponding to different mathematical structures. The first three levels correspond to noncommunicating parallel universes within the same mathematical structure: Level I simply means distant regions from which light hasn't yet had time to reach us, Level II covers regions that are forever unreachable because of the cosmological inflation of intervening space, and Level III, Everett's "Many Worlds," involves noncommunicating parts of the Hilbert space of quantum mechanics. Whereas all the parallel universes at Levels I, II and III obey the same fundamental mathematical equations (describing quantum mechanics, inflation, etc.), Level IV parallel universes dance to the tunes of different equations, corresponding to different mathematical structures. Figure 12.2 illustrates this four-level multiverse hierarchy, one of the core ideas of this book.

How the Mathematical Universe Hypothesis Implies the Level IV Multiverse

If the theory that the Level IV multiverse exists is correct, then since it has no free parameters whatsoever, all properties of all parallel universes (including the subjective perceptions of self-aware substructures in them) could in principle be derived by an infinitely intelligent mathe-

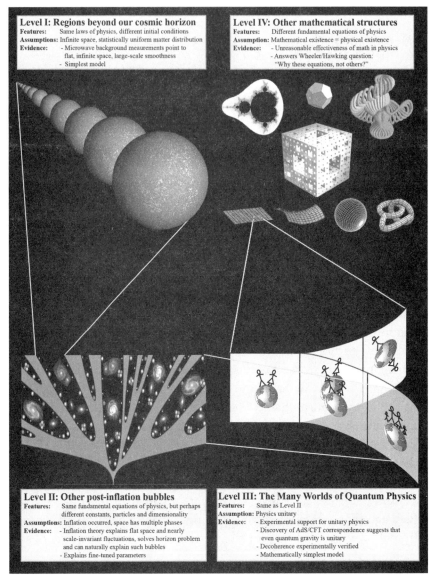

Figure 12.2: The parallel universes described in this book form a four-level hierarchy, where each multiverse is a single member among many at the level above it.

matician. But is this theory correct? Does the Level IV multiverse really exist?

Interestingly, in the context of the Mathematical Universe Hypothesis (MUH), the existence of the Level IV multiverse isn't optional. As we discussed in detail in the previous chapter, the MUH says that a mathematical structure *is* our external physical reality, rather than being merely a *description* thereof. This equivalence between physical and mathematical existence means that if a mathematical structure contains a self-aware substructure, it will perceive itself as existing in a physically real universe, just as you and I do (albeit generically a universe with different properties from ours). Stephen Hawking famously asked, "What is it that breathes fire into the equations and makes a universe for them to describe?" In the context of the MUH, there's thus no fire-breathing required, since the point isn't that a mathematical structure *describes* a universe, but that it *is* a universe. Moreover, there's no *making* required either. You can't *make* a mathematical structure—it simply exists. It doesn't exist in space and time—space and time may exist in it. In other words, all structures that exist mathematically have the same ontological status, and the most interesting question isn't which ones exist physically (they all do), but which ones contain life—and perhaps us. Many mathematical structures—the dodecahedron, for example—lack the complexity to support any kind of self-aware substructures, so it's likely that the Level IV multiverse resembles a vast and mostly uninhabitable desert, with life confined to rare oases, bio-friendly mathematical structures such as the one we inhabit. Analogously, we saw evidence in Chapter 6 that the Level II multiverse is mostly barren wasteland, with self-awareness confined to the tiny "Goldilocks" fraction of space where the value of the dark-energy density and other physical parameters are just right for life. In the Level I multiverse, the story appears to repeat itself, with life flourishing mainly in the tiny fraction of space that lies near planetary surfaces. So we humans are in a very privileged place indeed!

Exploring the Level IV Multiverse: What's Out There?

Our Local Neighborhood

Let's spend some time exploring the Level IV multiverse and the diverse zoo of mathematical structures that it contains, beginning in our local neighborhood. Although we still don't know exactly which mathematical structure we inhabit, it's not hard to imagine many small modifications that might give other valid mathematical structures. For example, the standard model of particle physics involves certain symmetries that mathematicians denote $SU(3) \times SU(2) \times U(1)$, and if we replace them by different symmetries, we'll end up with a mathematical structure with different kinds of particles and forces, where quarks, electrons and photons are replaced by other entities with novel properties. In some mathematical structures, there's no light. In others, there's no gravity. In Einstein's mathematical description of spacetime, the numbers 1 and 3 that respectively specify the number of time and space dimensions can be replaced by different values of your choice.

Although we discussed in Chapter 6 how inflation in a single mathematical structure with its single set of *fundamental* laws of physics can give rise to different *effective* laws of physics in different parts of space, forming a Level II multiverse, we're now talking about something more radical, where even the fundamental laws are different—where there's no quantum mechanics, say. If string theory can be rigorously defined mathematically, then there's a mathematical structure where string theory is the "correct Theory of Everything" in that structure, but everywhere else in the Level IV multiverse, it's not.

When contemplating the Level IV multiverse, we need to let our imagination fly, unencumbered by our preconceptions of what laws of

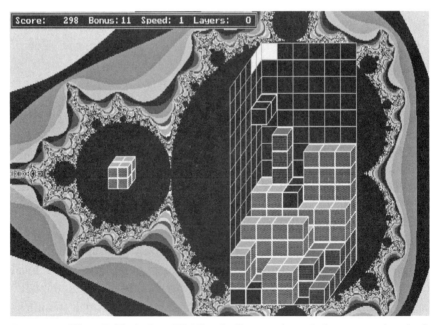

Figure 12.3: The 3-D Tetris clone FRAC embodies a mathematical structure where both space and time are discrete rather than continuous.

physics are supposed to be like. Consider space and time: rather than being continuous as our world suggests, they can be discrete, as in the computer games PAC-MAN and Tetris, or in John Conway's Game of Life, where motion can occur only in jerky jumps. As long as all user input is turned off so that the time evolution can be deterministically computed, these games all correspond to valid mathematical structures. For example, Figure 12.3 shows the 3-D Tetris clone called FRAC mentioned in Chapter 3, which I wrote with my friend Per Bergland in 1990, and if you play it without touching the keyboard (which isn't the best high-score strategy . . .), then the entire game from start to finish is determined by simple mathematical rules in the program, which makes it a mathematical structure that's part of the Level IV multiverse. It's been widely speculated that even our own Universe may exhibit some form of spacetime discreteness that's hidden away on such small scales that we haven't yet noticed.

Going still more radical, there are many mathematical structures that do away with space and time altogether, so that there's no meaningful sense in which anything is happening in them. Most of the structures exemplified in Figure 12.4 are of this type; there's nothing going on inside the abstract dodecahedron, say, because this mathematical structure contains no time.

Encoding	Mathematical structure
100	Empty set
105	Set of 5 elements
113100120	Triangle rotations
11220000110	The group C_2
11220001110	Boolean algebra
1132000012120201	The group C_3
12410002311003102	Tetrahedron rotations
126100024351100510243	Cube rotations
21422001010011110111	"Kite" graph
12810012305674100156204 73	Octahedron rotations

Figure 12.4: A computer program can automatically generate an ordered master list of all finite mathematical structures, where each one is encoded as a sequence of numbers. The table above shows a few examples, using the encoding scheme from my 2007 mathematical universe paper. The words and diagrams in the second column are redundant baggage, reflecting some ways in which we humans name and illustrate these structures.

Our Postal Code in the Level IV Multiverse

As we discussed in Chapter 10, a mathematical structure is simply a set of abstract elements with relations between them. To explore the Level IV multiverse more systematically, we can write a computer program that automatically generates a list of the mathematical structures that exist, starting with the simplest and progressing to ever more complex ones. Figure 12.4 shows ten entries from such a list, using an encoding scheme that I described in my 2007 mathematical-universe paper (http://arxiv.org/pdf/0704.0646.pdf). The details of the encoding don't matter for this discussion, except that it has the nice property that *every* mathematical structure with a finite number of elements will appear somewhere on the list. So every one of these mathematical structures can be identified by a single number: its line number on this master list.

For finite mathematical structures, all relations can be described by finite tables of numbers, generalizing the idea of a multiplication table to other kinds of relations. For structures with very large numbers of elements, these tables get big and bulky, which generates large encodings that end up far down on the list. However, a small fraction of these very large structures embody an elegant simplicity that make them quite simple to describe. Consider, for example, the mathematical structure where the elements are the whole numbers 0, 1, 2, 3, . . . and the relations are addition and multiplication. It would be hugely wasteful to specify how multiplication works by writing out an enormous multiplication table for all pairs of numbers: even if we limit ourselves to the first million numbers, we'd need a table with a million rows and a million columns, containing a trillion entries. Instead, we teach our school kids merely a multiplication table for the first ten numbers, and then a simple algorithm for how to use this table to multiply bigger multi-digit numbers. We describe multiplication to our computers even more efficiently than to our kids: by representing the numbers in binary, we need to specify only a 2×2 multiplication table for zeros and ones and a short computer program that specifies how to use this to multiply arbitrarily large numbers.

A computer program is stored as simply a finite string of zeros and ones (a *bit string*), which can be interpreted as a single whole number written out in binary. This gives us an alternative way of encoding and enumerating the mathematical structures from Figure 12.4: we let each mathematical structure be represented by the number whose bit string is the shortest computer program whose functions define all the relations of the structure. Now structures will appear near the top of the list as long as they're simple to describe, even if they're huge structures in terms of their number of elements. The complexity-theory pioneers Ray Solomonoff, Andrey Kolmogorov and Gregory Chaitin have defined the *algorithmic complexity* (or just *complexity*, for short) of a bit string as the length in bits of its shortest self-contained description, for example a computer program that outputs the string. This means that our alternative master list ranks mathematical structures in order of increasing complexity.

A nice feature of this list is that it can also handle mathematical structures with infinitely many elements. For example, to define the mathematical structure of all whole numbers with addition and multiplication, we need to specify just the shortest computer program that can read in arbitrarily large numbers and add and multiply them—*Mathematica* and other computer algebra–software packages have precisely such algorithms. Mathematical structures involving infinitely many points on a continuum, such as spacetime, electromagnetic fields, and wavefunctions, can often be well approximated by finite structures that can be processed on a computer as well—indeed, this is how my colleagues and I perform most of our theoretical-physics calculations in practice.

In summary, the Level IV multiverse can be systematically mapped

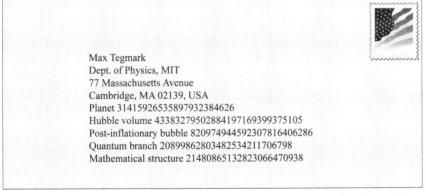

Max Tegmark
Dept. of Physics, MIT
77 Massachusetts Avenue
Cambridge, MA 02139, USA
Planet 31415926535897932384626
Hubble volume 43383279502884197169399375105
Post-inflationary bubble 82097494445923078164406286
Quantum branch 20899862803482534211706798
Mathematical structure 21480865132823066470938

Figure 12.5: To specify my address in the full physical reality, I need to list my location in the Level IV multiverse (my mathematical-structure number), in the Level III multiverse (my branch of the quantum wavefunction), in the Level II multiverse (my post-inflationary bubble), in the Level I multiverse (my horizon volume), and in our Universe. I've listed only finite numbers in this example, although there may be infinitely many members at all four levels, making my actual address involve numbers too huge to fit on an envelope.

by enumerating mathematical structures with a computer and studying their properties. If we one day succeed in identifying which mathematical structure we inhabit, then we can refer to it by its number on the master list, and for the first time specify our address in the full physical reality, as whimsically illustrated in Figure 12.5. Different countries on Earth have different schemes of specifying addresses: for example, some use postal codes with numbers, some use postal codes with letters and some use no postal codes at all. Similarly, the way you write the most local part of your address will vary between mathematical structures: most have neither quantum mechanics nor inflation, and therefore lack Levels III, II and I, as well as planets, whereas others may contain other types of parallel universes that we haven't even dreamed of.

The Structure of the Level IV Multiverse

It's interesting to study the Level IV multiverse. If we adopt the popular formalist definition of mathematics as "the study of mathematical structures," then studying the Level IV multiverse is what mathematicians do for a living. For a physicist like myself who believes in the Mathematical Universe Hypothesis, studying it is also tantamount to exploring the ultimate physical reality and searching for our place in it. And, conveniently, it's easier to explore the Level IV multiverse than any of the lower multiverses or even our own Universe, because it doesn't require rockets or telescopes—merely computers and ideas! I've therefore had lots of fun over the years writing computer software to perform the sort of mathematical-structure tabulation and classification that we have just talked about.

When doing this in practice, one encounters an enormous redundancy. There are very many different ways of writing a computer program that performs any given calculation, and there are similarly huge numbers of equivalent ways of describing finite mathematical structures by tables of numbers—corresponding to, for example, different ways of ordering/labeling the elements. As we discussed in Chapter 10, a mathematical structure is an equivalence class of descriptions, so the master list should contain each mathematical structure only once, specified only by that one of its many equivalent descriptions that is the shortest.

For any two mathematical structures, you can define a new one by combining all their elements and relations. Many structures on the master list are of this composite type, and when studying the Level IV multiverse, it makes sense to ignore them. This is because there are no relations connecting the two parts, which means that a self-aware observer in one of its parts will be forever unaware of and unaffected by the existence of the other part, so she can just as well act as though the other part didn't exist—or wasn't part of her mathematical structure. The only way in which composite structures may perhaps matter is if they enter into the solution of the measure problem, altering the likelihood you'd assign to living in different mathematical structures. Because a composite structure is more complicated to describe, it will typically occur much farther down the master list than its parts, which might give it a lower "measure." Indeed, for any finite number of structures in the Level IV multiverse, there's also a single composite structure extremely far down the master list that contains all of them.

Although the different mathematical structures in the Level IV multiverse aren't connected in any physically meaningful sense, there are many interesting relations between them at the meta-level. For example, we just discussed how one can be the combination of others. Another example is that one structure can in a sense describe another: the elements in the first structure can correspond to the relations in the second, and relations in the first can describe what happens when you combine relations in the second. In this sense, the twenty-four-relation "cube-rotations" structure from Figure 12.4 is described by what mathematicians call the "rotation group of the cube," a structure having twenty-four elements corresponding to all possible rotations that leave a perfect cube looking unchanged. Many different mathematical structures share these cube symmetries and thus have some claim to being *the* cube—for example, the structures whose elements correspond to cube faces, corners or edges and whose relations specify either how rotations rearrange these elements or who's neighboring whom.

Limits of the Level IV Multiverse: Undecidable, Uncomputable and Undefined

How big is the Level IV multiverse? For starters, there are infinitely many finite mathematical structures. As infinitely many as there are numbers 1, 2, 3, . . . to be precise, since we just saw that they can all be tabulated in a single numbered list. But how many infinite mathematical structures does the Level IV multiverse contain, each of which has an infinite number of elements? We saw that some infinite structures can also be defined and included in the master list together with the finite structures, by using computer programs to define their relations. However, embracing infinity opens a Pandora's box of ontological problems. To see this, consider a mathematical structure where the elements are the numbers 1, 2, 3, . . . which includes the three relations (functions) in the following list, rules that take numbers as input and compute a new number according to the definitions listed:

1. $P(n)$: Given a number n, $P(n)$ denotes the smallest prime that's greater than n.
2. $T(n)$: Given a number n, $T(n)$ denotes the smallest twin prime that's greater than n (a twin prime is a prime number that has a prime number as next-nearest neighbor; 11 and 13 are examples of twin primes).
3. $H(m,n)$: Given two numbers m and n, $H(m,n)$ equals 0 if the mth computer program on the master list of all computer programs will keep running forever when fed the bits of n as input, and $H(m,n)$ equals 1 if the program instead halts after a finite number of steps.

Does this structure qualify for membership in the Level IV multiverse, or is it not sufficiently well defined? The first function, $P(n)$, is a piece of cake: it's easy to write a program that starts checking whether the numbers following n are prime and stops as soon as it's found one, and we're guaranteed that this program will halt after a finite number of steps because we know that there are infinitely many primes—a fact that Euclid proved over two thousand years ago. So $P(n)$ is an example of what we call a *computable function*.

The second function, $T(n)$, is trickier: it's again easy to write a program that checks each number following n to see whether it's a twin prime, but if you plug in a number n greater than $3756801695685^{2666669}$ – 1 (the largest twin prime known as of my writing this), there's no guarantee that the program will ever stop and deliver an answer, because despite the best efforts of our most talented mathematicians, we still don't know whether there are infinitely many twin primes. So for now, we don't know whether $T(n)$ is a computable function and hence rigorously defined, and it's debatable whether a mathematical structure containing such a sloppily specified relation qualifies as well defined.

The third function, $H(m,n)$, is even more nefarious: the computer-science pioneers Alonzo Church and Alan Turing established that there's *no* program that can compute $H(m,n)$ for arbitrary input numbers m and n in a finite number of steps, so $H(m,n)$ is an example of what we call an *uncomputable function*. In other words, there's no program that can determine which other programs will eventually halt. Of course, any given program will either halt or it won't, but the catch is, just as with the twin primes, that you might have to wait forever to find out. The Church-Turing discovery of uncomputable functions is closely related to the discovery by the logician Kurt Gödel that some theorems of arithmetic are undecidable, meaning that they can neither be proved nor disproved in a finite number of steps.

Should a mathematical structure be considered well defined even if it contains a relation like H that can't be evaluated even on an arbitrarily powerful computer? If so, its structure can only be known to an oracle-like entity that in some sense has infinite powers and is capable of performing a truly infinite number of computational steps to get an answer. Such structures would never show up on the master list we discussed earlier, which covers only structures definable by normal computer programs, not ones requiring infinite oracle powers.

Finally, let's consider one of the most popular mathematical structures of our time: that of the so-called real numbers, such as $3.141592\ldots$, whose decimals go on forever. They form a continuum, and to specify even a single generic one, we need to list an infinite number of decimals, that is, an infinite amount of information. This means that conventional computer programs are hopelessly incapable of processing them: the problem isn't merely performing an infinite number of computational steps on a finite input as for the H-example, but that of inputting and outputting an infinite amount of information.

Alternatively, Kurt Gödel's work might make us worry that the MUH makes no sense with infinite mathematical structures because our Universe would be somehow inconsistent or undefined. If one accepts the mathematician David Hilbert's dictum that "mathematical existence is merely freedom from contradiction," then an inconsistent structure would not exist mathematically, let alone physically as in the MUH. Our standard model of physics includes everyday mathematical structures such as the whole numbers and real numbers. Yet Gödel's work leaves open the possibility that everyday mathematics is inconsistent, and that a finite-length proof exists within number theory itself, demonstrating that $0 = 1$. Using this shocking result, every other syntactically correct claim about whole numbers could in turn be proven to be true and mathematics as we know it would collapse like a house of cards.

So is the mathematical universe hypothesis ruled out by Gödel's incompleteness theorem? No, not as far as we know! Given any sufficiently powerful formal system, Gödel showed that we cannot use it to prove its own consistency, but this doesn't mean that it actually is inconsistent or that we have a problem. Indeed, our cosmos doesn't show any signs of being inconsistent or ill-defined, despite showing hints of being a mathematical structure. Moreover, what were we hoping for? Even if a formal system could be used to prove its own consistency, we'd remain unconvinced that it actually was consistent, since an inconsistent system can prove anything—including its consistency. We might be convinced if a simpler system that we have better reason to trust the consistency of could prove the consistency of a more powerful system—but unsurprisingly, that's impossible, as Gödel also proved. Of the many mathematicians with whom I'm friends, I've never heard anyone suggest that the mathematical structures that dominate modern physics (pseudo-Riemannian manifolds, Calabi-Yau manifolds, Hilbert spaces, etc.) are actually inconsistent.

All such uncertainties about undecidability and inconsistency apply only to mathematical structures with infinitely many elements. In the last chapter, we saw that the measure problem plaguing modern cosmology also applies only to mathematical structures with infinitely many elements, which begs a provocative question: Are infinities, undecidability, potential inconsistency, and the measure problem really inherent in the ultimate physical reality, or are they merely mirages, artifacts of our playing with fire and using powerful mathematical tools that are more convenient to work with than those that actually describe our Universe? More specifically, how well defined do mathematical structures need to be to be real, i.e., to be members of the Level IV multiverse? There's a range of interesting options for which structures qualify:

1. No structures (i.e., the Mathematical Universe Hypothesis is false).
2. Finite structures. These are trivially computable, since all their relations can be defined by finite look-up tables.
3. Computable structures (whose relations are defined by halting computations).
4. Structures with relations defined by computations that aren't guaranteed to halt (may require infinitely many steps), like our H-example.
5. Still more general structures, such as ones involving a continuum where typical elements require an infinite amount of information to describe.

The Computable Universe Hypothesis

An interesting possibility is the *Computable Universe Hypothesis* (here-after CUH) that option 3 is the limit and more general structures are disqualified:

> **Computable Universe Hypothesis (CUH):** *The mathematical structure that is our external physical reality is defined by comput-able functions.*

By this we mean that the relations (functions) that define the mathe-matical structure can all be implemented as computations that are guar-anteed to halt after a finite number of steps. If the CUH is false, then an even more conservative hypothesis is the Finite Universe Hypothesis (FUH); that option 2 is the limit: our external reality is a finite math-ematical structure.

I find it interesting that, as we discussed at the end of the last chapter, closely related issues have been hotly debated among mathematicians without any reference to physics. According to the finitist school of mathematicians, which included Leopold Kronecker, Hermann Weyl and Reuben Goodstein, a mathematical object doesn't exist unless it can be constructed from whole numbers in a finite number of steps. This leads directly to option 3.

According to the CUH, the mathematical structure that is our physi-cal reality has the attractive property of being computable and hence well defined in the strong sense that all its relations can be computed. There would thus be no physical aspects of our Universe that are uncomputable/undecidable, eliminating the concern that the work of Church, Turing and Gödel somehow makes our world incomplete or inconsistent. I don't know exactly what properties our physical reality has, but I'm confident that these properties exist in the sense of being well defined: I have no doubt that nature knows what it's doing!

Many authors have puzzled over why our physical laws appear rela-tively simple. For example, why does the standard model of particle physics have such simple symmetries as those we call $SU(3) \times SU(2) \times U(1)$, requiring only the 32 parameters from Chapter 10, when most alternatives are much more complicated? It's tempting to speculate that the CUH contributes to this relative simplicity by sharply limiting the complexity of nature. By banishing the continuum altogether, perhaps the CUH may also help downsize the inflationary landscape and resolve the cosmological measure problem, which as we discussed in the last chapter is in large part linked to the ability of a true continuum to undergo exponential stretching forever, producing infinite numbers of observers.

That was the good news. Although the CUH has attractive features such as ensuring that our Universe is rigorously defined and perhaps mitigating the cosmological measure problem by limiting what exists, it also poses serious challenges that need to be resolved.

A first concern about the CUH is that it may sound like a surrender of the philosophical high ground, effectively conceding that although all possible mathematical structures are "out there," some have privileged status. However, my guess is that if the CUH turns out to be correct, it will instead be because the rest of the mathematical landscape was a mere illusion, fundamentally undefined and simply not existing in any meaningful sense.

A more immediate challenge is that our current standard model (and virtually all historically successful theories) violate the CUH, and it's far from obvious whether a viable computable alternative exists. The main source of CUH violation comes from incorporating the continuum, usually in the form of real or complex numbers, which can't even comprise the input to a finite computation, since they generically require infinitely many bits to specify. Even approaches attempting to banish the classical spacetime continuum by discretizing or quantizing it tend to maintain continuous variables in other aspects of the theory, such as the strength of the electromagnetic field or the amplitude of the quantum wavefunction.

One interesting approach to this continuum challenge involves replacing real numbers by a mathematical structure that emulates the continuum while remaining computable—for example, what mathematicians refer to as algebraic numbers. Another approach that I feel is worth exploring is abandoning the continuum as fundamental and trying to recover it as an approximation. As mentioned, we've never measured anything in physics to more than about sixteen significant digits, and no experiment has been carried out whose outcome depends on the hypothesis that a true continuum exists, or hinges on nature computing something uncomputable. It's striking that many of the continuum models of classical mathematical physics (for example, the equations describing waves, diffusion or liquid flow) are known to be mere approximations of an underlying discrete collection of atoms. Quantum-gravity research suggests that even classical spacetime breaks down on very small scales. We therefore can't be sure that quantities that we still treat as continuous (such as spacetime, field strengths and quantum wavefunction amplitudes) aren't mere approximations of something discrete. Indeed, certain discrete computable structures (indeed, finite ones satisfying the FUH) can approximate our continuum physics models so well that we physicists use them when we need to compute things in practice, leaving open the question of whether the mathematical structure of our Universe is more like the former or more like the latter. Some authors such as Konrad Zuse, John Barrow, Jürgen Schmidhuber and Stephen Wolfram have gone as far as suggesting that the laws of nature are both computable and finite like a cellular automaton or computer simulation. (Note, however, that these suggestions differ from the CUH and FUH, by requiring the *time evolution* rather than the *description* [the relations] of the structure to be computable.)

Adding further twists, physics has also produced examples of how something continuous (like quantum fields) can produce a discrete solution (like a crystal lattice), which in turn appears like a continuous medium on large scales, which in turn has vibrations that behave like discrete particles called phonons. My MIT colleague Xiao-Gang Wen has shown that such "emergent" particles may even behave like ones in our standard model, raising the possibility that we may have multiple layers of effective continuous and discrete descriptions on top of what's ultimately a discrete computable structure.

The Transcendent Structure of Level IV

Above we explored how mathematical structures and computations are closely related, in that the former are defined by the latter. On the other hand, computations are merely special cases of mathematical structures. For example, the information content (memory state) of a digital computer is a string of bits, say, "1001011100111001 . . . " of great but finite length, equivalent to some large but finite whole number n written in binary. The information processing of a computer is a deterministic rule for changing each memory state into another (applied over and over again), so mathematically, it's simply a function f mapping the whole numbers onto themselves that gets iterated: $n \mapsto f(n) \mapsto f(f(n)) \mapsto \ldots$. In other words, even the most sophisticated computer simulation is merely

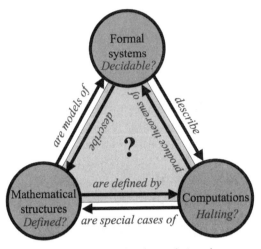

Figure 12.6: The arrows indicate the close relations between mathematical structures, formal systems and computations. The question mark suggests that these are all aspects of the same transcendent structure, whose nature we still haven't fully understood.

a special case of a mathematical structure, hence included in the Level IV multiverse.

Figure 12.6 illustrates how computations and mathematical structures are related not only to each other, but also to *formal systems*, the abstract symbolic systems of axioms and deduction rules that mathematicians use to prove theorems about mathematical structures. The boxes in Figure 12.1 correspond to such formal systems. If a formal system describes a mathematical structure, mathematicians say that the latter is a *model* of the former. Moreover, computations can generate theorems in formal systems (indeed, for certain classes of formal systems, there are algorithms that can compute all theorems).

Figure 12.6 also illustrates that there are potential problems at all three vertices of the triangle: mathematical structures may have relations that are undefined, formal systems may contain statements that are undecidable, and computations may fail to halt after a finite number of steps. The relations between the three vertices with their corresponding complications are illustrated by six arrows, explained in more detail in my 2007 mathematical-universe paper. Since different arrows are studied by different specialists in a range of fields from mathematical logic to computer science, the study of the triangle as a whole is somewhat interdisciplinary, and I think it deserves more attention.

I've drawn a question mark at the center of the triangle to suggest that the three vertices (mathematical structures, formal systems and computations) are simply different aspects of one underlying transcendent structure whose nature we still don't fully understand. This structure (perhaps restricted to the defined/decidable/halting part as per the CUH) exists "out there" in a baggage-free way, and is both the totality of what has mathematical existence and the totality of what has physical existence.

Implications of the Level IV Multiverse

So far in this chapter, we've argued that the ultimate physical reality is the Level IV multiverse, and started exploring its *mathematical* properties. Now let's explore its *physical* properties as well as other implications of the Level IV idea.

Symmetries and Beyond

If we turn our attention to some particular mathematical structure on the master list that serves as our atlas of the Level IV multiverse, how can we derive the physical properties that a self-aware observer in it would perceive it to have? In other words, how would an infinitely intelligent mathematician start with its mathematical definition and derive the physics description that we called the "consensus reality" in Chapter 9?*

We argued in Chapter 10 that her first step would be to calculate what symmetries the mathematical structure has. Symmetry properties are among the very few types of properties that every mathematical structure possesses, and they can manifest themselves as physical symmetries to the structure's inhabitants.

The question of what she should calculate next when exploring an arbitrary structure is largely uncharted territory, but I find it striking that in the particular mathematical structure that we inhabit, further study of its symmetries has led to a gold mine of further insights. The

---

* In the philosophy of science, the conventional approach holds that a theory of mathematical physics can be broken down into (i) a mathematical structure, (ii) an empirical domain and (iii) a set of correspondence rules that link parts of the mathematical structure with parts of the empirical domain. If the MUH is correct, then (ii) and (iii) are redundant in the sense that they can, at least in principle, be derived from (i). Instead, they can be viewed as a handy user's manual for the theory defined by (i).

German mathematician Emmy Noether proved in 1915 that each continuous symmetry of our mathematical structure leads to a so-called conservation law of physics, whereby some quantity is guaranteed to stay constant—and thereby has the sort of permanence that might make self-aware observers take note of it and give it a "baggage" name. All the conserved quantities that we discussed in Chapter 7 correspond to such symmetries: for example, energy corresponds to time-translation symmetry (that our laws of physics stay the same for all time), momentum corresponds to space-translation symmetry (that the laws are the same everywhere), angular momentum corresponds to rotation symmetry (that empty space has no special "up" direction) and electric charge corresponds to a certain symmetry of quantum mechanics. The Hungarian physicist Eugene Wigner went on to show that these symmetries also dictated all the quantum properties that particles can have, including mass and spin. In other words, between the two of them, Noether and Wigner showed that, at least in our own mathematical structure, studying the symmetries reveals what sort of "stuff" can exist in it. As I mentioned in Chapter 7, some physics colleagues of mine with a penchant for math jargon like to quip that a particle is simply "an element of an irreducible representation of the symmetry group." It's become clear that practically all our laws of physics originate in symmetries, and the physics Nobel laureate Philip Warren Anderson has gone even further, saying, "It is only slightly overstating the case to say that physics is the study of symmetry."

Why do symmetries play such an important role in physics? The MUH provides the answer that our physical reality has symmetry properties because it's a mathematical structure, and mathematical structures have symmetry properties. The deeper question of why the particular structure that we inhabit has so much symmetry then becomes equivalent to asking why we find ourselves in this particular structure, rather than in another one with less symmetry. Part of the answer may be that symmetries appear to be more the rule than the exception in mathematical structures, especially in large ones not too far down the master list, where simple algorithms can define relations for a vast number of elements precisely because they all have properties in common. An anthropic-selection effect may be at work as well: as pointed out by Wigner himself, the existence of observers able to spot regularities in the world around them probably requires symmetries, so given that we're observers, we should expect to find ourselves in a highly symmetric mathematical structure. For example, imagine trying to make sense of a world where experiments were never repeatable because their outcome depended on exactly where and when you performed them. If dropped rocks sometimes fell down, sometimes fell up and sometimes fell sideways, and everything else around us similarly behaved in a seemingly random way, without any discernible patterns or regularities, then there might have been no point in evolving a brain.

The way modern physics is usually presented, symmetries are treated as an input rather than an output. For example, Einstein founded special relativity on what's called Lorentz symmetry (the postulate that you can't tell whether you're standing still because all laws of physics, including those governing the speed of light, are the same for all uniformly moving observers). Similarly a symmetry called $SU(3) \times SU(2) \times U(1)$ is usually taken as a starting assumption for the standard model of particle physics. Under the Mathematical Universe Hypothesis, the logic is reversed: the symmetries aren't an assumption, but simply properties of the mathematical structure that can be calculated from its definition on the master list.

The Illusion of Initial Conditions

Compared to how we usually teach physics at MIT, the Level IV multiverse provides a very different starting point for the subject, and this causes most traditional physics concepts to be reinterpreted. As we just saw, some concepts such as symmetries retain their central status. In contrast, other concepts, such as initial conditions, complexity and randomness, get reinterpreted as mere illusions, existing only in the mind of the beholder and not in the external physical reality.

Let's first examine initial conditions, which we briefly encountered in Chapter 6. Nobody captures the traditional view of initial conditions better than Eugene Wigner: "Our knowledge of the physical world has been divided into two categories: initial conditions and the laws of nature. The state of the world is described by the initial conditions. These are complicated and no accurate regularity has been discovered in them. In a sense, the physicist isn't interested in the initial conditions, but leaves their study to the astronomer, geologist, geographer, etc." In other words, we physicists traditionally call the regularities that we understand "laws" and dismiss much of what we don't understand as "initial conditions." The laws let us predict how these conditions will change over time, but give no information about why they started out the way they did.

In contrast, the Mathematical Universe Hypothesis leaves no room for such arbitrary initial conditions, eliminating them altogether as a fundamental concept. This is because our physical reality is a mathematical structure that is *completely* specified in all respects by its mathematical definition in the master list. A purported Theory of Everything saying that everything just "started out" or "was created" in some not fully specified state would constitute an incomplete description, thus violating the MUH. A mathematical structure isn't allowed to be partly undefined. So traditional physics embraces initial conditions, while the MUH rejects them: what are we to make of this?

The Illusion of Randomness

Because of this requirement that everything be defined, the MUH also banishes another concept that has played a central role in physics: randomness. Regardless of whether anything *seems* random to an observer, it must ultimately be an illusion, not existing at the fundamental level, because there's nothing random about a mathematical structure. Yet the physics textbooks on my office bookshelves are full of that word: quantum measurements are said to produce random outcomes, and the heat in a cup of coffee is alleged to be caused by random motion of its molecules. Again traditional physics embraces something that the MUH rejects: what are we to make of this?

The initial-condition puzzle and the randomness puzzle are linked, and raise a pressing question. By a crude estimate, it takes almost a googol (10^{100}) bits of information to specify the actual state of every particle in our Universe right now. What's the origin of this information? The traditional answer involves a combination of initial conditions and randomness: lots of bits are needed to describe how our Universe started out, since traditional laws of physics don't specify this, and then we need additional bits to describe the outcome of various random processes that happened between then and now. Now that the MUH requires everything to be specified and banishes both initial conditions and randomness, how are we to account for all this information? If the mathematical structure is simple enough to be described by equations on a T-shirt, this at face value appears downright impossible! Let's now tackle these questions.

The Illusion of Complexity

How much information does our Universe really contain? As we have discussed, the information content (algorithmic complexity) of something is the length in bits of its shortest self-contained description. To appreciate the subtlety of this, let's first ask how much information each of the six different patterns in Figure 12.7 contains. At first glance, the two leftmost ones look very similar, like seemingly random patterns of 128 × 128 = 16,384 tiny black and white pixels. This suggests that we need about 16,384 bits to describe either of them, one bit to specify the color of each pixel. But whereas this is probably true for the upper pattern, which I created with a quantum–random number generator, there's a hidden simplicity in the lower pattern: it's just the binary digits of the square root of two! That simple description is enough to calculate the whole pattern: $\sqrt{2} \approx 1.414213562\ldots$, which is written as

How many bits of information are needed to describe each pattern?

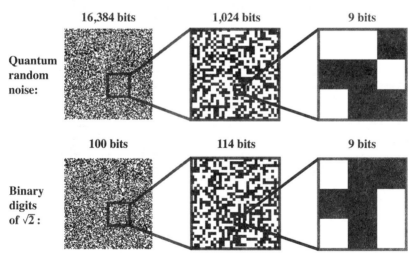

Figure 12.7: The complexity of a pattern (how many bits of information are needed to describe it) isn't always obvious. The upper left panel shows 128 × 128 = 16,384 squares that are randomly colored black or white, which typically can't be described using less than 16,384 bits. The smaller pieces of this pattern (top middle and right) consist of fewer random squares and therefore require fewer bits to describe. The lower left pattern, on the other hand, can be generated by a very short (100-bit, say) program, because it's simply the binary digits of $\sqrt{2}$ (0 = black square, 1 = white square). Describing the bottom middle pattern requires an additional 14 bits to specify which digit of $\sqrt{2}$ it starts at. Finally, the lower right pattern requires 9 bits just as the one above it; the pattern is so short that it doesn't help to specify that it's part of $\sqrt{2}$.

1.0100001010000110 . . . in binary. For argument's sake, let's say that this pattern of zeros and ones can be generated by a computer program that's 100 bits long. Then the apparent complexity of the lower left pattern is an illusion: we're looking not at 16,384 bits of information, but merely 100!

Things get even more intriguing when we start asking about the information content in small parts of these patterns. In the upper row of Figure 12.7, things are as we'd expect: smaller patterns are simpler and require less information to describe: we simply require one bit to describe each little black/white pixel. But in the bottom row, we see an example of the exact opposite! Here less is more, in the sense that the middle pattern is *more* complex than the left one, requiring more bits to describe. That's because it's no longer enough to simply say that it's the binary digits of $\sqrt{2}$: we also need to specify at which of the digits our pattern starts—which in this case requires another 14 bits of information. In summary, we've seen that *the whole can contain less information than the sum of its parts—and sometimes even less than one of its parts!*

Finally, the two rightmost patterns in Figure 12.7 each require 9 bits to describe. You and I know that the lower right pattern is hidden in the first 16,384 digits of $\sqrt{2}$, but for such a small pattern, this knowledge is no longer interesting or useful: there are only $2^9 = 512$ possible patterns of length 9, so you'll find our particular pattern hidden in *most* random-looking strings of thousands of zeros and ones.

Figure 12.8 shows the beautiful mathematical structure known as the Mandelbrot fractal, which further illustrates these ideas. It has the remarkable property of having intricate patterns down to arbitrarily tiny scales, and although many of these patterns look similar, no two are identical. How complex are the two images shown? They each contain about a million pixels, which in turn are represented by three bytes of information (a byte is eight bits), suggesting that each image requires a few megabytes to describe. However, the left image can in fact be computed from a program only a few hundred bytes long, implementing repeated use of the simple computation $z^2 + c$ described in the figure caption.

The right image is also simple, being merely a tiny part of the left one. But it's slightly more complex, requiring another 8 bytes to specify, with a 20-digit number, which of 10^{20} different parts it is. So once again, we see that less is more, in the sense that the apparent information content rises when we restrict our attention to a small part of the whole, thus losing the symmetry and simplicity that was inherent in the totality

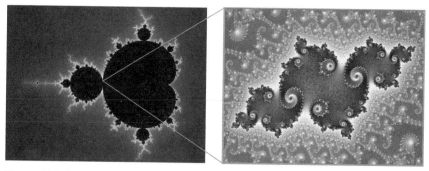

Figure 12.8: Despite its complex appearance with millions of elaborately colored pixels, the Mandelbrot fractal (left panel) has a very simple description: the points in the image correspond to what mathematicians call complex numbers c, and the colors encode how rapidly the complex number z blows up toward infinite size when you start with $z = 0$ and repeatedly keep squaring and adding c, i.e., repeatedly applying the simple equation $z \mapsto z^2 + c$. Paradoxically, the right image requires more information to describe even though it's simply a small part of the left one: if you cut the Mandelbrot fractal into about a hundred trillion trillion pieces, this is one of them, and the information contained in the right image basically tells you its address within the larger image, because the most economical way to specify it is as something like "Piece 31415926535897932384 of the Mandelbrot fractal."

of all parts taken together. For an even simpler example of this, consider that the algorithmic information content of a typical trillion-digit number is substantial, since the shortest program that prints it can't do much better than simply store all its trillion digits. Nonetheless, the list of *all* numbers 1, 2, 3, . . . can be generated by quite a trivial computer program, so the complexity of the whole set is smaller than that of a typical member.

Let's now return to our physical Universe and the near googol bits that appear required to specify it. Some scientists, such as Stephen Wolfram and Jürgen Schmidhuber, have wondered whether much of this complexity might also be a mere illusion, just as for the Mandelbrot fractal and the lower left pattern of Figure 12.7, resulting from a yet-to-be-discovered mathematical rule that's very simple. Although I find this to be an elegant idea, I'd bet against it: I view it as unlikely that all the numbers that characterize our Universe, from the patterns in the WMAP cosmic microwave–background maps to the positions of grains of sand on a beach, can be reduced to almost nothing by a simple data-compression algorithm. Indeed, as we saw in Chapter 5, cosmological inflation explicitly predicts that the cosmic seed fluctuations, from which much of this information ultimately originates, are distributed like random numbers for which such dramatic data compression is impossible.

These seed fluctuations specify all ways in which our early Universe differed from an easy-to-describe perfectly uniform plasma. Why do the patterns of cosmic seed fluctuations appear so random? We saw in Chapter 5 that, according to the cosmological standard model, inflation generates *all possible patterns* in different parts of space (in different universes throughout the Level I multiverse), and since we find ourselves in a rather typical part of this multiverse, we'll see a random-looking pattern without any hidden regularities to help us compress the information. The situation is a lot like the bottom row of Figure 12.7, where our Universe (corresponding to the right panel) corresponds to just a small random-looking part of the Level I multiverse (corresponding to the left panel) with its simple description. In fact, if you flip back to Chapter 6, you'll see that Figure 6.2 becomes equivalent to the bottom row of Figure 12.7 if we simply extend it to include more than a googolplex binary digits of $\sqrt{2}$, and expand the right panel to contain about a googol bits as our Universe does. It's widely believed among mathematicians (albeit still unproven) that the digits of $\sqrt{2}$ behave like random numbers, so that any possible pattern appears somewhere, just as universes with any possible initial conditions appear somewhere in the Level I multiverse. This means that a sequence of a googol digits from the digits of $\sqrt{2}$ actually tells us nothing at all about $\sqrt{2}$, merely about where in the digit sequence we're looking. Similarly, observing a googol bits of information about typical random-looking inflation-generated cosmic seed fluctuations only gives us information about where in the vast post-inflationary space we're looking.

Initial Conditions Reinterpreted

Above we worried about how to think about our initial conditions, and we now have a radical answer: *this information isn't fundamentally about our physical reality, but about our place in it.* The vast complexity we observe is an illusion in the sense that the underlying reality is quite simple to describe, and what requires close to a googol bits to specify is just our particular address in the multiverse. We discussed in Chapter 6 how our Galaxy contains many solar systems with different numbers of planets, so that when we say ours has eight, we're saying nothing fundamental about our Galaxy but something about our Galactic address. Because the Level I multiverse contains other Earths whose skies show all possible variations of cosmic microwave–background patterns or stellar constellations, the information contained in the WMAP map or a photo of the Big Dipper similarly tells us about our multiversal address. Analogously, the 32 physical constants from Chapter 10 tell us about our place in the Level II multiverse, if it exists. Although we thought all this information was about our physical reality, it was about us. The complexity is an illusion, existing only in the eye of the beholder.

I first thought of these ideas while biking to work through Munich's Englischer Garten back in 1995, and published them in a paper with the provocative title "Does our Universe in fact contain almost no information?" Now I realize that I should have dropped the word *almost*! Let me explain why. Our Level III multiverse reminds me more of the Mandelbrot fractal (Figure 12.8) than of our $\sqrt{2}$ example (Figure 12.7), because its pieces exhibit lots of regularity. Whereas all possible patterns occur equally often in the digits of $\sqrt{2}$, many patterns (pictures of your friends, say) don't occur anywhere in the Mandelbrot fractal. Just as most pieces of the Mandelbrot fractal seem to share a certain artistic style, dictated by that formula $z^2 + c$, most of the post-inflationary universes in the Level III multiverse share regularities in their time development that follow from quantum mechanics. When I referred to "almost no information," I meant the small amount of information needed to describe these regularities, specifying the mathematical structure that is the Level III multiverse. But in the light of the Mathematical Universe Hypothesis, not even this information tells us anything about the ultimate physical reality—rather, it simply tells us our address in the Level IV multiverse.

Randomness Reinterpreted

Okay, now that we've figured out how to interpret initial conditions, what about randomness? Here too the answer lies in the multiverse. We saw in Chapter 8 how the completely deterministic Schrödinger equation of quantum mechanics can give rise to apparent randomness from the subjective perspective of an observer in the Level III multiverse, and how the core process was a more general one having nothing to do with quantum mechanics: cloning. Specifically, randomness is simply how it feels when you're cloned: you can't predict what you'll perceive next if there'll be two copies of you perceiving different things. In Chapter 8 we saw that apparent randomness is caused by observer cloning in *some* cases. Now we see that it's in fact caused by cloning in *all* cases, since the MUH banishes fundamental randomness, the other logically possible explanation.

In other words, whereas apparently arbitrary initial conditions are caused by multiple universes, apparent randomness is caused by multiple yous. These two ideas merge into one and the same if we consider those parallel universes that contain a subjectively indistinguishable copy of you, so that there's both multiple universes and multiple yous. Then when you measure the initial conditions of your Universe, this information will appear random to all your copies, and it doesn't matter whether you interpret this information as coming from initial conditions or randomness—the information is the same. Observing which universe you're in reveals which copy of you is doing the observing.

How Complexity Suggests a Multiverse

We've talked a lot about the complexity of our Universe, but what about the complexity of our mathematical structure?

The MUH doesn't specify whether the complexity of the mathematical structure in the bird perspective is low or high, so let's consider both possibilities. If it's extremely high, then our quest to figure out its specification is clearly doomed. In particular, if describing the structure requires more bits than describing our observable Universe, then we can't even store the information about the structure in our Universe—it won't fit. An example of such a high-complexity theory would be the standard model with its 32 parameters from Chapter 10 explicitly specified as real numbers, such as $1/\alpha = 1/137.035999\ldots$, with infinitely many decimals lacking any simplifying pattern. Because even if one such parameter would require an infinite amount of information to store, the mathematical structure would be infinitely complex and impossible to specify in practice.

Most physicists hope for a complete Theory of Everything that's much simpler than this and can be specified by few enough bits to fit in a book, if not on a T-shirt—vastly fewer than the near googol bits needed to describe our Universe. Such a simple theory must predict a multiverse, regardless of whether the MUH is true or not. Why? Because this theory is by definition a complete description of reality: if it lacks enough bits to completely specify our Universe, then it must instead describe all possible combinations of stars, sand grains and such—so that the extra bits that describe our Universe simply encode which universe we're in, like a multiversal postal code would. The address written on the envelope in Figure 12.5 would then have a relatively short bottom line, specifying the theory, but the address lines right above would contain nearly a googol characters.

Are We Living in a Simulation?

We've just seen how the Mathematical Universe Hypothesis changes our perspective on many fundamental questions. Let's now turn to another such topic: that of simulated realities. Long a staple of science fiction, the idea that our external reality is some form of computer simulation has gained prominence with blockbuster movies such as *The Matrix*. Scientists such as Eric Drexler, Ray Kurzweil and Hans Moravec have argued that simulated minds are both possible and imminent, and some (for example Frank Tipler, Nick Bostrom and Jürgen Schmidhuber) have gone as far as discussing the probability that this has already happened—that we're simulated.

Why should you think you're simulated? Well, many science-fiction authors have explored scenarios where future space colonization transforms much of the matter in our Universe into ultra-advanced computers that simulate huge numbers of observer moments subjectively indistinguishable from yours. Nick Bostrom and others have argued that in this case, it's most likely that your current observer moment is in fact one of the simulated ones, since they're more numerous. However, I think that this argument logically self-destructs: if the argument is valid, then your indistinguishable simulated copies can make it, too, implying that there are way more doubly simulated copies, and that you're probably a simulation within a simulation. Making this argument repeatedly, you conclude that you're probably a simulation within a simulation within a simulation, and so on, arbitrarily many levels down—a *reductio ad absurdum*. I think the logical mistake happens at the very first step: if you're willing to assume that you're simulated, then as emphasized by Phillip Helbig, the computational resources of your own (simulated) universe are irrelevant: what matters are the computational resources in the universe where the simulation is taking place, about which you know essentially nothing.

Others have argued that it's fundamentally impossible for our reality to be a simulation. Seth Lloyd has advanced the intermediate possibility that we live in an analog simulation performed by a quantum computer, albeit not a computer designed by anybody—rather, because the structure of quantum field theory is mathematically equivalent to that of a spatially distributed quantum computer. In a similar spirit, Konrad Zuse, John Barrow, Jürgen Schmidhuber, Stephen Wolfram and others have explored the idea that the laws of physics correspond to a classical computation. Let's explore these ideas in the context of the Mathematical Universe Hypothesis.

The Time Misconception

Suppose that our Universe is indeed some form of computation. A common misconception in the universe-simulation literature is that our physical notion of a one-dimensional time must then necessarily be equated with the step-by-step one-dimensional flow of the computation. I'll argue below that if the MUH is correct, then computations don't need to *evolve* our Universe, but merely *describe* it (defining all its relations).

The temptation to equate time steps with computational steps is understandable, given that both form a one-dimensional sequence where (at least for the non-quantum case) the next step is determined by the current state. However, this temptation stems from an outdated classical description of physics: there's generically no natural and well-defined global time variable in Einstein's general relativity, and even less so in quantum gravity where time is known to emerge only as an approximate property of certain "clock" subsystems. Indeed, linking frog-perspective time with computer time is unwarranted even within the context of classical physics. The rate of time flow perceived by an observer in the simulated universe is completely independent of the rate at which a computer runs the simulation, a point emphasized in Greg Egan's science-fiction novel *Permutation City*. Moreover, as we discussed in the last chapter and as stressed by Einstein, it's arguably more natural to view our Universe not from the frog perspective as a three-dimensional space where things happen, but from the bird perspective as a four-dimensional spacetime that merely is. There should therefore be no need for the computer to compute anything at all—it could simply store all the four-dimensional data, that is, encode all properties of the mathematical structure that is our Universe. Individual time slices could then be read out sequentially if desired, and the "simulated" world should still feel as real to its inhabitants as in the case where only three-dimensional data is stored and evolved. In conclusion: *the role of the simulating computer isn't to compute the history of our Universe, but to specify it.*

How specify it? The way in which the data are stored (the type of computer, the data format, etc.) should be irrelevant, so the extent to which the inhabitants of the simulated universe perceive themselves as real should be independent of whatever method is used for data compression. The physical laws that we've discovered provide great means of data compression, since they make it sufficient to store the initial data at some time together with the equations and a program computing the future from these initial data. As emphasized on pages 340–344, the initial data might be extremely simple: popular initial states from quantum field theory with intimidating names such as *the Hawking-Hartle wavefunction* or *the inflationary Bunch-Davies vacuum* have very low algorithmic complexity, since they can be defined in brief physics papers, yet simulating their time evolution would simulate not merely one universe like ours, but a vast decohering collection of parallel ones. It's therefore plausible that our Universe (and even the whole Level III multiverse) could be simulated by quite a short computer program.

A Different Sort of Computation

The previous example referred to our particular mathematical structure, with its quantum mechanics and so on. More generally, as we've discussed, a complete description of an arbitrary mathematical structure is by definition a specification of the relations between its elements. We saw earlier in this chapter that for these relations to be well defined, all these functions must be *computable*: there must exist a computer program that can compute the relations in a finite number of computational steps. Each relation of the mathematical structure is thus defined by a computation. In other words, if our world is a well-defined mathematical structure in this sense, then it's indeed inexorably linked to computations, albeit computations of a different sort than those usually associated with the simulation hypothesis: these computations don't *evolve* our Universe, but *describe* it by evaluating its relations.*

Does a Simulation Really Need to Be Run?

A deeper understanding of the relations between mathematical structures, formal systems and computations (the triangle in Figure 12.6)

---

* Indeed, as pointed out by Ken Wharton in his paper "The Universe Is Not a Computer," at http://arxiv.org/pdf/1211.7081.pdf, our laws of physics may be such that the past doesn't uniquely determine the future, so the idea that our Universe can be simulated even in principle is a hypothesis that shouldn't be taken for granted.

would shed light on many of the thorny issues we've encountered in this book. One such issue is the measure problem that plagued us in the last chapter, which is in essence the problem of how to deal with annoying infinities and predict probabilities for what we should observe. For example, since every universe simulation corresponds to a mathematical structure, and therefore already exists in the Level IV multiverse, does it in some meaningful sense exist "more" if it is also run on a computer? This question is further complicated by the fact that eternal inflation predicts an infinite space with infinitely many planets, civilizations and computers, some of which may be running universe simulations, and that the Level IV multiverse also includes an infinite number of mathematical structures that can be interpreted as computer simulations.

The fact that our Universe (together with the entire Level III multiverse) may be simulatable by a quite short computer program calls into question whether it makes any ontological difference whether simulations are "run" or not. If, as I have argued, the computer need only describe and not compute the history, then the complete description would probably fit on a single memory stick, and no CPU power would be required. It would appear absurd that the existence of this memory stick would have any impact whatsoever on whether the multiverse it describes exists "for real." Even if the existence of the memory stick mattered, some elements of this multiverse will contain an identical memory stick that would "recursively" support its own physical existence. This wouldn't involve any Catch-22, chicken-or-the-egg problem regarding whether the stick or the multiverse was created first, since the multiverse elements are four-dimensional spacetimes, whereas "creation" is of course only a meaningful notion *within* a spacetime.

So are we simulated? According to the MUH, our physical reality is a mathematical structure, and as such, it exists regardless of whether someone here or elsewhere in the Level IV multiverse writes a computer program to simulate/describe it. The only remaining question is then whether a computer simulation could make our mathematical structure in any meaningful sense exist even more than it already did. If we solve the measure problem, perhaps we'll realize that simulating it would increase its measure slightly, by some fraction of the measure of the mathematical structure within which it's simulated. My guess is that this would be a tiny effect at best, so if asked, "Are we simulated?," I'd bet my money on "No!"

Relation Between the MUH, the Level IV Multiverse and Other Hypotheses

An interesting variety of ultimate-reality proposals have been put forth by various researchers at the interfaces between philosophy, information theory, computer science and physics, and for excellent recent overviews, I recommend Brian Greene's book *The Hidden Reality* and Russell Standish's book *Theory of Nothing*.

On the philosophy side, the proposal closest to the Level IV multiverse is the theory of *modal realism* by the late philosopher David Lewis, which posits that "all possible worlds are as real as the actual world." His late philosophy colleague Robert Nozick made a similar proposal termed *the principle of fecundity*. One common criticism of modal realism asserts that because it posits that all imaginable universes exist, it makes no testable predictions at all. The Level IV multiverse can be thought of as a smaller and more rigorously defined reality by virtue of replacing Lewis's "all possible worlds" by "all mathematical structures." The Level IV multiverse does *not* imply that all imaginable universes exist. We humans can imagine many things that are mathematically undefined and hence don't correspond to mathematical structures. Mathematicians publish papers with existence proofs that demonstrate the mathematical consistency of various mathematical-structure descriptions precisely because to do this is difficult and not possible in all cases.

On the computer-science side, the most closely related proposals are that our physical reality is some form of computer simulation or simulations, as we discussed earlier in this chapter. The relation is most clearly seen in Figure 12.6 where the two ideas correspond to two different vertices of the triangle: our reality is a computation according to the simulation hypothesis, as opposed to a mathematical structure according to the MUH. Under the simulation hypothesis, computations *evolve* our Universe, but under the MUH they merely *describe* it by evaluating its relations. According to the computational-multiverse theories explored by Jürgen Schmidhuber, Stephen Wolfram and others, the *time evolution* needs to be computable, while according to the Computable Universe Hypothesis (CUH), it's the *description* (the relations) that must be computable. John Barrow and Roger Penrose have suggested that only structures complex enough for Gödel's incompleteness theorem to apply can contain self-aware observers. Earlier, we saw that the CUH in a sense postulates the exact opposite.

Testing the Level IV Multiverse

We have argued that the External Reality Hypothesis (ERH), which says that there is an external physical reality completely independent of us humans, implies the Mathematical Universe Hypothesis (MUH), which says that our external physical reality is a mathematical structure, which in turn implies the existence of the Level IV multiverse. Therefore, the most straightforward way to strengthen or weaken our evidence for the Level IV multiverse is to further study and test the ERH. While the jury is still out on the ERH, I think it's fair to say that most of my physics colleagues subscribe to it, and that the recent successes of the standard models of particle physics and cosmology do little to suggest that our ultimate physical reality, whatever it is, fundamentally revolves around us humans and can't exist without us. That said, let's nonetheless explore two ways of potentially testing the MUH and the Level IV multiverse more directly.

The Typicality Prediction

As we saw in Chapter 6, the discovery that a physical parameter seems fine-tuned to allow life can be interpreted as evidence of a multiverse where the parameter takes a broad range of values, because this interpretation makes it unsurprising that a habitable universe like ours exists, and predicts that this is where we'll find ourselves. In particular, we saw that some of the strongest evidence for a Level II multiverse comes from the observed fine-tuning of the dark-energy density. Could there be fine-tuning evidence even for Level IV, at least in principle?

At a 2005 physics conference in Cambridge, while my friend Anthony Aguirre and I were taking a late-evening walk through the quaint courtyards of Trinity College, I suddenly realized that the answer was *yes*. Here's why.

Suppose you're getting out of your friend's car after she's driven you to a town you know nothing about, and you notice a confusing zoo of signs (see Figure 12.9) banning parking everywhere on the street except for the one place where she parked. She explains that, as part of

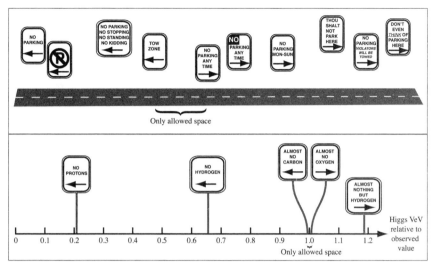

Figure 12.9: If a street has lots of randomly placed signs, each banning parking on the whole street on either the left or right side of the sign, it's quite unlikely that parking will be allowed *anywhere* on it: this happens only if all left arrows end up to the left of all right arrows, as in the top panel example. Similarly, if a universe has a physical parameter that must satisfy lots of constraints for life to be allowed (bottom panel), it's a priori unlikely that there's any habitable range of parameter values. Situations such as those illustrated here can therefore be interpreted as evidence for the existence of many streets or many mathematical structures in a Level IV multiverse, respectively.

an antipollution campaign, the new mayor has ordered ten signs to be randomly placed on each street, each one banning parking on the whole street either to the left or to the right side of the sign. After doing some math, you realize that this crazy random process will typically ban parking everywhere on a street, with only about a 1% chance that there's an allowed space;* this happens only if all signs with left arrows get placed to the left of all signs with right arrows.

What are you to make of this? Is it just a lucky coincidence? If you abhor unexplained coincidences as a typical scientist does, then you'll lean toward the one interpretation that doesn't require a wild stroke of luck: that there are many streets in this strange town, probably in the ballpark of a hundred or more. This makes it likely that there's legal parking on *some* street, and since your friend knows the town, it's totally unsurprising that this is where your friend has chosen to park. This fine-tuning example differs from those of Chapter 6 because

---

* If there are n random signs, the probability that any parking is allowed is $(n + 1)/2^n$: once the signs have been placed, there are 2^n ways of orienting the left/right arrows, and only $n + 1$ of these ways corresponds to all the left arrows being to the left of all the right arrows.

what appears to be fine-tuned isn't something continuous, such as the dark-energy density, but rather something discrete: all the directions of the left- or right-pointing arrows conspire in a surprising way.

My parking example was admittedly silly, but as the lower panel of Figure 12.9 illustrates, we observe something rather similar in our Universe. The horizontal axis shows a parameter related to the recently discovered Higgs particle, and recent work by John Donoghue, Craig Hogan, Heinz Oberhummer and their collaborators has shown that, much like the dark-energy density, it appears highly fine-tuned: it's about sixteen orders of magnitude smaller than one might naturally expect, yet changing it by even a percent up or down dramatically reduces the amount of either carbon or oxygen produced by stars. Increasing it by 18% radically reduces fusion of hydrogen into *any* other atoms by stars, while reducing it by 34% makes hydrogen atoms decay into neutrons as their proton gobbles up their electron. Reducing it fivefold makes even lone protons decay to neutrons, guaranteeing a universe with no atoms at all.

How should we interpret this? Well, first of all, this looks like further evidence for a Level II multiverse across which some physical parameters vary. Just as this can explain why we find a dark-energy density that's just right to allow galaxies to form, this can clearly also explain why we find Higgs properties that are just right to allow more complex atoms than hydrogen to form—and it's not surprising that we're in one of the relatively few universes with both interesting atoms and interesting galaxies if life requires at least a minimal level of complexity.

But Figure 12.9 raises a second question as well: why do the five arrows in the bottom panel conspire to allow *any* habitable range of Higgs properties? This could well be a fluke: five random arrows would allow some range with 19% probability, so we need only invoke a small amount of luck. Moreover, because of how nuclear physics works, these five arrows aren't independent, so I don't view this particular five-arrow example as strong evidence of anything. However, it's perfectly plausible that further physics research could uncover more striking fine-tuning of this discrete type with, say, ten or more arrows conspiring to allow a habitable range for some physical parameter or parameters.* And if this happens, then we can argue just as for the top panel: that this is evidence for the existence not of other streets, but of other *universes* where the

---

* It's easy to generalize this discrete fine-tuning definition to the case where more than one parameter can vary.

laws of physics are different, giving quite different requirements for life! In some cases, these universes might exist in the Level II multiverse, in a region where the same fundamental laws of physics give rise to a different phase of space with other effective laws. In other instances, however, this might be proven to be impossible, in which case these other universes would have to obey different *fundamental* laws, corresponding to different mathematical structures in the Level IV multiverse. In other words, while we currently lack direct observational support for the Level IV multiverse, it's possible that we may get some in the future.

The Mathematical-Regularity Prediction

We've mentioned Wigner's famous 1960 essay where he argued that "the enormous usefulness of mathematics in the natural sciences is something bordering on the mysterious," and that "there is no rational explanation for it." The Mathematical Universe Hypothesis provides this missing explanation. It explains the utility of mathematics for describing the physical world as a natural consequence of the fact that the latter *is* a mathematical structure, and we're simply uncovering this bit by bit. The various approximations that constitute our current physics theories are successful because simple mathematical structures can provide good approximations of certain aspects of more-complex mathematical structures. In other words, our successful theories aren't mathematics approximating physics, but mathematics approximating mathematics.

One of the key testable predictions of the Mathematical Universe Hypothesis is that physics research will uncover further mathematical regularities in nature. This predictive power of the mathematical-universe idea was expressed by Paul Dirac in 1931: "The most powerful method of advance that can be suggested at present is to employ all the resources of pure mathematics in attempts to perfect and generalize the mathematical formalism that forms the existing basis of theoretical physics, and after each success in this direction, to try to interpret the new mathematical features in terms of physical entities."

How successful has this prediction been so far? Two millennia after the Pythagoreans promulgated the basic idea of a mathematical universe, further discoveries made Galileo describe nature as being "a book written in the language of mathematics." Then much more far-reaching mathematical regularities were uncovered, ranging from the motions of planets to the properties of atoms, prompting those awestruck endorsements by Dirac and Wigner. After this, the standard models of particle physics and cosmology revealed new "unreasonable" mathematical order to a spectacular extent, from the microcosm of elementary particles to the macrocosm of the early Universe, arguably enabling all physics measurements ever made to be successfully calculated from the 32 numbers listed in Table 10.1. I know of no other compelling explanation for this trend than that the physical world really is completely mathematical.

Looking toward the future, there are two possibilities. If I'm wrong and the MUH is false, then physics will eventually hit an insurmountable roadblock beyond which no further progress is possible: there would be no further mathematical regularities left to discover even though we still lacked a complete description of our physical reality. For example, a convincing demonstration that there's such a thing as fundamental randomness in the laws of nature (as opposed to deterministic observer cloning that merely *feels* random subjectively) would therefore refute the MUH. If I'm right, on the other hand, then there'll be no roadblock in our quest to understand reality, and we're limited only by our imagination!

THE BOTTOM LINE

- The Mathematical Universe Hypothesis implies that mathematical existence equals physical existence.
- This means that all structures that exist mathematically exist physically as well, forming the Level IV multiverse.
- The parallel universes we've explored form a nested four-level hierarchy of increasing diversity: Level I (unobservably distant regions of space), Level II (other post-inflationary regions), Level III (elsewhere in quantum Hilbert space) and Level IV (other mathematical structures).
- Intelligent life appears to be rare, with most of Levels I, II and IV being uninhabitable.
- Exploring the Level IV multiverse doesn't require rockets or telescopes, merely computers and ideas.
- The simplest mathematical structures can be listed by a computer in telephone-book fashion, with each one having its own unique number.
- Mathematical structures, formal systems and computations are closely related, suggesting that they're all aspects of the same transcendent structure whose nature we still haven't fully understood.
- The MUH is compatible with Gödel incompleteness theorem: the mathematical structure we inhabit can be inconsistent even if we can't prove it.
- The Computable Universe Hypothesis (CUH) that the mathematical structure that is our external physical reality is defined by computable functions would make everything provably consistent.
- The Finite Universe Hypothesis (FUH) that our external physical reality is a finite mathematical structure implies the CUH and eliminates all concerns about reality being undefined.
- The CUH/FUH may help solve the measure problem and explain why our Universe is so simple.
- The MUH implies that there are no undefined initial conditions: initial conditions tell us nothing about physical reality, merely about our address in the multiverse.
- The MUH implies that there's no fundamental randomness: randomness is simply the way cloning feels subjectively.
- The MUH implies that most of the complexity we observe is an illusion, existing only in the eye of the beholder, being merely information about our address in the multiverse.
- A collection of things can be simpler to describe than one of its parts.
- Our multiverse is simpler than our Universe, in the sense that it can be described with less information, and the Level IV multiverse is simplest of all, requiring essentially no information to describe.
- We probably don't live in a simulation.
- The MUH is in principle testable and falsifiable.

13
Life, Our Universe and Everything

Figure 13.1: When we ask what everything is made of and zoom in to ever smaller scales, we find that the ultimate building blocks of matter are mathematical structures, objects whose properties are mathematical properties. When we ask how big everything is and zoom out to ever-larger scales, we end up at the same place: in the realm of mathematical structures, indeed a Level IV multiverse of all mathematical structures.

How Big Is Our Physical Reality?

I feel honored that you, my dear reader, have joined my reality-exploration journey all the way to this last chapter. We've traveled far, from the extra-galactic macrocosm to the subatomic microcosm, encountering a grander reality than I ever dreamed of in my wildest childhood dreams, with four different levels of parallel universes.

How does this all fit together? Figure 13.1 shows how I think about it. In the first part of the book, we pursued the question "How big is everything?" and explored ever-larger scales: we're on a planet in a galaxy in a universe that I think is in a doppelgänger-laden Level I multiverse in a more diverse Level II multiverse in a quantum-mechanical Level III multiverse in a Level IV multiverse of all mathematical structures. In the second part of the book, we pursued the question "What's everything made of?" and explored ever smaller scales: we're made of cells made of molecules made of atoms made of elementary particles, which are purely mathematical structures in the sense that their only properties are mathematical properties. Although we don't yet know what if anything these particles are made of, string theory and its leading competitors all suggest that any more fundamental building blocks are purely mathematical as well. In this sense, even though our two intellectual expeditions set off in opposite directions, toward the large and the small, respectively, they ended up in the same place: in the realm of mathematical structures. Whereas all roads were said to lead to Rome, our two roads to reality both lead to mathematics. This elegant confluence reflects the fact that one mathematical structure can contain others within it, explaining all the mathematical regularities that physics has uncovered as aspects or approximations of the grand mathematical structure that is our full external reality. On the largest and smallest scales, the mathematical fabric of reality becomes evident, while it remains easy to miss on the intermediate scales that we humans are usually aware of.*

---

* This expansion of our ontology in physics is reminiscent of the expansion of our ontology in mathematics over the past centuries. Mathematicians call this *generalization*: the insight that what we're studying is part of a larger structure.

360 LIFE, OUR UNIVERSE AND EVERYTHING

The Case for a Smaller Reality

I've painted a picture for you of our ultimate physical reality as I see it. Personally, I find this reality breathtakingly beautiful and inspiringly grand. But is it real? Or could it be that the picture is misleading, with much of the grandeur being mere mirages? Do you really live in a multiverse? Or is the whole question a silly one, lying beyond the pale of science? Let me give you my two cents.

Multiverse ideas have traditionally received short shrift from the establishment: we've seen that Giordano Bruno with his infinite-space multiverse got burned at the stake in 1600 and Hugh Everett with his quantum multiverse got burned on the physics job market in 1957. As I mentioned, I've even felt some of the heat firsthand, with senior colleagues suggesting that my multiverse-related publications were nuts and would ruin my career. There's been a sea change in recent years, however. Parallel universes are now all the rage, cropping up in books, movies and even jokes: "You passed your exam in many parallel universes—but not in this one."

This airing of ideas certainly hasn't led to a consensus among scientists, but it's made the multiverse debate much more nuanced and, in my opinion, more interesting, with scientists moving beyond shouting sound bites past each other and genuinely trying to understand opposing points of view. A nice example of this is a recent anti-multiverse article in *Scientific American* by the relativity pioneer George Ellis, which I highly recommend reading (see http://tinyurl.com/antiverse).

As we discussed in Chapter 6, we use the term *our Universe* to mean the spherical region of space from which light has had time to reach us during the 14 billion years since our Big Bang. When talking about parallel universes, we distinguished between four different levels: Level I (other such regions far away in space where the apparent laws of physics are the same, but where history played out differently because things started out differently), Level II (regions of space where even the apparent laws of physics are different), Level III (parallel worlds elsewhere in Hilbert space where quantum reality plays out) and Level IV (totally disconnected realities governed by different mathematical equations). In his critique, George Ellis classifies many of the arguments in favor of these multiverse levels and argues that they all have problems. Here's my summary of his main anti-multiverse arguments:

1. Inflation may be wrong (or not eternal).
2. Quantum mechanics may be wrong (or not unitary).
3. String theory may be wrong (or lack multiple solutions).
4. Multiverses may be unfalsifiable.
5. Some claimed multiverse evidence is dubious.
6. Fine-tuning arguments may assume too much.
7. It's a slippery slope to even bigger multiverses.

(George didn't actually mention argument 2 in his article, but I'm adding it here because I think he would have if the editor had allowed him more than six pages.)

What's my take on this critique? Interestingly, I agree with all of these seven statements—and nonetheless, I'll still happily bet my life savings on the existence of a multiverse!

Let's start with the first four. As we saw in Chapter 6, inflation naturally produces the Level I multiverse, and if you add in string theory with a landscape of possible solutions, you get Level II as well. As we saw in Chapter 8, quantum mechanics in its mathematically simplest collapse-free ("unitary") form gives you Level III. So if these theories are ruled out, then key evidence for these multiverses collapses. Remember: *Parallel universes are not a theory—they're predictions of certain theories.*

To me, the key point is that if theories are scientific, then it's legitimate science to work out and discuss all their consequences even if they involve unobservable entities. For a theory to be falsifiable, we need not be able to observe and test all its predictions, merely at least one of them. My answer to argument 4 is therefore that what's scientifically testable are our mathematical theories, not necessarily their implications, and that this is quite okay. As we discussed in Chapter 6, because Einstein's theory of general relativity has successfully predicted many things that we can observe, we also take seriously its predictions for things we can't observe—for example, what happens inside black holes. Likewise, if we're impressed by the successful predictions of inflation or quantum mechanics so far, then we also need to take seriously their other predictions, including the Level I and Level III multiverses. George even mentions the possibility that eternal inflation may one day be ruled out: to me, this is simply an argument that eternal inflation is a scientific theory.

String theory certainly hasn't come as far as inflation and quantum mechanics in terms of establishing itself as a testable scientific theory. However, I suspect that we'll be stuck with a Level II multiverse even if string theory turns out to be a red herring. It's quite common for mathematical equations to have multiple solutions, and as long as the fundamental equations describing our reality do, then eternal inflation generically creates huge regions of space that physically realize each of these solutions, as we saw in Chapter 6. For example, the equations governing water molecules, which have nothing to do with string theory, permit the three solutions corresponding to steam, liquid water and ice, and if space itself can similarly exist in different phases, inflation will tend to realize them all.

George lists a number of observations purportedly supporting multiverse theories that are dubious at best, such as evidence that certain constants of nature aren't really constant, and evidence in the cosmic microwave–background radiation of collisions with other universes or strangely connected space. I totally share his skepticism of these claims. In all these cases, however, the controversies have been about the analysis of the data, much as it was in the 1989 cold-fusion debacle. To me, the very fact that scientists are making these measurements and arguing about data details is further evidence that this is within the pale of science: this is precisely what separates a scientific controversy from a nonscientific one!

We saw in Chapter 6 that our Universe appears surprisingly fine-tuned for life in the sense that if you tweaked many of our constants of nature by just a tiny amount, life as we know it would be impossible. Why? If there's a Level II multiverse where these "constants" take all possible values, it's not surprising that we find ourselves in one of the rare universes that are inhabitable, just as it's not surprising that we find ourselves living on Earth rather than Mercury or Neptune. George objects to the fact that you need to assume a multiverse theory to draw this conclusion, but that's how we test any scientific theory: we assume that it's true, work out the consequences, and discard the theory if the predictions fail to match the observations. Some of the fine-tuning appears extreme enough to be quite embarrassing—for example, we saw that we need to tune the dark energy to about 123 decimal places to make habitable galaxies. To me, an unexplained coincidence can be a telltale sign of a gap in our scientific understanding. Dismissing it by saying, "We just got lucky—now stop looking for an explanation!" is not only unsatisfactory, but also tantamount to ignoring a potentially crucial clue.

George argues that if we take seriously that anything that could happen does happen, we're led down a slippery slope toward even larger multiverses, such as the Level IV one. Since this is my favorite multiverse level, and I'm one of the very few proponents of it, this is a slope that I'm happy to slide down!

George also mentions that multiverses may fall foul of Occam's razor by introducing unnecessary complications. As a theoretical physicist, I judge the elegance and simplicity of a theory not by its ontology, but by the elegance and simplicity of its mathematical equations—and it's quite striking to me that the mathematically simplest theories tend to give us multiverses. It's proven remarkably hard to write down a theory that produces exactly the universe we see and nothing more.

Finally, there's an anti-multiverse argument that I commend George for avoiding, but which is in my opinion the most persuasive one of all for most people: parallel universes just seem too weird to be real. But as we discussed in Chapter 1, this is exactly what we should expect: evolution endowed us with intuition only for those everyday aspects of physics that had survival value for our distant ancestors, leading to the prediction that whenever we use technology to glimpse reality beyond the human scale, our evolved intuition should break down. We've seen this happen again and again with counterintuitive features of relativity theory, quantum mechanics, etc., and should expect the ultimate theory of physics, whatever it turns out to be, to feel weirder still.

The Case for a Greater Reality

Having looked at anti-multiverse arguments, let's now analyze the pro-multiverse case a bit more closely. I'm going to argue that all the controversial issues melt away if we accept the External Reality Hypothesis from Chapter 10: *There exists an external physical reality completely independent of us humans.* Suppose that this hypothesis is correct. Then most multiverse critique rests on some combination of the following three dubious assumptions:

> **Omnivision Assumption:** *Physical reality must be such that at least one observer can in principle observe all of it.*

> **Pedagogical-Reality Assumption:** *Physical reality must be such that all reasonably informed human observers feel they intuitively understand it.*

> **No-Copy Assumption:** *No physical process can copy observers or create subjectively indistinguishable observers.*

Assumptions 1 and 2 appear to be motivated by little more than human hubris. The Omnivision Assumption effectively redefines the word *exists* to be synonymous with what's observable by us humans, which is akin to being an ostrich with its head in the sand. Those who insist on the Pedagogical-Reality Assumption will typically have rejected comfortingly familiar childhood notions such as Santa Claus, Euclidean space, the Tooth Fairy and creationism—but have they really worked hard enough to free themselves from comfortingly familiar notions that are more deeply rooted? In my personal opinion, our job as scientists is to try to figure out how the world works, not to tell it how to work based on our philosophical preconceptions.

If the Omnivision Assumption is false, then there are by definition things that exist despite being unobservable even in principle. Because our Universe definition includes everything that's in principle observable, this means that our Universe isn't all that exists, so we live in a multiverse. If the Pedagogical-Reality Assumption is false, then the objection that multiverses are too weird makes no logical sense. If the No-Copy Assumption is false, then there's no fundamental reason why there can't be copies of you elsewhere in the external reality—indeed, we've seen in Chapters 6 and 8 how both eternal inflation and collapse-free quantum mechanics provide mechanisms for creating them.

Moreover, we argued in Chapter 10 that the External Reality Hypothesis implies the *Mathematical Universe Hypothesis*: that our external physical reality is a mathematical structure. In Chapter 12 we saw how this in turn implies the Level IV multiverse, which contains all other multiverse levels within it. In other words, we basically get stuck with all these parallel universes as soon as we accept that there's an external reality independent of us.

In summary, we've seen throughout this book how humanity's self-image has evolved. We humans have long had a tendency toward hubris, arrogantly imagining ourselves at center stage, with everything revolving around us, but we've repeatedly been proven wrong: it is instead we who are revolving around the Sun, which is itself revolving around the center of one galaxy among countless others in a universe that may in turn be but one in a four-level multiverse hierarchy. I hope this makes us humbler. However, whereas we humans had overestimated our physical powers in the grand scheme of things, we had underestimated our mental powers! Our ancestors thought they were forever grounded, and could never truly understand the nature of the stars and what lay beyond. Then they realized how far they could get without flying into space to examine celestial objects—by letting their human minds fly. Thanks to breakthroughs in physics, we're gaining ever-deeper insights into the very nature of reality. We've found ourselves inhabiting a reality far grander than our ancestors ever dreamed of, and this means that our future potential for life is much grander than we thought. With physical resources nearly limitless, it's our future ingenuity that will make the key difference; so our destiny is in our own hands.

The Future of Physics

If I'm wrong and the Mathematical Universe Hypothesis is false, it means that fundamental physics is doomed to eventually hit a roadblock beyond which we can't understand our physical reality any better, because it lacks a mathematical description. If I'm right, then there's no roadblock, and everything is in principle understandable to us. I think that would be wonderful, because then we'll be limited only by our own imagination.

By our imagination and our willingness to do hard work, to be more specific. As we mentioned in Chapter 10, the answer Douglas Adams gave to his ultimate question of life, the Universe, and everything was hardly an answer that laid all questions to rest. Similarly, the answer I'm proposing to the question about the ultimate nature of reality ("It's all math," or more specifically, "It's the Level IV multiverse") leaves most of our traditional big questions unanswered. Instead of getting answered, most questions get rephrased. For example, "What are the equations of quantum gravity?" turns into "Where in the Level IV multiverse are we?"—a question that appears as difficult to answer as the original. So the ultimate question about physical reality would change. We'd abandon as misguided the question of which particular mathematical equations describe all of reality, and instead ask how to compute the frog's view of our Universe—our observations—from the bird's view. That would determine whether we've uncovered the true structure of our particular Universe, and help us figure out which corner of the mathematical cosmos is our home.

This situation where fundamental questions can be easier to answer than less-fundamental ones is actually typical in physics: if we find the correct equations describing quantum gravity, they'll provide a deeper understanding about what space, time and matter are, but they won't help us model global climate change more accurately, even though they in principle explain all the relevant physics of weather. The devil is in the details, and figuring out these details often requires hard work that's rather independent of the ultimate underlying theory.

In this spirit, we're going to devote the rest of this book to exploring some specific big questions that bring us farther and farther from fundamental physics and closer and closer to home. Since the earlier parts of the book focused heavily on our past, it's fitting to end our journey by focusing on our future.

The Future of Our Universe—How Will It End?

If the Mathematical Universe Hypothesis is correct, then there isn't much to say about the future of our physical reality as a whole: since it exists outside of space and time, it can't end or disappear any more than it can get created or change. However, if we head closer to home and zoom in to the particular mathematical structure that we inhabit, which contains space and time within it, then things get much more interesting. Here in our neck of the woods, things are such that they appear to change from the vantage point of observers such as us, and it's natural to ask what will ultimately happen.

So how's our Universe going to end, billions of years from now? I have five main suspects for our upcoming cosmic apocalypse or "cosmochalypse," illustrated in Figure 13.2 and summarized in Table 13.1: the Big Chill, the Big Crunch, the Big Rip, the Big Snap and Death Bubbles.

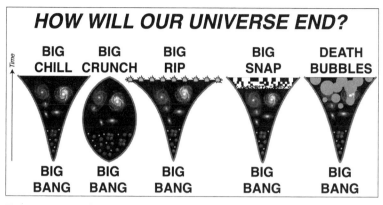

Figure 13.2: We know that our Universe began with a hot Big Bang 14 billion years ago, expanded and cooled, and merged its particles into atoms, stars and galaxies. But we don't know its ultimate fate. Proposed scenarios include a Big Chill (eternal expansion), a Big Crunch (recollapse), a Big Rip (an infinite expansion rate tearing everything apart), a Big Snap (the fabric of space revealing a lethal granular nature when stretched too much) and Death Bubbles (space "freezing" in lethal bubbles that expand at the speed of light).

Future of space	Big Chill	Big Crunch	Big Rip	Big Snap	Death Bubbles
Lasts forever?	Yes	No	No	No	No
Approaches infinite size?	Yes	No	Yes	No	No
Approaches infinite density?	No	Yes	Yes	No	No
Stable?	Yes	Yes	Yes	No	No
Infinitely stretchable?	Yes	Yes	Yes	No	Yes

Table 13.1: The future of space in five cosmic-doomsday scenarios

As we saw in Chapter 3, our Universe has now been expanding for about 14 billion years. The Big Chill is when our Universe keeps expanding forever, diluting our cosmos into a cold, dark and ultimately dead place. I think of it as the T. S. Eliot option: "This is the way the world ends / Not with a bang but a whimper." If you as Robert Frost prefer the world ending in fire rather than ice, then cross your fingers for the Big Crunch, where the cosmic expansion is eventually reversed and everything comes crashing back together in a cataclysmic collapse akin to a backwards Big Bang. Finally, the Big Rip is like the Big Chill for the impatient, where our galaxies, planets and even atoms get torn apart in a grand finale a finite time from now. Which of these three should you bet on? That depends on what the dark energy from Chapter 4, which makes up about 70% of the mass of our Universe, will do as space continues to expand. It can be any one of the Chill, Crunch or Rip depending on whether the dark energy sticks around unchanged, dilutes to negative density or anti-dilutes to higher density, respectively. Since we still have no clue what dark energy is, I'll just tell you how I'd bet: 40% on the Big Chill, 9% on the Big Crunch and 1% on the Big Rip.

What about the other 50% of my money? I'm saving it for the "none of the above" option, because I think we humans need to be humble and acknowledge that there are basic things we still don't understand. The nature of space, for example. The Chill, Crunch and Rip endings all assume that space itself is stable and infinitely stretchable.

We used to think that space was just the boring static stage upon which the cosmic drama unfolds. Then Einstein taught us that space is really one of the key actors: it can curve into black holes, it can ripple as gravitational waves, and it can stretch as an expanding universe. Perhaps it can even freeze into a different phase much like water can, as we explored in Chapter 6, with lethal fast-expanding bubbles of the new phase offering another wild-card cosmochalypse candidate. We also used to think that you can't get more space without taking it away from someone else. However, as we saw in Chapter 3, Einstein's gravity theory says the exact opposite: more volume can be created in a particular region between some galaxies without this new volume expanding into other regions—the new volume simply stays between those same galaxies. Moreover, Einstein's theory says that space stretching can always continue, allowing our Universe to approach infinite volume as in the Big Chill and Big Rip scenarios. This sounds a bit too good to be true, which makes me wonder: is it?

A rubber band looks nice and continuous, just like space, but if you stretch it too much, it snaps. Why? Because it's made of atoms, and with enough stretching, this granular atomic nature of the rubber becomes important. Could it be that space too has some sort of granularity on a scale that's simply too small for us to have noticed? Mathematicians like to model space as an idealized continuum without any granularity, where it makes sense to talk about arbitrarily short distances. We use this continuous space model in most of the physics classes we teach at MIT, but do we really know that it's correct? Certainly not! In fact, there's mounting evidence against it, as we discussed in Chapter 11. In a simple continuous space, you'd need to write out infinitely many decimal places just to specify the exact distance between two random points, but physics titan John Wheeler showed that quantum effects probably make any digits after the thirty-fifth decimal place meaningless, because our whole classical notion of space breaks down on smaller scales, perhaps being replaced by a strange foamy structure. It's a bit like when you keep zooming a photo on your screen and discover that what looked smooth and continuous is actually granular like a rubber band, in this case made up of pixels that can't be further subdivided (see Figure 11.3).

Because that photo is pixelized, it contains only a finite amount of information and can be conveniently transmitted over the Internet. Similarly, there's mounting evidence that our observable Universe contains only a finite amount of information, which would make it easier to understand how nature can compute what to do next. The holographic principle we mentioned in Chapter 6 suggests that our Universe contains at most ten to the power 124 bits of information, which averages to about 10 terabytes for each volume that can fit an atom.

Now here's what bothers me. The Schrödinger equation of quantum mechanics that we encountered in Chapter 7 implies that information can't be created or destroyed. Which means that the number of gigabytes per liter of space keeps dropping as our Universe expands. This expansion continues forever in the Big Chill scenario (the front-runner cosmochalypse candidate based on polling my astrophysics colleagues), so what happens when the information content gets diluted down to a megabyte per liter, which is less than a cell phone can store? To a byte per liter? We can't say specifically what will happen until we have a detailed model to replace continuous space, but I think it's a safe bet that it will be something bad that will gradually alter the laws of physics as we know them and make our form of life impossible—welcome to what I call the "Big Snap."

Here's what bothers me even more: a simple calculation suggests that this will happen within a few billion years, even before our Sun runs out of fuel and engulfs Earth. Our best theory for what put the bang into our Big Bang, the inflation theory from Chapter 5, says that there was an awful lot of rapid space-stretching going on in our early Universe, with some regions getting much more stretched than others. If space can get stretched only by a maximum amount before suffering a Big Snap, then most of the volume (and consequently most of the galaxies, stars, planets and observers) will be found in the regions that have stretched the most and are close to snapping.

What would an impending Big Snap be like? If the granularity of space gradually grows, then the smallest-scale structures would get messed up first. We might first notice the properties of nuclear physics starting to change, for example by previously stable atoms undergoing radioactive decay. Then atomic physics would start changing, messing up all of chemistry and biology. Fortunately, our Universe has provided gamma-ray bursts as a convenient early-warning system which, like a canary in a coal mine, might alert us long before a Big Snap could harm us. Gamma-ray bursts are cataclysmic cosmic explosions blasting out detectable short-wavelength gamma rays from halfway across our Universe. In continuous space, all wavelengths move at the same speed, the speed of light, but in the simplest kinds of granular space, shorter wavelengths move slightly slower. Yet we've recently observed gamma rays of quite different wavelengths race for billions of years through space from a distant explosion, arriving in a photo finish within a hundredth of a second of each other. Taken at face value, this rules out an impending Big Snap for billions and billions of years to come, flying in the face of what we predicted in the last paragraph.

In fact, the problem is even worse. Our space isn't expanding uniformly: indeed, some regions, such as our Galaxy, aren't expanding at all. One could therefore imagine galaxy-dwelling observers happily surviving long after intergalactic space has undergone a Big Snap, as long as deleterious effects from these faraway regions don't propagate into the galaxies. But this scenario saves only the observers, not the underlying theory! Indeed, the discrepancy between theory and observation merely gets worse: repeating the previous argument now predicts that we're most likely to find ourselves alive and well in a galaxy *after* the Big Snap has taken place throughout most of space, so the lack of any strange gamma-ray time delays becomes even harder to explain.

So we've concocted a strange brew by blending together some of the most cherished ingredients of cosmology and quantum physics, adding some experimental data, and stirring it up. The result? The ingredients don't mix well, suggesting that there's something wrong with at least one of them. I love mysteries, and find paradoxes to be nature's best gifts to us physicists, often providing clues to future breakthroughs. I think we're due for a breakthrough on the nature of space, and that the Big Snap paradox is an interesting hint.

The Future of Life

Starting with the full physical reality of the Level IV multiverse, we've now zoomed in on our particular Universe and discussed its long-term fate. Let's continue even closer to home and consider the future of life. Out of all the awe-inspiring properties that our Universe has, the one I find the most inspiring is that it's come alive, containing self-aware entities such as us who can enjoy it and ponder its mysteries.

So what are the future prospects for life? Are we humans alone in our Universe, or are there other civilizations out there that might interact with us or destroy us? Will our human life spread throughout our Universe, perhaps in some evolved form? We'll explore these fascinating questions below, but first let's tackle some more pressing ones: what are some of the main threats to the future survival of life on our planet, and what can we do to mitigate them?

Existential Risk

When I was fifteen, I had a thought that really shocked me. I was well aware that we humans worried a lot. We worried about personal challenges such as health, relationships, money and career, and we also worried about threats to our family, our friends and our society. But what about the greatest threats of all, that could potentially destroy all human life—were we really worrying enough about this? No, we weren't!

I realized that I'd lived my life lulled into a false sense of security, naively believing that everything that needed to be worried about was being taken care of by someone else. As a toddler, I never worried about dinner because I knew my parents had a plan for that. I didn't worry about my safety because I knew that the fire and police departments had a plan for that. Gradually, I realized that the grown-ups around me weren't as omniscient and omnipotent as I'd thought, and that there were many small problems that I had to solve myself. But the really big and most important problems facing humankind, they were given top priority by our political leaders. Surely?

I never questioned this until the frightening truth hit me like a brick when I was fifteen. My personal wake-up call was learning details about the nuclear-arms race. It really astonished me to realize that here we were together, billions of us, on this precious and beautiful blue planet, and even though essentially none of us wanted a full-scale nuclear war, there was significant risk we'd have one in my lifetime, most likely by accident. Perhaps the risk was 1% per year, perhaps 100 times less, perhaps 10 times more—in any case, the risk was absurdly high given the stakes. And yet it wasn't even considered the number-one election issue in any country. Moreover, this is just one example among many of what Nick Bostrom has termed *existential risk*, something that could either annihilate Earth-originating intelligent life or permanently and drastically curtail its potential.*

The American futurist Buckminster Fuller has described this basic predicament much more poetically than I did in my teens, as our collective voyage on "Spaceship Earth." As it blazes through cold and barren space, our spaceship both sustains and protects us. It's stocked with major but limited supplies of water, food and fuel. Its atmosphere keeps us warm and (via its ozone layer) shielded from the Sun's harmful ultraviolet rays, and its magnetic field shelters us from lethal cosmic rays. Surely any responsible spaceship captain would make it a top priority to safeguard its future existence by avoiding asteroid collisions, onboard explosions, overheating, ultraviolet-shield destruction and premature depletion of supplies? Well, our spaceship crew hasn't made *any* of these issues a top priority, devoting (by my estimate) less than a millionth of its resources to them. In fact, our spaceship doesn't even have a captain!

Later, we'll explore why we humans are so bad at managing the greatest threats to our long-term survival, and what we can do about it. First, however, let's briefly survey what some of these threats are. Figure 13.3 summarizes some of the existential risks I find most relevant. Let's start on the right end of the timeline, in the distant future, and work our way back home toward the present.

Our Dying Sun

Let's begin with astronomical and geological threats, and then turn to human-created threats. Earlier, we discussed five "cosmochalypse" scenarios for the end of our Universe: the Big Chill, Big Crunch, Big Rip, Big Snap and Death Bubbles. Although we don't know which of them, if any, will actually happen, my guess is that there's no need to panic, and that our Universe will avoid wholesale destruction for tens of billions of years.

What we know with certainty, however, is that our 4.5-billion-year-old Sun will cause us problems much sooner. It keeps shining progressively more brightly, because of the complex dynamics of the fusion reactions

---

* For good introductions to existential risk, I recommend http://www.existential-risk .org and Martin Rees's book *Our Final Hour*.

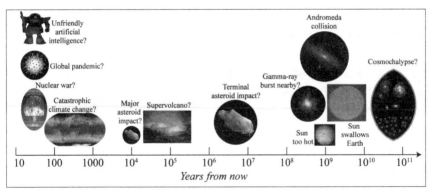

Figure 13.3: Examples of what could destroy life as we know it or permanently curtail its potential. Whereas our Universe itself will likely last for at least tens of billions of years, our Sun will scorch Earth in about a billion years and then swallow it unless we move it to a safe distance, and our Galaxy will collide with its neighbor in about 3.5 billion years. Although we don't know exactly when, we can predict with near certainty that long before this, asteroids will pummel us and supervolcanoes will cause yearlong sunless winters. In the immediate term, we may face self-inflicted problems such as climate change, nuclear war, global pandemics and unfriendly superhuman artificial intelligence.

in its core as the hydrogen fuel gradually gets depleted. Forecasts suggest that about a billion years from now, this solar brightening will start having a catastrophic effect on Earth's biosphere, and that a runaway greenhouse effect will eventually boil off our oceans, much like what has already happened on Venus. Unless we do something about it, that is.

Interestingly, there may be something that can be done. The astronomers Donald Korycansky, Greg Laughlin and Fred Adams have shown that, by clever use of asteroids, Earth can be kept at constant temperature by gradually moving it out to a larger orbit around the warming Sun. Their basic idea is to nudge a large asteroid to fly very close to Earth every 6,000 years or so and give us a gravitational tug in the right direction. Each such close encounter would be fine-tuned to send the asteroid passing near Jupiter and Saturn to get its energy and angular momentum reset to the required values for the next Earth encounter—we've successfully used such "gravitational assists" before, to send spacecraft such as NASA's Voyager probes into the outer Solar System. If successful, this scheme could extend Earth's habitability from about 1 billion to about 6 billion years. After that, our Sun will end its life as we know it, bloating into a red giant, and more radical measures may be required both to prevent it from engulfing Earth and to keep our atmosphere at a reasonable temperature.

Around the same time, a few billion years from now, our entire Milky Way Galaxy will collide and merge with its nearest big neighbor, the Andromeda galaxy. This isn't quite as bad as it sounds, because their constituent stars are so far apart relative to their size that they'll mostly miss each other: if our Sun were the size of an orange in Boston, then its nearest neighbor star, Proxima Centauri, would be in my native Stockholm. Instead of colliding, most stars will intermingle to form a single new galaxy, "Milkomeda." However, as we'll see next, this may exacerbate problems with supernovae and asteroid impacts.

Asteroids, Supernovae and Supervolcanoes

Our fossil record reveals five major extinction events during the last 500 million years, each killing off more than 50% of all animal species. Although the details are actively debated, it's widely believed that they were all triggered by various astronomical and geological events. The most recent of these "big five" extinctions appears to have been triggered by a Mount Everest–sized asteroid crashing into the Mexican coastline about 65 million years ago, whose most famous casualties were the non-avian dinosaurs. With an impact energy equivalent to many millions of hydrogen-bomb explosions, it blasted out a 180-kilometer crater and engulfed our planet in a dark dust cloud that blocked sunlight for years, causing widespread ecosystem collapse.

Earth regularly gets hit by objects from space of various sizes and compositions, so the question isn't *if* we'll suffer another similarly deadly collision, but *when*. The answer is largely up to us: a good network of robotic telescopes should be able to give us decades of advance warning of dangerous inbound asteroids, which is ample time to develop, launch and execute a mission to deflect them. If this is done sufficiently far in advance, only a gentle nudge is needed, which can be applied for example with a "gravity tractor" (a satellite whose gravitational pull nudges the asteroid toward it), a satellite-based laser (which ablates material from the asteroid's surface and sends the asteroid recoiling in the opposite direction), or even by painting the asteroid so that the radiation pressure corresponding to solar heating will push it differently. If time is short, a riskier approach is required, such as a kinetic impactor (a satellite tackling the asteroid off course like a football player) or nuclear explosion.

As a warm-up, we can practice deflecting the smaller and more numerous asteroids that strike Earth more frequently. For example, the 1908 Tunguska event was caused by an object weighing about as much as an oil tanker, which didn't pose any existential risk, but whose roughly 10-megaton blast would have killed millions had it hit a large city. Once we've mastered the art of deflecting small asteroids for our protection, we'll be prepared when the next big one arrives, and we'll also be able to use this same technical know-how for the longer-term engineering project we discussed earlier: harnessing asteroids to enlarge Earth's orbit away from the gradually brightening Sun.

Asteroids certainly didn't cause all mass extinctions. Another astronomical suspect, a gamma-ray burst from a supernova explosion, has been blamed for the second-largest recorded extinction, which took place about 450 million years ago. Although the forensic evidence is currently too weak for a guilty verdict, the suspect certainly had the means and a plausible opportunity. When some massive and fast-rotating stars explode as supernovae, they fire off part of their enormous explosion energy as a beam of gamma rays. If such a killer beam hit Earth, it would deliver a one-two punch: it would both zap us directly and destroy our ozone layer, after which our Sun's ultraviolet light would start sterilizing Earth's surface.

There are interesting links between the different astronomical threats. Occasionally, a random star will stray close enough to our Solar System that it will perturb the orbits of distant asteroids and comets, sending a swarm of them into the inner Solar System where some might collide with Earth. For example, the star Gliese 710 is predicted to pass within a light-year of us in about 1.4 million years, four times closer than our current nearest neighbor, Proxima Centauri.

Moreover, today's orderly traffic flow where most stars orbit around the center of our Milky Way Galaxy in the same direction, as in a roundabout, will be replaced by a chaotic mess when our Galaxy merges with Andromeda, significantly increasing the frequency of disruptive close encounters with other stars that could trigger a hail of asteroids or ultimately even eject Earth from our Solar System. This galactic collision will also cause gas clouds to collide, triggering a burst of star formation, and the heaviest newborn stars will soon explode as supernovae, which may be too close for comfort.

Returning closer to home, we also face "the enemy within": events caused by our own planet. Supervolcanoes and massive floods of basalt lava are prime suspects in many extinction events. They have the potential to create "volcanic winter" by enveloping Earth in a dark dust cloud, blocking sunlight for years much as a major asteroid impact would. They may also disrupt ecosystems globally by infusing the atmosphere with gases that produce toxicity, acid rain or global warming. Such a super-eruption in Siberia is widely blamed for the greatest recorded extinction of all, the "Great Dying," which wiped out 96% of all marine species about 250 million years ago.

Self-Inflicted Problems

In summary, we humans face many existential risks involving astronomical or geological effects; I've summarized only those I personally take most seriously. When I think about all such risks, the conclusion I draw is actually rather optimistic:

1. It's likely that future technologies can help life flourish for billions of years to come.
2. We and our descendants should be able to develop these technologies in time if we have our act together.

By first eliminating the most urgent problems, on the left side in Figure 13.3, we'll buy ourselves time to tackle the remaining ones.

Ironically, these most-urgent problems are largely self-inflicted. Whereas most geological and astronomical disasters loom thousands, millions or billions of years from now, we humans are radically changing things on time scales of decades, opening up a Pandora's box of new existential risks. By transforming water, land and air with fishing, agriculture and industry, we're driving about 30,000 species to extinction each year, in what some biologists are calling "the Sixth Extinction." Is it soon our turn to go extinct, too?

You've undoubtedly followed the acrimonious debate about human-caused risks, ranging from global pandemics (accidental or deliberate) to climate change, pollution, resource depletion and ecosystem collapse. Let me tell you a bit more about the two human-caused risks that concern me the most: accidental nuclear war and unfriendly artificial intelligence.

Accidental Nuclear War

A serial killer is on the loose! A suicide bomber! Beware the bird flu! Although headline-grabbing scares are better at generating fear, boring old cancer is more likely to do you in. Although you have less than a 1% chance per year to get it, live long enough, and it has a good chance of getting you in the end. As does accidental nuclear war.

During the half century that we humans have been tooled up for nuclear Armageddon, there has been a steady stream of false alarms that could have triggered all-out war, with causes ranging from computer malfunction, power failure and faulty intelligence to navigational error, bomber crash and satellite explosion. This bothered me so much when I was seventeen that I volunteered as a freelance writer for the Swedish peace magazine *PAX*, whose editor-in-chief Carita Andersson kindly nurtured my enthusiasm for writing, taught me the ropes and let me pen a series of news articles. Gradual declassification of records has revealed that some of these nuclear incidents carried greater risk than was appreciated at the time. For example, it became clear only in 2002 that during the Cuban Missile Crisis, the USS *Beale* had depth-charged an unidentified submarine that was in fact Soviet and armed with nuclear weapons, and whose commanders argued over whether to retaliate with a nuclear torpedo.

Despite the end of the Cold War, the risk has arguably grown in recent years. Inaccurate but powerful ICBMs undergirded the stability of "mutually assured destruction," because a first strike couldn't prevent massive retaliation. The shift toward more accurate missile navigation, shorter flight times and better enemy submarine tracking erodes this stability. A successful missile-defense system would complete this erosion process. Both Russia and the United States retain their launch-on-warning strategies, requiring launch decisions to be made on five- to fifteen-minute time scales where complete information may be unavailable. On January 25, 1995, Russian president Boris Yeltsin came within minutes of initiating a full nuclear strike on the United States because of an unidentified Norwegian scientific rocket. Concern has been raised over a U.S. project to replace the nuclear warheads on two of the twenty-four D5 ICBMs carried by Trident submarines with conventional warheads, for possible use against Iran or North Korea:

Russian early-warning systems would be unable to distinguish them from nuclear missiles, expanding the possibilities for unfortunate mis-understandings. Other worrisome scenarios include deliberate malfeasance by military commanders triggered by mental instability and/or fringe political/religious agendas.

But why worry? Surely, if push came to shove, reasonable people would step in and do the right thing, just as they have in the past? Nuclear nations do indeed have elaborate countermeasures in place, just as our body does against cancer. Our body can normally deal with isolated deleterious mutations, and it appears that fluke coincidences of as many as four mutations may be required to trigger certain cancers. Yet if we roll the dice enough times, shit happens—Stanley Kubrick's dark nuclear war comedy *Dr. Strangelove* illustrates this with a triple coincidence.

Accidental nuclear war between two superpowers may or may not happen in my lifetime, but if it does, it will obviously change everything. The climate change we're currently worrying about pales in comparison with nuclear winter, where a global dust cloud blocks sunlight for years, much like when an asteroid or supervolcano caused a mass extinction in the past. The 2008 economic turmoil was of course nothing compared to the resulting global crop failures, infrastructure collapse and mass starvation, with survivors succumbing to hungry armed gangs systematically pillaging from house to house. Do I expect to see this in my lifetime? I'd give it about 30%, putting it roughly on par with my getting cancer. Yet we devote way less attention and resources to reducing the risk of nuclear disaster than we do for cancer. And whereas humanity as a whole survives even if 30% get cancer, it's less obvious to what extent our civilization would survive a nuclear Armageddon. There are concrete and straightforward steps that can be taken to slash this risk, as spelled out in numerous reports by scientific organizations, but these never become major election issues and tend to get largely ignored.

An Unfriendly Singularity

The Industrial Revolution has brought us machines that are stronger than us. The information revolution has brought us machines that are smarter than us in certain limited ways. In what ways? Computers used to outperform us only on simple, brute-force cognitive tasks such as rapid arithmetic or database searching, but in 2006, a computer beat the world chess champion Vladimir Kramnik, and in 2011, a computer dethroned Ken Jennings on the American quiz show *Jeopardy!* In 2012, a computer was licensed to drive cars in Nevada after being judged safer than a human driver. How far will this development go? Will computers eventually beat us at *all* tasks, developing superhuman intelligence? I have little doubt that this *can* happen: our brains are a bunch of particles obeying the laws of physics, and there's no physical law precluding particles from being arranged in ways that can perform even-more-advanced computations. But *will* it actually happen, and would that be a good thing or a bad thing? These questions are timely: although some think machines with superhuman intelligence can't be built in the foreseeable future, others such as the American inventor and author Ray Kurzweil predict their existence by 2030, making this arguably the single most urgent existential risk to plan for.

The singularity idea

In summary, it's unclear whether the development of ultra-intelligent machines will or should happen, and artificial-intelligence experts are divided. What I think is quite clear, however, is that, if it happens, the effects will be explosive. The British mathematician Irving Good explained why in 1965, two years before I was born: "Let an ultra-intelligent machine be defined as a machine that can far surpass all the intellectual activities of any man however clever. Since the design of machines is one of these intellectual activities, an ultraintelligent machine could design even better machines; there would then unquestionably be an 'intelligence explosion,' and the intelligence of man would be left far behind. Thus the first ultraintelligent machine is the last invention that man need ever make, provided that the machine is docile enough to tell us how to keep it under control."

In a thought-provoking and sobering 1993 paper, the mathematician and sci-fi author Vernor Vinge called this intelligence explosion "the Singularity," arguing that it was a point beyond which it was impossible for us to make reliable predictions.

I suspect that if we can build such ultra-intelligent machines, then the first one will be severely limited by the software we've written for it, and that we'll have compensated for our lack of understanding about how to optimally program intelligence by building hardware with significantly more computing power than our brains have. After all, our neurons are no better or more numerous than those of dolphins, just differently connected, suggesting that software can sometimes be more important than hardware. This situation would probably enable this first machine to radically improve itself over and over and over again simply by rewriting its own software. In other words, whereas it took us humans millions of years of evolution to radically transcend the intelligence of our apelike ancestors, this evolving machine could similarly soar beyond the intelligence of its ancestors, us humans, in a matter of hours or seconds.

After this, life on Earth would never be the same. Whoever or whatever controls this technology would rapidly become the world's wealthiest and most powerful, outsmarting all financial markets and out-inventing and out-patenting all human researchers. By designing radically better computer hardware and software, such machines would enable their power and their numbers to rapidly multiply. Soon technologies beyond our current imagination would be invented, including any weapons deemed necessary. Political, military and social control of the world would soon follow. Given how much influence today's books, media and web content have, I suspect that machines able to outpublish billions of ultra-talented humans could win our hearts and minds even without outright buying or conquering us.

Who controls the singularity?

If a singularity occurs, how would it affect our human civilization? We obviously don't know for sure, but I think it will depend on who/what initially controls it, as illustrated in Figure 13.4. If the technology is initially developed by academics or others who make it open source, I think the resulting free-for-all situation will be highly unstable and lead to control by a single entity after a brief period of competition. If this entity is an egoistic human or for-profit corporation, I think government control will soon follow as the owner takes over the world and becomes the government. An altruistic human might do the same. In this case, the human-controlled artificial intelligences (AIs) would effectively be like enslaved gods, entities with understanding and ability vastly beyond us humans, but nonetheless doing whatever their owner told them to do. Such AIs might be as superior to today's computers as we humans are to ants.

It may prove impossible to keep such superintelligent AIs enslaved even if we try our utmost to keep them "boxed in," disconnected from the Internet. As long as they can communicate with us, they could come

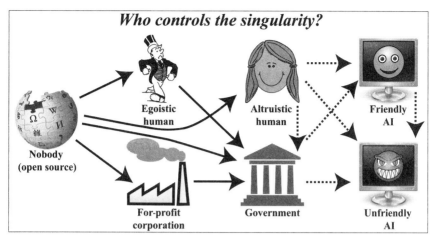

Figure 13.4: If the singularity does occur, it will make a huge difference who controls it. I suspect that the "nobody" option is totally unstable and would, after a brief period of competition, lead to control by a single entity. I think control by an egoistic human or a for-profit corporation would lead to government control, as the owner effectively takes over the world and becomes the government. An altruistic human might do the same, or choose to cede control to a friendly artificial intelligence (AI) that can better protect human interests. However, an unfriendly AI could become the ultimate controller by outwitting its creator and rapidly developing traits entrenching its power.

to understand us well enough to figure out how to sweet-talk us into doing something seemingly innocuous that allows them to "break out," go viral, and take over. I very much doubt that we could contain such a breakout given how we struggle to eradicate even the vastly simpler human-made computer viruses of today.

To forestall a breakout, or to serve human interests better, its owner may choose to voluntarily cede power to what AI researcher Eliezer Yudkowsky terms a "friendly AI," which, no matter how advanced it eventually gets, retains the goal of having a positive rather than negative effect on humanity. If this is successful, then the friendly AIs would act as benevolent gods, or zookeepers, keeping us humans fed, safe and fulfilled while remaining firmly in control. If all human jobs get replaced by machines under friendly-AI control, humanity could still remain reasonably happy if the products we need were given to us effectively for free. In contrast, the scenario in which an egoistic human or for-profit corporation controls the singularity would probably result in the greatest income disparity that our planet has ever seen, since history suggests that most humans prefer amassing personal wealth over spreading it around.

Even the best-laid plans often fail, however, and a friendly-AI situation might be unstable, eventually transforming into one controlled by an unfriendly AI, whose goals don't coincide with those of us humans, and whose actions end up destroying both humanity and everything we care about. Such destruction could be incidental rather than purposeful: the AI may simply want to use Earth's atoms for other purposes that are incompatible with our existence. The analogy with how we humans treat lower life-forms isn't encouraging: if we want to build a hydroelectric dam and there happens to be some ants in the area that would drown as a result, we'll build the dam anyway—not out of any particular antipathy toward ants, but merely because we're focused on goals we consider more important.

The internal reality of ultra-intelligent life

If there's a singularity, would the resulting AI, or AIs, feel conscious and self-aware? Would they have an internal reality? If not, they're for all practical purposes zombies. Of all traits that our human form of life has, I feel that consciousness is by far the most remarkable. As far as I'm concerned, it's how our Universe gets meaning, so if our Universe gets taken over by life that lacks this trait, then it's meaningless and just a huge waste of space.

As we discussed in Chapters 9 and 11, the nature of life and consciousness is a hotly debated subject. My guess is that these phenomena can exist much more generally than in the carbon-based examples we know of. As mentioned in Chapter 11, I believe that *consciousness is the way information feels when being processed*. Since matter can be arranged to process information in numerous ways of vastly varying complexity, this implies a rich variety of levels and types of consciousness. The particular type of consciousness that we subjectively know is then a phenomenon that arises in certain highly complex physical systems that input, process, store and output information. Clearly, if atoms can be assembled to make humans, the laws of physics also permit the construction of vastly more advanced forms of sentient life.

If we humans eventually trigger the development of more intelligent entities through a singularity, I therefore think it's likely that they, too, would feel self-aware, and should be viewed not as mere lifeless machines but as conscious beings like us. However, their consciousness may subjectively feel quite different from ours. For example, they'd probably lack our strong human fear of death: as long as they've backed themselves up, all they stand to lose are the memories

THE FUTURE OF LIFE 383

they've accumulated since their most recent backup. The ability to readily copy information and software between AIs would probably reduce the strong sense of individuality that's so characteristic of our human consciousness: there would be less of a distinction between you and me if we could trivially share and copy all our memories and abilities, so a group of nearby AIs may feel more like a single organism with a hive mind.

If this is true, then it can reconcile long-term survival of life with the doomsday argument from Chapter 11: what's about to end is not life itself, but our reference class, self-aware observer moments that subjectively feel approximately like our human minds. Even if a multitude of sophisticated hive minds colonize our Universe during billions of years, we shouldn't be any more surprised that we aren't them than we should be that we aren't ants.

Reactions to the singularity

People's reactions to the possibility of a singularity vary dramatically. The friendly-AI vision has a venerable history in the science-fiction literature, undergirding Isaac Asimov's famous three laws of robotics that were intended to ensure a harmonious relationship between robots and humans. Stories in which AIs outsmart and attack their creators have been popular as well, as in the *Terminator* movies. Many dismiss the singularity as "the rapture of the geeks," and view it as a far-fetched science-fiction scenario that won't happen, at least not for the foreseeable future. Others think that it's likely to happen, and that if we don't plan for it carefully, it will probably destroy not only our human species, but also everything we ever cared about, as we explored earlier. I serve as an advisor to the Machine Intelligence Research Institute (http://intelligence .org), and many of its researchers fall into this category, viewing the singularity as the most serious existential risk of our time. Some of them feel that if the friendly-AI vision of Yudkowsky and others can't be guaranteed, then the best approach is to keep future AIs locked in under firm human control or not to develop advanced AIs at all.

Although we've so far focused our discussion on negative consequences of a singularity, others, such as Ray Kurzweil, feel that a singularity would be something hugely positive, indeed the best thing that could happen to humanity, solving all our current human problems.

Does the idea of humankind getting replaced by more advanced life sound appealing or appalling to you? That probably depends strongly on the circumstances, and in particular on whether you view the future beings as our descendants or our conquerors.

If parents have a child who's smarter than them, who learns from them, and then goes out and accomplishes what they could only dream of, they'll probably feel happy and proud even if they know they can't live to see it all. Parents of a highly intelligent mass murderer feel differently. We might feel that we have a similar parent-child relationship with future AIs, regarding them as the heirs of our values. It will therefore make a huge difference whether future advanced life retains our most cherished goals.

Another key factor is whether the transition is gradual or abrupt. I suspect that few are disturbed by the prospects of humankind gradually evolving, over thousands of years, to become more intelligent and better adapted to our changing environment, perhaps also modifying its physical appearance in the process. On the other hand, many parents would feel ambivalent about having their dream child if they knew it would cost them their lives. If advanced future technology doesn't replace us abruptly, but rather upgrades and enhances us gradually, eventually merging with us, then this might provide both the goal retention and the gradualism required for us to view post-singularity life-forms as our descendants. Mobile phones and the Internet have already enhanced the ability of us humans to achieve what we want, arguably without significantly eroding our core values, and singularity optimists believe that the same can be true of brain implants, thought-controlled devices and even wholesale uploading of human minds to a virtual reality.

Moreover, this could open up space, the final frontier. After all, extremely advanced life capable of spreading throughout our Universe can probably only come about in a two-step process: first intelligent beings evolve through natural selection, then they choose to pass on the torch of life by building more advanced consciousness that can further improve itself. Unshackled by the limitations of our human bodies, such advanced life can rise up and eventually inhabit much of our observable Universe, an idea long explored by science-fiction writers, AI aficionados and trans-humanist thinkers.

In summary, will there be a singularity within a few decades? And is this something we should work for or against? I think it's fair to say that we're nowhere near consensus on either of these two questions, but that doesn't mean it's rational for us to do nothing about the issue. It could be the best or worst thing ever to happen to humankind, so if there's even a 1% chance that there'll be a singularity in our lifetime, I think a reasonable precaution would be to spend at least 1% of our GDP studying the issue and deciding what to do about it. So why don't we?

Human Stupidity: A Cosmic Perspective

My career has given me a cosmic perspective in which existential risk management feels more urgent, as summarized in Figure 13.5. We professors are often forced to hand out grades, and if I were teaching Risk Management 101 and had to give us humans a midterm grade based on our existential risk management so far, you could argue that I should give a B– on the grounds that we're muddling by and still haven't dropped the course. From my cosmological perspective, however, I find our performance pathetic, and can't give more than a D: the long-term potential for life is literally astronomical, yet we humans have no convincing plans for dealing with even the most-urgent existential risks, and we devote a minuscule fraction of our attention and resources to developing such plans. Compared with the roughly twenty million U.S. dollars spent last year on the Union of Concerned Scientists, one of the largest organizations focused on at least some existential risks, the United States alone spent about five hundred times more on cosmetic surgery, about a thousand times more on air-conditioning for troops, about five thousand times more on cigarettes, and about thirty-five thousand times more on its military, not counting military health care, military retirement costs or interest on military debt.

How can we humans be so shortsighted? Well, given that evolution has prepared us mainly for technologies like sticks and rocks, perhaps

	Standard perspective	Cosmic perspective	
Humans	Pinnacle of evolution	You ain't seen nothin' yet!	
Space	Obsession about our planet	10^{57} times more volume available	*Huge potential!*
Time	Obsession about next 50 years	Billions of years available	*Extinction probability/decade* $\sim 10^{-1} - 10^{-4}$?
Midterm grade	B–	D	

Figure 13.5: The importance of managing existential risk in a reasonable way becomes more obvious in a cosmic perspective, highlighting the huge future potential that we stand to lose if we mess up and destroy our human civilization.

we shouldn't be surprised that we're dealing with modern technology so poorly, but rather that we're not doing even worse. Here I am sitting in a large wood-and-stone box repeatedly pressing little black squares while staring at a glowing rectangle in front of me. I haven't met a single living organism today, and I've been sitting here for hours, illuminated by a strange growing spiral above me. The fact that I'm nonetheless feeling happy is testament to how remarkably adaptable the brains evolution has endowed us humans with are. As is the fact that I've learned to interpret the squiggly black patterns on my glowing rectangle as words telling a story, and that I know how to calculate the age of our Universe, even though none of these specific abilities had any survival value to my cave-dwelling ancestors. But just because we can do a lot doesn't mean we can do everything necessary. External forces have changed our environment slowly over the past 100,000 years of human history, and evolution has gradually helped us adapt. But recently, we ourselves have changed our environment way too fast for evolution to keep up, and we've made it so complex that it's hard even for the world's leading experts to fully understand the limited aspects they focus on. So it's no wonder that we sometimes lose sight of the big picture and prioritize short-term gratification over the long-term survival of our spaceship. For example, that glowing spiral above my head gets powered by burning coal into carbon dioxide, which contributes to overheating our spaceship, and now that I think of it, I really should have turned it off long ago.

Human Society: A Scientific Perspective

So here we are on Spaceship Earth, heading into an asteroid belt of existential risks without a plan or even a captain. We clearly need to do something about this, but what should our goals be, and how can we best accomplish them? The *what* question is ethical, whereas the *how* question is scientific. Both are clearly crucial. To paraphrase Einstein, "science without ethics is blind; ethics without science is lame." However (and this is a point that my friend Geoff Anders likes to emphasize), there are some ethical conclusions that we have nearly universal agreement on (such as "not having a global nuclear war is better than having one"), which we're nonetheless doing a dismal job turning into practical goals that we effectively advance. This is why I gave us a D grade in existential-risk mitigation, and I think it's unfair to blame this failure mainly on difficulties with ethics and the *what* question. Rather, I think that we should start with the problems where we humans have broad agreement on what our goals are, such as the long-term survival of our civilization, and use a scientific approach to tackling the question of *how* to achieve these goals (I'm using the word *scientific* in a broad sense of emphasizing the use of logical reasoning). I don't feel that it's enough to simply say things like "a change of heart on a vast scale has to be achieved"—we need more concrete strategies. So how should we pursue our goals? How can we help humanity become less shortsighted when it charts out its future course? In essence, how can we make reason play a greater role in decision making?

Changes in our human society result from a complex set of forces pushing in different directions, often working against each other. From a physics perspective, the easiest way to change a complex system is to find an instability, where the effect of pushing with a small force gets amplified into a major change. For example, we saw that a gentle nudge to an asteroid can prevent it from hitting Earth a decade later. Analogously, the easiest way for a single person to affect society is by exploiting an instability, as captured by numerous physics-based metaphors: an idea can be a "spark in a powder keg," "spread like wildfire," have a "domino effect," or "snowball out of control."* For example, if you want to tackle the existential risk from killer asteroids, the hard way is to build an asteroid deflector–rocket system. The easier way is to spend much less money building an early-warning system, knowing that once you have information about an incoming asteroid, raising money for the rocket system will be easy.

I think that for making our planet a better place, many of the easiest instabilities to utilize involve spreading correct information. For reason to play a role in decision making, the relevant information needs to be in the heads of those making the decisions. As illustrated in Figure 13.6, this typically involves three steps, all of which frequently fail: the information must be created/discovered, disseminated by the discoverer and learned by the decision maker. Once discoveries have propagated around the triangle into the heads of others, they enable further discoveries, fueling the growth of human knowledge in a virtuous cycle. Some

---

* Most instabilities involve some form of runaway self-reproduction/chain reaction: for example, burning trees in a forest produce more burning trees, free neutrons in a nuclear bomb produce more free neutrons, a carrier of the bubonic plague produces more carriers, and one buyer of a disruptively successful product produces more buyers.

discoveries have the added advantage of making the triangle itself more efficient: the printing press and the Internet have radically facilitated both dissemination and learning, while better detectors and computers have greatly assisted researchers. Yet even today, there's room for major improvements to all three links of the information triangle.

Scientific research and other information creation is clearly a good investment for society, as are attempts to counter censorship and other impediments to information dissemination. In terms of utilizing instabilities, however, I think that the lowest-hanging fruit is on the bottom arrow in Figure 13.6: learning. Despite spectacular success in research, I feel that our global scientific community has been nothing short of a spectacular failure when it comes to educating the public and our decision makers. Haitians burned twelve "witches" in 2010. In the United States, polls have shown that 39% of Americans consider astrology sci-

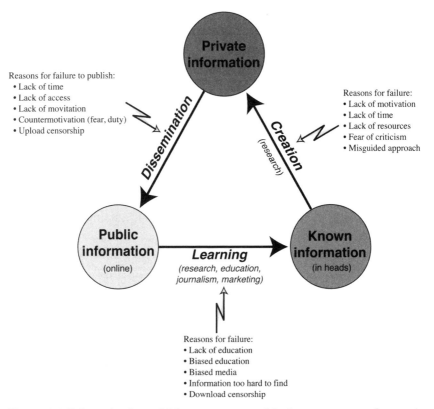

Figure 13.6: Information is crucial for reason to prevail in the management of our society. When important information is discovered, it needs to be made publicly available, then learned by those to whom it's relevant.

entific, and 46% believe that our human species is less than 10,000 years old. If everyone understood the concept of "scientific concept," these percentages would be zero. Moreover, the world would be a better place, since people with a scientific lifestyle, basing their decisions on correct information, maximize their chances of success. By making rational buying and voting decisions, they also strengthen the scientific approach to decision making in companies, organizations and governments.

Why have we scientists failed so miserably? I think the answers lie mainly in psychology, sociology and economics. A scientific lifestyle requires a scientific approach to both gathering information and using information, and both have their pitfalls. You're clearly more likely to make the right choice if you're aware of the full spectrum of arguments before making your mind up, yet there are many reasons why people don't get such complete information. Many lack access to it (97% of Afghans don't have Internet, and in a 2010 poll, 92% didn't know about the 9/11 attacks). Many are too swamped with obligations and distractions to seek it. Many seek information only from sources that confirm their preconceptions—for example, a 2012 poll showed 27% of Americans believing that Barack Obama was probably or definitely born in another country. The most valuable information can be hard to find even for those who are online and uncensored, buried in an unscientific media avalanche.

Then there's what we do with the information we have. The core of a scientific lifestyle is to change your mind when faced with information that disagrees with your views, avoiding intellectual inertia, yet many laud leaders who stubbornly stick to their views as "strong." Richard Feynman hailed "distrust of experts" as a cornerstone of science, yet herd mentality and blind faith in authority figures is widespread. Logic forms the basis of scientific reasoning, yet wishful thinking, irrational fears and other cognitive biases often dominate decisions.

So what can we do to promote a scientific lifestyle? The obvious answer is improving education. In some countries, having even the most rudimentary education would be a major improvement (less than half of all Pakistanis can read). By undercutting fundamentalism and intolerance, education would curtail violence and war. By empowering women, it would curb poverty and the population explosion. However, even countries that offer everybody education can make major improvements. All too often, schools resemble museums, reflecting the past rather than shaping the future. The curriculum should shift from one watered down by consensus and lobbying to skills our century needs for relationships, health, contraception, time management, critical thinking and recognizing propaganda. For youngsters, learning a global language and typing should trump long division and writing cursive. In the Internet age, my own role as a classroom teacher has changed. I'm no longer needed as a conduit of information, which my students can simply download on their own. Rather, my key role is inspiring a scientific lifestyle, curiosity and desire to learn more.

Now let's get to the most interesting question: how can we *really* make a scientific lifestyle take root and flourish? Reasonable people have been making similar arguments for better education since long before I was in diapers, yet rather than improving, education and adherence to a scientific lifestyle is arguably deteriorating further in many countries, including the United States. Why? Clearly because there are powerful forces pushing in the opposite direction, and they're pushing more effectively. Corporations concerned that a better understanding of certain scientific issues would harm their profits have an incentive to muddy the waters, as do fringe religious groups concerned that questioning their pseudo-scientific claims would erode their power.

So what can we do? The first thing we scientists need to do is get off our high horses, admit that our persuasive strategies have failed, and develop a better strategy. We have the advantage of having the better arguments, but the anti-scientific coalition has the advantage of better funding. However, and this is painfully ironic, it's also more scientifically organized! If a company wants to change public opinion to increase their profits, it deploys scientific and highly effective marketing tools. What do people believe today? What do we want them to believe tomorrow? Which of their fears, insecurities, hopes and other emotions can we take advantage of ? What's the most cost-effective way of changing their minds? Plan a campaign. Launch. Done. Is the message oversimplified or misleading? Does it unfairly discredit the competition? That's par for the course when marketing the latest smartphone or cigarette, so it would be naive to think that the code of conduct should be any different when this coalition fights science. Yet we scientists are often painfully naive, deluding ourselves that just because we think we have the moral high ground, we can somehow defeat this corporate-fundamentalist coalition by using obsolete unscientific strategies. Based on what scientific argument will it make a hoot of a difference if we grumble, "We won't stoop that low" and "People need to change" in faculty lunchrooms and recite statistics to journalists? We scientists have basically been saying, "Tanks are unethical, so let's fight tanks with swords."

To teach people what a scientific concept is and how a scientific lifestyle will improve their lives, we need to go about it scientifically: we need new science-advocacy organizations that use all the same scientific marketing and fund-raising tools as the anti-scientific coalition employ. We'll need to use many of the tools that make scientists cringe, from ads and lobbying to focus groups that identify the most effective sound bites. We won't need to stoop all the way down to intellectual dishonesty, however. Because in this battle, we have the most powerful weapon of all on our side: the facts.

The Future of You—Are You Insignificant?

After spending most of this book heading away to explore the most distant and abstract levels of our physical reality, we've devoted this last chapter to gradually returning homeward, discussing the future of our own Universe and the future of our human civilization. Let's finish by returning all the way home, to explore what this means for us personally—for you and me.

The Meaning of Life

As we've seen, the fundamental mathematical equations that appear to govern our physical reality make no reference to meaning, so a universe devoid of life would arguably have no meaning at all. Through us humans and perhaps additional life-forms, our Universe has gained an awareness of itself, and we humans have created the concept of meaning. So in this sense, *our Universe doesn't give life meaning, but life gives our Universe meaning*.

Although "What's the meaning of life?" can be interpreted in many different ways, some of which may be too vague to have a well-defined answer, one interpretation is very practical and down-to-Earth: "Why should I want to go on living?" The people I know who feel that their lives are meaningful usually feel happy to wake up in the morning and look forward to the day ahead. When I think about these people, it strikes me that they split into two broad groups based on where they find their happiness and meaning. In other words, the problem of meaning seems to have two separate solutions, each of which works quite well for at least some people. I think of these solutions as "top-down" and "bottom-up."

In the top-down approach, the fulfillment comes from the top, from the big picture. Although life here and now may be unfulfilling, it has meaning by virtue of being part of something greater and more meaningful. Many religions embody such a message, as do families, organizations and societies where individuals are made to feel part of something grander and more meaningful that transcends individuality.

In the bottom-up approach, the fulfillment comes from the little things here and now. If we seize the moment and get the fulfillment we need from the beauty of those little flowers by the roadside, from helping a friend or from meeting the gaze of a newborn child, then we can feel grateful to be alive even if the big picture involves less-cheerful elements such as Earth getting vaporized by our dying Sun and our Universe ultimately getting destroyed.

For me personally, the bottom-up approach provides more than enough of a raison d'être, and the top-down elements I'm about to argue for simply feel like an additional bonus. For starters, I find it utterly remarkable that it's possible for a bunch of particles to be self-aware, and that this particular bunch that's Max Tegmark has had the fortune to get the food, shelter and leisure time to marvel at the surrounding universe leaves me grateful beyond words.

Why We Should Care About Our Own Universe

In addition, I feel motivation and inspiration from top-down thinking, about the potential future of life in our Universe that we discussed at length earlier in this chapter. But if there are parallel universes where all physically possible futures play out, why should we care about our own Universe? If all outcomes will happen, why should we care about what choices we make? Indeed, why should we lift a finger or care about anything at all if the Level IV multiverse exists and even change itself is an illusion? We face a choice between two rational alternatives:

1. We care about at least something, and therefore go ahead and live life, making logical decisions reflecting the things we care about.

2. We care about nothing, and therefore do nothing at all or act completely randomly.

Both you and I have already made our choices, selecting option 1. It seems like the smart choice to me.

But this choice has logical consequences. When I think about people I care about, it feels logical to also care about the civilization, the planet, and the universe that they belong to. In contrast, it feels logical to care less about other universes, because my decisions here in our Universe by definition can't have any effect on them—they're therefore unaffected by what I care about. With this logic, let's limit our remaining discussion to our own Universe, and explore our role in it.

Are We Insignificant?

When gazing up on a clear night, it's easy to feel insignificant. For most of my life, the more I learned about the vastness of our cosmos and our place in it, the more insignificant I felt. But not anymore!

Since our earliest ancestors admired the stars, our human egos have suffered a series of blows. For starters, we're smaller than we thought. As we saw in Part I of the book, Eratosthenes showed that Earth was larger than millions of humans, and his Hellenic compatriots realized that the Solar System was thousands of times larger still. Yet for all its grandeur, our Sun turned out to be merely one rather ordinary star among hundreds of billions in a galaxy that in turn is merely one of hundreds of billions in our observable Universe, the spherical region from which light has had time to reach us during the 14 billion years since our Big Bang. Our lives are small temporally as well as spatially: if this 14-billion-year cosmic history were scaled to one year, then 100,000 years of human history would be 4 minutes and a 100-year life would be 0.2 seconds. Further deflating our hubris, we've learned that we're not that special either. Darwin taught us that we're animals; Freud taught us that we're irrational; machines now outpower us and outsmart us in chess and on the *Jeopardy!* quiz show. Adding insult to injury, cosmologists have found that we're not even made of the majority substance.

The more I learned about this, the less significant I felt. But I've suddenly changed my mind and turned more optimistic about our cosmic significance. Why? Because I've come to believe that advanced evolved life is very rare, yet has huge future potential, making our place in space and time remarkably significant.

Are We Alone?

When I give lectures about cosmology, I often ask the audience to raise their hands if they think there's intelligent life elsewhere in our Universe. Infallibly, almost everyone does, from kindergartners to college students. When I ask why, the basic answer I tend to get is that space is so huge that there's got to be life somewhere, at least statistically speaking. But is this argument really correct? I think it's wrong—let me explain why.

As the American astronomer Francis Drake pointed out, the probability of there being intelligent life in a given place can be calculated by multiplying together the probability of there being a habitable environment there (say an appropriate planet), the probability that life will evolve there, and the probability that this life will evolve to become intelligent. When I was a grad student, we had no clue about any of these three probabilities. After the past decade's dramatic discoveries of planets orbiting other stars, it now seems likely that habitable planets are abundant, with billions in our own Galaxy alone. The probability of evolving life and intelligence, however, remains extremely uncertain: some experts think that one or both are rather inevitable and occur on most habitable planets, whereas others think that one or both are extremely rare because of one or more evolutionary bottlenecks that require a wild stroke of luck to pass through. Some proposed bottlenecks involve chicken-or-the-egg problems at the earliest stages of self-reproducing life: for example, for a modern cell to build a ribosome, the highly complex molecular machine that reads our genetic code and builds our proteins, it needs another ribosome, and it's not obvious that the very first ribosome could evolve gradually from something simpler. Other proposed bottlenecks involve the development of higher intelligence. For example, although dinosaurs ruled Earth for over 100 million years, a thousand times longer than we modern humans have been around, evolution didn't seem to inevitably push them toward higher intelligence and inventing telescopes or computers.

In other words, I think it's fair to say that we still have no clue what fraction of all planets harbor intelligent life: a priori, before actually observing any other planets to check, any order-of-magnitude guess is about as good as any other. This is a standard way of modeling such extreme uncertainty in science, and goes by the geeky-sounding name *uniform logarithmic prior*; in plain English, it means that the fraction of planets with intelligent life is roughly equally likely to be one in a thousand, one in a million, one in a billion, one in a trillion, one in a quadrillion, and so on.

Given this, how far from us is our nearest-neighbor intelligent civilization? From our assumption, it follows that this distance also has a uniform logarithmic prior, so a priori, before looking, the answer is roughly equally likely to be 10^{10} meters, 10^{20} meters, 10^{30} meters, 10^{40} meters, and so on, as illustrated in Figure 13.7.

Now let's turn to what we know from observation. So far, direct astronomical searches have turned up no evidence for extraterrestrial intelligence, and there's no widely accepted evidence that aliens have visited Earth. My personal interpretation of this is that the fraction of planets harboring intelligence is minuscule, and that there's probably no highly intelligent life within about 10^{21} meters of us, i.e., in our own Galaxy or its immediate vicinity. I'm basing this conclusion on several assumptions:

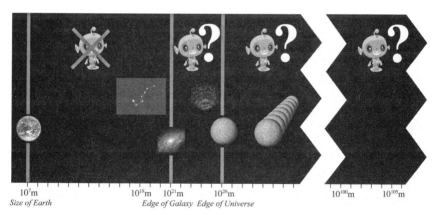

Figure 13.7: Are we alone? The huge uncertainties about how life and intelligence evolved suggest that our nearest neighbor civilization in space is roughly equally likely to be anywhere along the horizontal axis above, making it quite unlikely that it's between the edge of our Galaxy (about 10^{21} meters away) and the edge of our Universe (about 10^{26} meters away). If it were much closer than this range, there should be so many other advanced civilizations in our Galaxy that we'd probably have noticed, which suggests that we're in fact alone in our Universe.

1. Interstellar colonization is physically possible and can easily be accomplished if a civilization as advanced as ours has a million years to develop the required technology.
2. There are billions of habitable planets in our Galaxy, many of which formed not only millions but billions of years before Earth.
3. A non-negligible fraction of civilizations that can colonize space would choose to do so.

For assumption 1, I'm keeping an open mind about what technologies may be used. For example, rather than physically sending large human-sized organisms through space, it may be more efficient to send swarms of small self-assembling nanoprobes that build factories on landing and assemble any larger life-forms using "emailed" instructions transmitted at the speed of light via electromagnetic radiation.* Common objections to assumption 3 include the supposition of advanced civilizations being intrinsically kind or otherwise uninterested in colonization, perhaps because their advanced technology allows them to accomplish all their goals using the resources they already have. Alternatively, perhaps they keep a low profile for self-protection or other reasons, or colonize only in a way that we don't notice: this has been called the *zoo hypothesis* by the U.S. astronomer John A. Ball, and features in sci-fi classics such as Olaf Stapledon's *Star Maker*. Personally, I think we shouldn't underestimate the diversity of advanced civilizations by assuming that they all share the same goals: all it takes is *one* civilization deciding to overtly colonize all it can, and it will engulf our Galaxy and beyond. Faced with this risk, even civilizations otherwise uninterested in colonization may feel pressured to expand for self-protection.

If my interpretation is correct, then the closest civilization is about 1,000, . . . 000 meters away, where the total number of zeros is roughly equally likely to be 21, 22, 23, . . . 100, 101, 102, etc.—but not much

* The economist Robin Hanson has made an interesting point about assumption 1. The apparent incompatibility between the abundance of habitable planets in our Galaxy and the lack of extraterrestrial visitors, known as the *Fermi paradox*, suggests the existence of what Hanson calls a "Great Filter," an evolutionary/technological roadblock somewhere along the developmental path from nonliving matter to space-colonizing life. If we discover independently evolved primitive life in our Solar System, this would suggest that primitive life is not rare, and that the roadblock lies after our current human stage of development—perhaps because assumption 1 is false, or because almost all advanced civilizations self-destruct before they are able to colonize. I'm therefore crossing my fingers that all searches for life on Mars and elsewhere find nothing: this is consistent with the scenario where primitive life is rare but we humans got lucky, so that we have the roadblock behind us and have extraordinary future potential.

smaller than 21. However, for this civilization to be in our own Universe, whose radius is about 10^{26} meters, the number of zeros can't exceed 26, and the probability of the number of zeros falling in the narrow range between 22 and 26 is quite small. This is why I think we're alone in our Universe.

Are We Really Insignificant?

I've just argued that we're probably the most intelligent life-form in our entire Universe. This is a minority view,* and I may well be wrong, but it's at the very least a possibility that we can't currently dismiss. Let's therefore explore the implications of its being true and us being the only civilization in our Universe that has advanced to the point of building telescopes.

It was the cosmic vastness that made me feel insignificant to start with. Yet those grand galaxies are visible and beautiful to us—and only to us. It's only we who give them any meaning, making our small planet the most significant place in our entire observable Universe. If we didn't exist, all those galaxies would be just a meaningless and gigantic waste of space.

I also felt that my short lifespan appeared insignificant when compared with the vastness of cosmic time. However, this brief century of ours is arguably the most significant one in the history of our Universe: the one when its meaningful future gets decided. We'll have the technology to either self-destruct or seed our cosmos with life. The situation is so unstable that I doubt that we can dwell at this fork in the road for more than another century. If we end up going the life route rather than the death route, then in a distant future, our cosmos will be teeming with life that all traces back to what we do here and now. I have no idea how we'll be thought of, but I'm sure that we won't be remembered as insignificant.

In this book, we've explored our physical reality, seeing through the eyes of science a breathtakingly beautiful universe, which through us humans has come alive and started becoming aware of itself. We've seen that life's future potential in our Universe is grander than the wildest dreams of our ancestors, tempered by an equally real potential for intel-

* However, John Gribbin comes to a similar conclusion in his 2011 book *Alone in the Universe*. For a spectrum of intriguing perspectives on this question, I also recommend Paul Davies's 2011 book *The Eerie Silence*.

ligent life to go permanently extinct. Will life in our Universe fulfill its potential or squander it? I think this will be decided in our lifetime here on Spaceship Earth, by you, me and our fellow passengers. Let's make a difference!

THE BOTTOM LINE

- Even though our two intellectual expeditions set off in opposite directions, toward the large and the small, they ended up in the same place: in the realm of mathematical structures.
- On the largest and smallest scales, the mathematical fabric of reality becomes evident, while it remains easy to miss on the intermediate scales that we humans are usually aware of.
- If the ultimate fabric of reality really is mathematical, then everything is in principle understandable to us, and we'll be limited only by our own imagination.
- Although the Level IV multiverse is eternal, our particular Universe might end in a Big Chill, Big Crunch, Big Rip, Big Snap or with death bubbles.
- Evidence suggests that there's no other life-form as advanced as us humans in our entire Universe.
- From a cosmic perspective, the future potential of life in our Universe is vastly greater than anything we've seen so far.
- Yet we humans devote only meager attention and resources to existential risks that threaten life as we know it, including accidental nuclear war and unfriendly artificial intelligence.
- Although it's easy to feel insignificant in our vast cosmos, the entire future of life in our Universe will arguably be decided on our planet in our lifetime—by you, me and our fellow passengers on Spaceship Earth. Let's make a difference!

Acknowledgments

In addition to the people mentioned in the preface, I'm grateful to the organizations whose research grants have helped enable the research described in this book: NASA, the National Science Foundation, the Packard Foundation, the Research Corporation for Science Advancement, the Kavli Foundation, the John Templeton Foundation, the University of Pennsylvania and the Massachusetts Institute of Technology. I also wish to thank Jonathan Rothberg and an anonymous donor for their generous support of the Omniscope project.

Suggestions for Further Reading

This book has drawn on a huge corpus of work by the scientific community. Most of it is published in technical journal papers that you'll find cited in my own technical papers at http://space.mit.edu/home/tegmark/technical.html. However, there's also a rich literature of books aiming to explain the core ideas to non-experts. In addition to the references I've called out in footnotes, here's a small sample of such books, from the many wonderful ones that have been written, through which you can continue exploring topics that we've covered. I've tried to group them by their main focus, even though many cover other topics as well. One or more integral symbols ∫ indicate that a book is more technical/mathematical, akin to chili-pepper symbols indicating spiciness on restaurant menus.

COSMOLOGY (CHAPTERS 2–4)

Adams, Fred, and Greg Laughlin. *The Five Ages of the Universe.* New York: The Free Press, 1999.

Chown, Marcus. *The Magic Furnace: The Search for the Origins of Atoms.* New York: Oxford University Press, 2001.

de Grasse Tyson, Neil. *Death by Black Hole: And Other Cosmic Quandaries.* New York: W. W. Norton & Company, 2007.

Finkbeiner, Ann. *A Grand and Bold Thing: An Extraordinary New Map of the Universe Ushering in a New Era of Discovery.* New York: Free Press, 2010.

Greene, Brian. *The Fabric of the Cosmos.* New York: Knopf, 2004.

Hawking, Stephen. *A Brief History of Time.* New York: Touchstone, 1993.

Kirshner, Robert P. *The Extravagant Universe: Exploding Stars, Dark Energy, and the Accelerating Cosmos.* Princeton: Princeton Science Library, 2004.

Kragh, Helge. *Cosmology and Controversy: The Historical Development of Two Theories of the Universe.* Princeton: Princeton University Press, 1996.

Krauss, Lawrence. *A Universe from Nothing: Why There Is Something Rather than Nothing.* New York: Free Press, 2012.

Rees, Martin. *Just Six Numbers: The Deep Forces That Shape the Universe.* New York: BasicBooks, 2000.

Rees, Martin. *Our Cosmic Habitat.* Princeton: Princeton University Press, 2002.

Seife, Charles. *Alpha and Omega: The Search for the Beginning and End of the Universe.* New York: Penguin Books, 2004.

Singh, Simon. *Big Bang: The Origin of the Universe.* New York: HarperCollins, 2004.

Smolin, Lee. *Time Reborn: From the Crisis in Physics to the Future of the Universe.* Boston: Houghton Mifflin Harcourt, 2013.

Weinberg, Steven. *The First Three Minutes: A Modern View of the Origin of the Universe.* New York: BasicBooks, 1993.

INFLATION, MULTIVERSE LEVELS I–II (CHAPTERS 5–6)

Barrow, John. *The Book of Universes: Exploring the Limits of the Cosmos.* New York: W. W. Norton & Company, 2011.

Davies, Paul. *Cosmic Jackpot: Why Our Universe Is Just Right for Life.* New York: Houghton Mifflin, 2007.

Guth, Alan. *The Inflationary Universe.* New York: Perseus Books Group, 1997.

∫∫ Linde, Andrei D. *Particle Physics and Inflationary Cosmology.* Chur, Switzerland: Harwood Academic Publishers, 1990.

Steinhardt, Paul J., and Neil Turok. *Endless Universe: Beyond the Big Bang.* New York: Doubleday, 2007.

Susskind, Leonard. *The Cosmic Landscape: String Theory and the Illusion of Intelligent Design.* New York: Little, Brown and Company, 2005.

Vilenkin, Alexander. *Many Worlds in One: The Search for Other Universes.* New York: Hill and Wang, 2006.

QUANTUM MECHANICS, MULTIVERSE LEVEL III (CHAPTERS 7–8)

Byrne, Peter. *The Many Worlds of Hugh Everett III: Multiple Universes, Mutual Assured Destruction, and the Meltdown of a Nuclear Family.* New York: Oxford University Press, 2010.

Cox, Brian, and Jeff Forshaw. *The Quantum Universe (And Why Anything That Can Happen, Does).* Boston: Da Capo Press, 2012.

Deutsch, David. *The Beginning of Infinity: Explanations That Transform Our World.* New York: Allen Lane, 2012.

Deutsch, David. *The Fabric of Reality.* New York: Allen Lane, 1997.

∫∫ Everett, Hugh. "The Many-Worlds Interpretation of Quantum Mechanics." Ph.D. diss., Princeton University, 1957. Free download at http://www.pbs.org/wgbh/nova/manyworlds/pdf/dissertation.pdf.

∫∫ Everett, Hugh. *The Many-Worlds Interpretation of Quantum Mechanics*, edited by Bryce S. DeWitt and Neill Graham. Princeton: Princeton University Press, 1973.

∫∫ Giulini, Domenico, and Erich Joos, Claus Kiefer, Joachim Kupsch, Ion-Olimpiu Stamatescu and H. Dieter Zeh. *Decoherence and the Appearance of a Classical World in Quantum Theory.* Berlin: Springer, 1996.

Kaiser, David. *How the Hippies Saved Physics: Science, Counterculture, and the Quantum Revival.* New York: W. W. Norton & Company, 2011.

∫ Saunders, Simon, and Jonathan Barrett, Adrian Kent and David Wallace. *Many Worlds? Everett, Quantum Theory & Reality.* Oxford: Oxford University Press, 2010.

MULTIVERSES IN GENERAL (CHAPTERS 6 AND 8)

∫ Carr, Bernard J., ed. *Universe or Multiverse?* Cambridge, Mass.: MIT Press, 2007.

Carroll, Sean. *From Eternity to Here: The Quest for the Ultimate Theory of Time.* Oxford: Oneworld Publications, 2011.

Greene, Brian. *The Hidden Reality.* New York: Knopf, 2011.

Kaku, Michio. *Parallel Worlds: A Journey Through Creation, Higher Dimensions, and the Future of the Cosmos.* New York: Anchor Books, 2006.

Lewis, David. *On the Plurality of Worlds.* Oxford: Blackwell Publishing, 1986.

THE MIND (CHAPTERS 9 AND 11)

Blackmore, Susan. *Conversations on Consciousness: What the Best Minds Think about Free Will, and What It Means to Be Human.* New York: Oxford University Press, 2006.

Bostrom, Nick. *Anthropic Bias: Observation Selection Effects in Science and Philosophy.* New York: Routledge, 2002.

Damasio, Antonio. *The Feeling of What Happens.* New York: Harcourt Brace, 2000.

Damasio, Antonio. *Self Comes to Mind: Constructing the Conscious Brain.* New York: Pantheon Books, 2010.

Dennett, Daniel. *Consciousness Explained.* Boston: Little, Brown and Company, 1992.

Hawkins, Jeff, and Sandra Blakeslee. *On Intelligence.* New York: Henry Holt and Company, 2004.

Hut, Piet, Mark Alford and Max Tegmark. "On Math, Matter and Mind," *Foundations of Physics,* January 15, 2006, http://arxiv.org/pdf/physics/0510188.pdf.

Koch, Christof. "A 'Complex' Theory of Consciousness," *Scientific American,* August 18, 2009, http://www.scientificamerican.com/article.cfm?id=a-theory-of-consciousness.

Koch, Christof. *The Quest for Consciousness: A Neurobiological Approach.* Englewood, Col.: Roberts & Company Publishers, 2004.

Kurzweil, Ray. *How to Create a Mind: The Secret of Human Thought Revealed.* New York: Viking Penguin, 2012.

Penrose, Roger. *The Emperor's New Mind.* Oxford: Oxford University Press, 1989.

Pinker, Steven. *How the Mind Works.* New York: W. W. Norton and Company, 1997.

Tononi, Giulio. "Consciousness as Integrated Information: A Provisional Manifesto," *The Biological Bulletin,* 2008, http://www.biolbull.org/content/215/3/216.full.

Tononi, Giulio. *Phi: A Voyage from the Brain to the Soul.* New York: Pantheon Books, 2012.

Velmans, Max, and Susan Schneider, eds. *The Blackwell Companion to Consciousness.* Malden, Mass.: Blackwell Publishing, 2007.

MATHEMATICS, COMPUTATION, COMPLEXITY (CHAPTERS 10–12)

Barrow, John D. *Pi in the Sky.* Oxford: Clarendon Press, 1992.

Barrow, John D., *Theories of Everything.* New York: Ballantine Books, 1991.

Chaitin, Gregory J. *Algorithmic Information Theory.* (Cambridge: Cambridge University Press, 1987.

Davies, Paul. *The Mind of God.* New York: Touchstone, 1993.

Goodstein, Reuben L. *Constructive Formalism: Essays on the Foundations of Mathematics.* Leicester: Leister University College Press, 1951.

Hersh, Reuben, *What Is Mathematics, Really?* Oxford: Oxford University Press, 1999.

Levin, Janna. *A Madman Dreams of Turing Machines.* New York: Anchor Books, 2007.

Livio, Mario. *Is God a Mathematician?.* New York: Simon & Schuster, 2009.

Lloyd, Seth. *Programming the Universe: A Quantum Computer Scientist Takes on the Cosmos.* New York: Vintage Books, 2007.

Rucker, Rudy. *Infinity and the Mind.* Boston: Birkhäuser, 1982.

Standish, Russell K. *Theory of Nothing.* Charleston, S.C.: BookSurge, 2006.

Wolfram, Stephen. *A New Kind of Science.* New York: Wolfram Media, 2002.

FUTURE OF LIFE (CHAPTER 13)

Bostrom, Nick, and Milan Ćirković, eds. *Global Catastrophic Risks.* Oxford: Oxford University Press, 2008.

Davies, Paul. *The Eerie Silence: Renewing Our Search for Alien Intelligence.* New York: Houghton Mifflin Harcourt, 2011.

Drexler, K. Eric. *Engines of Creation: The Coming Era of Nanotechnology.* London: Fourth Estate, 1985.

Dyson, Freeman. *A Many-Colored Glass: Reflections on the Place of Life in the Universe.* Charlottesville: University of Virginia Press, 2007.

Fuller, R. Buckminster. *Operating Manual for Spaceship Earth.* Buckminster Fuller Institute, http://bfi.org/about-bucky/resources/books/operating-manual-spaceship-earth.

Gribbin, John R. *Alone in the Universe: Why Our Planet Is Unique.* Hoboken, N.J.: John Wiley & Sons, 2011.

Kurzweil, Ray. *The Age of Spiritual Machines: When Computers Exceed Human Intelligence.* New York: Viking, 1999.

Kurzweil, Ray. *The Singularity Is Near: When Humans Transcend Biology.* New York: Viking, 2005.

Kurzweil, Ray, and Terry Grossman. *Transcend: Nine Steps to Living Well Forever.* New York: Viking, 2010.

Moravec, Hans. *Robot: Mere Machine to Transcendent Mind.* Oxford: Oxford University Press, 1999.

Rees, Martin. *Our Final Hour: A Scientist's Warning.* New York: Perseus Books, 1997.

Sagan, Carl. *Pale Blue Dot: A Vision of the Human Future in Space.* New York: Random House, 1997.

FUNDAMENTAL PHYSICS, STRING THEORY, QUANTUM GRAVITY

Barbour, Julian. *The End of Time: The Next Revolution in Physics.* Oxford: Oxford University Press, 1999.

Barrow, John D., and Frank J. Tipler. *The Anthropic Cosmological Principle.* Oxford: Clarendon Press, 1986.

Carroll, Sean. *The Particle at the End of the Universe: How the Hunt for the Higgs Boson Leads Us to the Edge of a New World.* New York: Dutton, 2012.

∫ Einstein, Albert. *Relativity: The Special and General Theory.* London: Really Simple Media, 2011.

∫∫ Feynman, Richard, and Robert Leighton and Matthew Sands. *The Feynman Lectures on Physics.* 3 vols. New York: Addison-Wesley, 1964.

Gamow, George. *Mr. Tompkins in Paperback.* Cambridge: Cambridge University Press, 1940.

Greene, Brian. *The Elegant Universe.* New York: W. W. Norton and Company, 2003.

Musser, George. *The Complete Idiot's Guide to String Theory.* New York: Penguin Group, 1998.

∫∫ Penrose, Roger. *The Road to Reality: A Complete Guide to the Laws of the Universe.* New York: Knopf, 2005.

Randall, Lisa. *Warped Passages: Unraveling the Mysteries of the Universe's Hidden Dimensions.* New York: Ecco, 2005.

Smolin, Lee. *Three Roads to Quantum Gravity.* New York: BasicBooks, 2001.

Smolin, Lee. *The Trouble with Physics: The Rise of String Theory, the Fall of a Science, and What Comes Next.* Boston: Houghton Mifflin, 2006.

Susskind, Leonard. *The Black Hole War: My Battle with Stephen Hawking to Make the World Safe for Quantum Mechanics.* New York: Little, Brown and Company, 2008.

Weinberg, Steven L. *Dreams of a Final Theory: The Scientist's Search for the Ultimate Laws of Nature.* New York: Pantheon, 1992.

Wigner, Eugene P. *Symmetries and Reflections.* Cambridge, Mass.: MIT Press, 1967.

Wilczek, Frank. *The Lightness of Being: Mass, Ether and the Unification of Forces.* New York: BasicBooks, 2008.

∫ Zeh, H. Dieter. *The Physical Basis of the Direction of Time.* 4th ed. Berlin: Springer, 2002.

Index

[Page numbers followed by *f*, *n*, or *t* refer to figures, footnotes or tables, respectively.]